Springer Series in Reliability Engineering

Series Editor

Professor Hoang Pham
Department of Industrial and Systems Engineering
Rutgers, The State University of New Jersey
96 Frelinghuysen Road
Piscataway, NJ 08854-8018
USA

Other titles in this series

The Universal Generating Function in Reliability Analysis and Optimization
Gregory Levitin

Warranty Management and Product Manufacture
D.N.P. Murthy and Wallace R. Blischke

Maintenance Theory of Reliability
Toshio Nakagawa

System Software Reliability
Hoang Pham

Reliability and Optimal Maintenance
Hongzhou Wang and Hoang Pham

Applied Reliability and Quality
B.S. Dhillon

Shock and Damage Models in Reliability Theory
Toshio Nakagawa

Risk Management
Terje Aven and Jan Erik Vinnem

Satisfying Safety Goals by Probabilistic Risk Assessment
Hiromitsu Kumamoto

Offshore Risk Assessment (2nd Edition)
Jan Erik Vinnem

The Maintenance Management Framework
Adolfo Crespo Márquez

Human Reliability and Error in Transportation Systems
B.S. Dhillon

Complex System Maintenance Handbook
D.N.P. Murthy and Khairy A.H. Kobbacy

Recent Advances in Reliability and Quality in Design
Hoang Pham

Product Reliability
D.N.P. Murthy, Marvin Rausand and Trond Østerås

Mining Equipment Reliability, Maintainability, and Safety
B.S. Dhillon

Advanced Reliability Models and Maintenance Policies
Toshio Nakagawa

Justifying the Dependability of Computer-based Systems
Pierre-Jacques Courtois

Poong Hyun Seong
Editor

Reliability and Risk Issues in Large Scale Safety-critical Digital Control Systems

With Additional Contributions by
Poong Hyun Seong
Hyun Gook Kang
Han Seong Son
Jong Gyun Choi
Man Cheol Kim
Jong Hyun Kim
Jae Whan Kim
Seo Ryong Koo
Seung Jun Lee
Jun Su Ha

Springer

Professor Poong Hyun Seong, PhD
Department of Nuclear
and Quantum Engineering
Korea Advanced Institute of Science
and Technology (KAIST)
373-1, Guseong-dong, Yuseong-gu
Daejeon, 305-701
Republic of Korea

ISBN 978-1-84996-797-6 e-ISBN 978-1-84800-384-2

DOI 10.1007/978-1-84800-384-2

Springer Series in Reliability Engineering ISSN 1614-7839

British Library Cataloguing in Publication Data
Reliability and risk issues in large scale safety-critical
digital control systems. - (Springer series in reliability
engineering)
1. Digital control systems 2. Digital control systems -
Reliability
I. Seong, Poong Hyun
629.8'312

Cover design: deblik, Berlin, Germany

Printed on acid-free paper

9 8 7 6 5 4 3 2 1

springer.com

Preface

Reliability and risk issues for safety-critical digital control systems are associated with hardware, software, human factors, and the integration of these three entities. The book is divided into four parts. Each part, consisting of three chapters, deals with all entities.

Component level digital hardware reliability, existing hardware reliability theories, and related digital hardware reliability issues (Chapter 1), digital system reliability and risk including hardware, software, human factors, and integration (Chapter 2), and countermeasures using cases from nuclear power plants (Chapter 3) are presented in Part I.

Existing software reliability models and associated issues (Chapter 4), software reliability improvement techniques as countermeasures of software reliability modeling (Chapter 5), and a CASE tool called "NuSEE (nuclear software engineering environment)" which was developed at KAIST (Chapter 6) are presented in Part II.

Selected important existing human reliability analysis (HRA) methods including first- and second-generation methods (Chapter 7), human factors considered in designing and evaluating large-scale safety-critical digital control systems (Chapter 8), and a human performance evaluation tool, called "HUPESS (human performance evaluation support system)", which was developed at KAIST as a countermeasure to human-factors-related issues (Chapter 9) are presented in Part III.

The integrated large-scale safety-critical control system, which consists of hardware and software that is usually handled by humans, is presented in Part IV. This book emphasizes the need to consider hardware, software, and human factors, not separately, but in an integrated manner. Instrument failure significantly affecting human operator performance was demonstrated in many cases, including the TMI-2 incidents. These issues are discussed in Chapter 10. An analytical HRA method for safety assessment of the integrated digital control systems including human operators, which is based on Bayes' theorem and information theory, is discussed in Chapter 11. Using this method, it is concluded that human operators are crucial in reliability and risk issues for large-scale safety-critical digital control systems. An operator system which supports human cognitive behavior and actions, "INDESCO (integrated decision support system to aid the cognitive activities of operators)" which was developed at KAIST is discussed in Chapter 12.

This book can be read in different ways. If a reader wants to read only the current issues in any specific entity, he/she can read the first two chapters of either Part I, II, or III, or the first chapter of Part IV. If a reader wants to read only countermeasures developed at KAIST in any specific entity, he/she may read either Chapter 3, 6, or 9, or Chapters 11 and 12.

There are many co-authors of this book. Part I was mainly written by Drs. Jong Gyun CHOI and Hyun Gook KANG from KAERI (Korea Atomic Energy Research Institute). Part II was mainly written by Professor Han Seong SON from Joongbu University and Dr. Seo Ryong KOO from Doosan Heavy Industries and Construction Co., Ltd.. The main writers of Part III are Mr. Jae Whan KIM from KAERI, Dr. Jong Hyun KIM from KHNP (Korea Hydro and Nuclear Power) Co., Ltd., and Dr. Jun Su HA from KAIST. The integration part, Part IV, was mainly written by Drs. Man Cheol KIM and Seung Jun LEE from KAERI.

Last but not least, I would like to thank Mrs. Shirley Sanders and Professor Charles Sanders for their invaluable support for English editing of this entire book. Without their help, this book might have not been published.

Republic of Korea
May 2008

Poong Hyun Seong

Professor
Department of Nuclear
and Quantum Engineering
Korea Advanced Institute
of Science and Technology (KAIST)

Contents

Part II Software-related Issues and Countermeasures

List of Contributors

Poong Hyun Seong
Department of Nuclear and Quantum Engineering,
Korea Advanced Institute of Science and Technology (KAIST)

Hyun Gook Kang
Integrated Safety Assessment Division,
Korea Atomic Energy Research Institute (KAERI)

Han Seong Son
Department of Game Engineering, Joongbu University

Jong Gyun Choi
I&C and Human Factors Division, KAERI

Man Cheol Kim
Integrated Safety Assessment Division, KAERI

Jong Hyun Kim
MMIS Team, Nuclear Engineering and Technology Institute (NETEC),
Korea Hydro and Nuclear Power (KHNP) Co., Ltd.

Jae Whan Kim
Integrated Safety Assessment Division, KAERI

Seo Ryong Koo
Nuclear Power Plant BG, Doosan Heavy Industries and Construction Co., Ltd.

Seung Jun Lee
Integrated Safety Assessment Division, KAERI

Jun Su Ha
Center for Advanced Reactor Research, KAIST

List of Figures

List of Tables

Part I

Hardware-related Issues and Countermeasures

1

Reliability of Electronic Components

Jong Gyun Choi[1] and Poong Hyun Seong[2]

[1]I&C/Human Factors Division
Korea Atomic Energy Research Institute
1045 Daedeok-daero, Yuseong-gu, Daejeon, 305-353, Korea
choijg@kaeri.re.kr

[2]Department of Nuclear and Quantum Engineering
Korea Advanced Institute of Science and Technology
373-1 Guseong-dong, Yuseong-gu, Daejeon, 305-701, Korea
phseong@kaist.ac.kr

Electronics is the study of charge flow through various materials and devices, such as semiconductors, resistors, inductors, capacitors, nano-structures, and vacuum tubes [1]. The electronic component is any indivisible electronic building block packaged in a discrete form with two or more connected leads or metallic pads. The components are intended to be connected together, usually by soldering to a printed circuit board, to create an electronic circuit with a particular function. The representative electronic components are integrated circuits (microprocessors, RAM), resistors, capacitors, and diodes. These electronic components are the major hardware components making up digital systems.

Digital system developers generally consider the various factors for selecting and purchasing the proper electronic components among various alternatives. These factors include cost, market share, maturity, dimension, and performance. Performance involves capability, efficiency, and reliability. Capability is the ability of a component to satisfy its required or intended function. Efficiency means how easily and effectively the component can realize its required function or objectives. Reliability is defined as the ability for a component to continue operating without failure.

Reliability is one of the essential attributes which determines quality of electronic components for safety-critical application. Both manufacturers and customers of electronic components need to define and predict the reliability in a common way. Reliability prediction is the process that estimates the component's ability to perform its required function without failure during its life.

Reliability prediction is performed during concept, design, development, operation, and maintenance phases. It is used for other purposes [2]:

- To assess whether reliability goals can be reached
- To compare alternative designs

- To identify potential design weakness and potential improvement opportunities
- To plan logistics support strategies, such as spare parts provisioning and calculation of warranty and lifecycle costs
- To provide data for system reliability and safety analysis
- To estimate mission reliability
- To predict the field reliability performance
- To establish objectives for reliability tests

Failure needs first to be defined in order to understand the concept of reliability. Failure is defined as the occurrence of an event whereby the component does not perform a required function. The component is in one of two states, good or failure state, within the operating environment (Figure 1.1).

The root causes of failure are summarized in [2]:

- Causes related to design process, such as design rule violation, design errors resulting from overstressed parts, timing faults, reverse current paths, documentation or procedural errors, and non-tested or latent failures
- Causes related to manufacturing process, such as workmanship defects caused by manual or automatic assembly or rework operations, test errors, and test equipment faults
- Causes related to physical environment, such as excessive operating temperature, humidity or vibration, exceeding electromagnetic threshold, and foreign object or mishandling damage
- Causes related to humans, such as operator errors, incorrectly calibrated instruments, and maintenance errors

The failure of electronic components can be categorized into three classes according to the length of time that the failure is active [3]: permanent failure, intermittent failure, and transient failure. Permanent failure is caused by a physical defect or an inadequacy in the design of the component. Permanent failure is persistent, consistent, and reproducible, and continues to indefinitely exist. Intermittent failure appears for a specified time interval and disappears for a

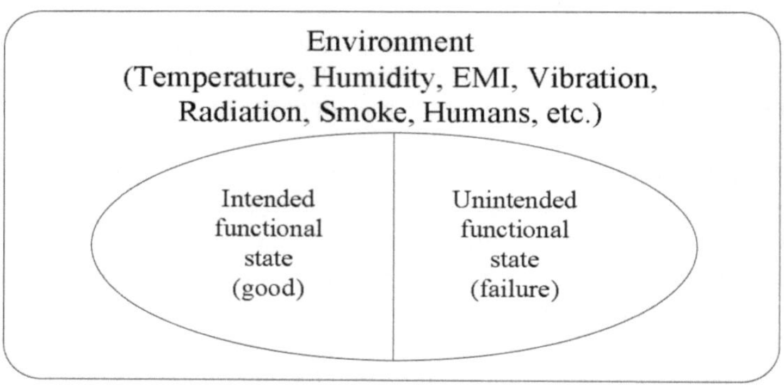

Figure 1.1. Functional state of the component

specified time interval repeatedly. Transient failure is reversible and not associated with any persistent physical damage to the component.

Methods that predict the reliability of the electronic components and some important issues are described in this chapter. Mathematical background related to reliability prediction is described in Section 1.1. Reliability prediction models of the permanent failures are introduced in Section 1.2. Reliability prediction models of the intermittent failures are dealt with in Section 1.3. Reliability prediction models of the transient failures are treated in Section 1.4. The chapter is summarized in Section 1.5.

1.1 Mathematical Reliability Models

When $F(t)$, a cumulative distribution function (CDF), is defined as the probability that the component will fail at time less than or equal to time t, the mathematical form of the $F(t)$ is:

$$F(t) = \Pr(T \le t), \quad t \ge 0 \tag{1.1}$$

where T is a random variable meaning time to failure and t represents the particular time of interest.

When $f(t)$ is defined as the probability density function of $F(t)$, $F(t)$ is expanded as:

$$F(t) = \Pr(T \le t) = \int_0^t f(\tau)d\tau, \quad t \ge 0 \tag{1.2}$$

Reliability is defined as the conditional probability that the component will perform its required function for a given time interval under given conditions, and is mathematically expressed as [4]:

$$R(t) = \Pr(T \ge t \mid C_1, C_2, ...) \tag{1.3}$$

where C_1, C_2, ... are given conditions, such as environmental conditions.

When the C_1, C_2, ... are implicitly considered, reliability is expressed simply as:

$$R(t) = \Pr(T \ge t) = 1 - \Pr(T < t) = 1 - F(t) = 1 - \int_0^t f(\tau)d\tau \tag{1.4}$$

If the hazard rate function (or, failure rate function) is defined by:

$$h(t) = \frac{f(t)}{1 - F(t)} \tag{1.5}$$

The reliability is rewritten by:

$$R(t) = 1 - F(t) = 1 - \int_0^t f(\tau)d\tau = \exp\left[-\int_0^t h(\tau)d\tau\right] \tag{1.6}$$

Reliability is mathematically calculated only if the accurate failure distribution is known (Equation 1.6).

Another important concept related to reliability is the mean time to failure (MTTF) or the expected lifetime of the component. This is expressed mathematically as:

$$\text{MTTF} = E(t) = \int_0^\infty \tau f(\tau)d\tau = \int_0^\infty R(\tau)d\tau \tag{1.7}$$

The reliability, failure rate, and the MTTF are calculated, assuming the component failure is exponentially distributed, as:

$$R(t) = 1 - \int_0^t \lambda e^{-\lambda\tau} d\tau = e^{-\lambda t},$$

$$h(t) = \frac{f(t)}{1 - F(t)} = \lambda = \text{const},$$

$$\text{and MTTF} = \int_0^\infty \tau\lambda e^{-\lambda\tau} d\tau = \frac{1}{\lambda}$$

The mathematical relationship between several reliability-related measures, $F(t)$, $f(t)$, $R(t)$, and $h(t)$ is shown in Table 1.1. Three representative statistical distribution functions, exponential, Weibull, and lognormal distribution, which are most commonly used for reliability prediction are shown in Table 1.2. The failure of electronic components is categorized into three classes according to the length of time that the failure is active: permanent failure, intermittent failure, and transient failure.

Table 1.1. Mathematical relationship between the representative reliability measures

	$f(t)$	$F(t)$	$R(t)$	$h(t)$
$f(t) =$	$f(t)$	$F'(t)$	$-R'(t)$	$h(t)\exp\left[-\int_0^t h(\tau)d\tau\right]$
$F(t) =$	$\int_0^t f(\tau)d\tau$	$F(t)$	$1 - R(t)$	$1 - \exp\left[-\int_0^t h(\tau)d\tau\right]$
$R(t) =$	$\int_t^\infty f(\tau)d\tau$	$1 - F(t)$	$R(t)$	$\exp\left[-\int_0^t h(\tau)d\tau\right]$
$h(t) =$	$f(t)/\int_t^\infty f(\tau)d\tau$	$F'(t)/[1 - F(t)]$	$-[\ln R(t)]$	$h(t)$

Table 1.2. Mathematical reliability measures about three representative failure distribution models

Failure distribution	Exponential	Weibull	Lognormal
$f(t)=$	$\lambda e^{-\lambda t}$	$\frac{\beta t^{\beta-1}}{\alpha^{\beta}} e^{-\left(\frac{t}{\alpha}\right)^{\beta}}$	$\frac{1}{\sigma_t t\sqrt{2\pi}} e^{\left[-\frac{1}{2\sigma_t^2}(\ln t-\mu_t)^2\right]}$
$F(t)=$	$1-e^{-\lambda t}$	$1-e^{-\left(\frac{t}{\alpha}\right)^{\beta}}$	$\frac{1}{\sigma_t t\sqrt{2\pi}}\int_0^t \frac{1}{\theta} e^{\left[-\frac{1}{2}\frac{(\ln\theta-\mu_t)^2}{\sigma_t^2}\right]} d\theta$
$R(t)=$	$e^{-\lambda t}$	$e^{-\left(\frac{t}{\alpha}\right)^{\beta}}$	$1-\frac{1}{\sigma_t t\sqrt{2\pi}}\int_0^t \frac{1}{\theta} e^{\left[-\frac{1}{2}\frac{(\ln\theta-\mu_t)^2}{\sigma_t^2}\right]} d\theta$
$h(t)=$	λ	$\frac{\beta}{\alpha} e^{-\left(\frac{t}{\alpha}\right)^{\beta-1}}$	$\frac{f(t)}{1-F(t)}$
MTTF =	$\frac{1}{\lambda}$	$\alpha\Gamma\left(\frac{1+\beta}{\beta}\right)$	$e^{\left(\mu_t+\frac{1}{2}\sigma_t^2\right)}$

The failure rate function of electronic components is modeled mathematically by the following equation:

$$h_{\text{component}}(t) = h_{\text{permanent}}(t) + h_{\text{intermittent}}(t) + h_{\text{transient}}(t) \tag{1.8}$$

The reliability of a component is calculated from Equations 1.6 and 1.8, if failure rate functions for these three classes of failures ($h_{\text{permanent}}$, $h_{\text{intermittent}}$, and $h_{\text{transient}}$) are accurately identified since reliability measures in Table 1.1 are mathematically interrelated with each other.

1.2 Permanent Failure Models of the Electronic Components

The permanent failure rate curve is generally known as a bathtub curve (Figure 1.2). The vertical axis is the failure rate function and the horizontal axis is the time. A bathtub curve is divided by three regions in time; early failure period region, random failure period region, and finally wear-out period region. The failure in the early failure period region is due to defects in materials that occurred during design, manufacturing, handling ,and installation. The failure rate in this region decreases as time goes on. The interval of early failure regions can be several weeks to months. The failure in the random failure period region is due to unstable environmental stress and accidents. The failure rate in this region is relatively constant. The component spends most of its lifetime in this random failure period region. The failure rate of the component increases due to material degradation and other mechanisms in the wear-out failure period region.

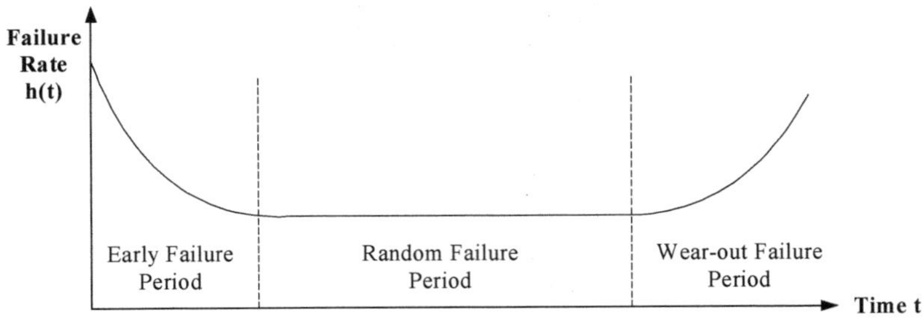

Figure 1.2. Bathtub curve

A component will experience a random failure period while operating in the field if sufficient screening and testing has been done to prevent the occurrence of early failure, and only the component which has survived an early failure period is delivered to the customer. In this case, the component failure rate will be constant while operating in the field and independent of time:

$$h_{\text{permanent}}(t) = \lambda_{\text{permanent}} \tag{1.9}$$

The permanent failure has exponential distribution and the reliability due to permanent failure is easily calculated as $e^{-\lambda t}$ (Table 1.2, Equation 1.9). MIL-HDBK-217 proposes a representative reliability prediction method that deals with permanent failures of electronic components, assumming a constant failure rate [5]. It contains failure rate models for nearly every type of electronic component used in modern military systems from microcircuits to passive components, such as integrated chips, resistors, and capacitors. It provides two methods, the part stress method and the part count method, to obtain the constant failure rate of components. A part stress analysis method is applicable when detailed information regarding the component is available, such as pin number and electrical stress. This method is used only if the detailed design is completed. A part count analysis method is useful during the early design phase, when insufficient information is given regarding the component. The failure rate calculated by a part count analysis method is a rough estimation.

For example, the constant failure rate of DRAM based on the part stress method proposed by MIL-HDBK-217F N2 is determined from:

$$\lambda_{\text{permanent,DRAM}} = (C_1\pi_T + C_2\pi_E)\pi_Q\pi_L \tag{1.10}$$

Where:
C_1 = Die complexity failure rate
C_2 = Package failure rate
π_T = Temperature factor
π_E = Environmental factor
π_Q = Quality factor
π_L = Learning factor

C_1 depends on circuit complexity and C_2 depends on packing type and package pin number. The values of π_T, π_Q, and π_L are determined by the operating temperature, quality, and production history, respectively.

MIL-HDBK-217 was updated several times and became the standard by which reliability predictions were performed after the first version was published by the US Navy in 1962 [6]. The last version of the MIL-HDBK-217 was revision F Notice 2, which was released on February 28, 1995. The Reliability Information Analysis Center (RIAC) published RIAC-HDBK-217Plus with expectation that 217Plus will eventually become a *de facto* standard to replace MIL-HDBK-217 [7]. The failure rate models in this handbook are called empirical models because they are based on historical field failure data to estimate the failure rate of the electronic device.

Various agencies and industries have proposed empirical models dedicated to their own industry and have published the specific industrial handbooks. The representative models are summarized as [8]:

- RAC's PRISM
- SAE's HDBK
- Telecordia SR-332
- CNET's HDBK

The common assumption of these models is that electric components have a constant failure rate and the failure is exponentially distributed. The models are based on field failure data and in some cases, from laboratory testing and extrapolation from similar components. Empirical models have been widely used in military and industry because they are relatively simple and there is no alternative for reliability practitioners.

Some reliability professionals have criticized these empirical models in that they cannot accurately predict the reliability of the components. The inaccuracy of these empirical models is due to:

1. *The constant failure rate assumption*
   The constant failure rate assumption means that the component failure is exponentially distributed and is independent of time. This assumption makes the mathematical calculation of reliability simple because the integration of probability density function is easier than other distributions. The failure distribution is described by a single parameter, λ (Table 1.2). Investigation and analysis of the failure data of semiconductors and electronic systems collected by various researchers advocated that the hazard rate of semiconductors has a shape like a rollercoaster curve over the operating lifetime [9, 10]. They showed that the hazard rate was not constant, but decreased with age during its useful life. Therefore, reliability obtained from the constant failure rate assumption may result in erroneous decisions about logistics support strategies, such as spare parts provisioning and calculation of warranty and lifecycle costs. They also showed that the hazard rate function is always increasing in the neighborhood of zero, and therefore, cannot exhibit a traditional bathtub curve [11].

2. *The lack of accuracy*
   Empirical models predict failure rates more conservatively than actual failure rates in the field of the components [12, 13]. The predicted failures of components cannot represent field failures because the reliability prediction models are based upon industry-average values of failure rates, which are neither vendor- nor device-specific [14, 15]. Some empirical models use false factors to calculate the failure rate of components. For example, some empirical models include the temperature factor, π_T, to consider the influence of temperature on electronic component failure rate. The temperature factor is based on the Arrhenius equation which deals with temperature dependency on chemical reaction rate. But there is no statistical correlation between temperature and observed component failure rate for bipolar logic ICs [16], and the Arrhenius model does not apply to electronic component reliability [17–19]. Empirical models having these inadequate factors can cause the system developer to focus on an erroneous design.

3. *Out of date information*
   The approach of empirical models is to fit the curve based on historical field failure data. This approach requires sufficient time (a few years or more) to collect field failure data and update the failure rate model. Component manufacturers not only continually improve the design and manufacturing process, but adapt advanced technology for improving reliability. Therefore, empirical models always lag behind up-to-date technology and do not reflect emerging technology.

4. *Transfer of primary failure causes*
   Primary system failures are caused not by component failures but by non-component factors, such as system design error, system manegement error, user handling error, and interconnection error between components due to improvement of component design, manufacturing, and process control. Empirical models are not effective in predicting system reliability and safety because they mainly treat component failures as the primary cause of system failure [17].

5. *No information about failure modes and mechanisms*
   General failure mechanisms of electronic components are electro-migration, corrosion, stress-migration, temperature cycling, and thermal shock. A good understanding of these failure mechanisms is necessary for preventing, detecting, and correcting the failures associated with design, manufacture, and operation of components. Empirical models do not provide detailed information about component failures, such as failure site, mechanisms, and modes. Therefore, empirical models are not effective in identifying potential design weaknesses and improvement opportunities, since the designer and manufacture cannot obtain information about cause-and-effect relationships for failures from empirical models.

6. *The difference of failure rate calculated by empirical models*
Six empirical models, such as MIL-HDBK-217, Bellcore RPP, NTT procedure, British Telecom procedure, CNET procedure, and Siemens procedure, showed that the failure rate of 64K DRAM ranged from 8 FIT (result from British Telecom procedure) to 1950 FIT (result from CNET procedure) under the same physical and operating characteristics [20]. Failure rate results of the same component differed widely between the empirical models, MIL-HDBK-217, HRD4, Siemens, CNET, and Bellcore [20, 21].

Other models for reliability prediction that were contrary to empirical models were stress and damage models based on the physics-of-failure (PoF) approach [22]. These models generally predict the time to failure of components as a reliability measure by analyzing root-cause failure mechanisms, which are governed by fundamental mechanical, electrical, thermal, and chemical processes. This approach starts from the fact that various failure mechanisms of the component are well known and that the failure models for these failure mechanisms are available.

Failure mechanisms of semiconductor components have been classified into wear-out and overstress mechanisms [23, 24]. Wear-out mechanisms include fatigue, crack growth, creep rupture, stress-driven diffusive voiding, electro-migration, stress migration, corrosion, time-dependent dielectric breakdown, hot carrier injection, surface inversion, temperature cycling, and thermal shock. Overstress mechanisms include die fracture, popcorning, seal fracture, and electrical overstress.

A failure model for each failure mechanism has been established. These models generally provide the time to failure of the component for each identified failure mechanism based on information of the component geometry, material properties, its environmental stress, and operating conditions. Representative failure models are the Black model [25] for electro-migration failure mechanism, the Kato and Niwa model [26] for stress-driven diffusive voiding failure mechanism, the Fowler–Nordheim tunneling model [27] for time-dependent dielectric breakdown failure mechanism, the Coffin–Manson model [28] for temperature cycling failure mechanism, and the Peck model [29] for corrosion failure mechanism. For example, the Black model for electro-migration mechanism proposed the mean time to failure as:

$$TF = \frac{w_{met} t_{met}}{j^n A_{para} e^{-E_a / K_B T}} \tag{1.10}$$

where:
TF = mean time to failure (h)
w_{met} = metallization width (cm)
t_{met} = metallization thickness (cm)
A_{para} = parameter depending on sample geometry, physical characteristics of the film and substrate, and protective coating
j = current density (A/cm^2)

K_B = the Boltzmann constant
n = experimentally determined exponent
E_a = the activation energy (eV)
T = steady-state temperature (kelvin)

The failure of an electronic component is caused by multiple failure mechanisms at several sites. All potential failure mechanisms which cause component failure are identified and time to the failure due to each identified failure mechanism is calculated using relevant failure models. The minimum time to failure of the component is determined to predict the reliability of the component [30]. The steps shown in Figure 1.3 are generally taken to predict the reliability of the component, based on stress and damage models [2]. Although stress and damage models have many advantages, compared with empirical models, they also have problems that are categorized as:

1. *Difficulty in obtaining input parameters*
   Each failure model for failure mechanism requires many input data that contribute to the failure, such as component geometry, material properties, operating conditions, and environmental conditions. Some of these input data are known only to the manufacturer of the component. Some of these input data are obtained only from dedicated testing which is costly and requires special expertise. For example, the Black model requires input variables, such as metallization width, metallization thickness, a parameter obtained from experiment, current density, and steady-state temperature (Equation 1.10). System designers and reliability engineers may have difficulty in using failure models because input data for each failure model may not be easily available or impossible to obtain if the manufacturer does not provide them.

2. *Difficulty in applying multiple failure models for one component reliability prediction*
   An electronic component can fail due to multiple failure mechanisms, such as die fracture, popcorn, seal fracture, fatigue, crack growth, creep rupture, stress-driven diffusive voiding, electro-migration, stress migration, corrosion, time-dependent dielectric breakdown, hot carrier injection, surface inversion, temperature cycling, and thermal shock. Dozens of failure models need to be applied and combined to predict the reliability of only one component. This approach is costly.

3. *Limit of failure models*
   Component failures generally occur due to incorrect design, manufacturing defects, mishandling, and component defects. Stress and damage models generally deal with component design but do not address external causes, such as manufacturing defects and mishandling errors. Therefore, this approach cannot be used to estimate field reliability [6], which may be an important measure for system developers and reliability engineers. This approach also is not practical for assessing system reliability.

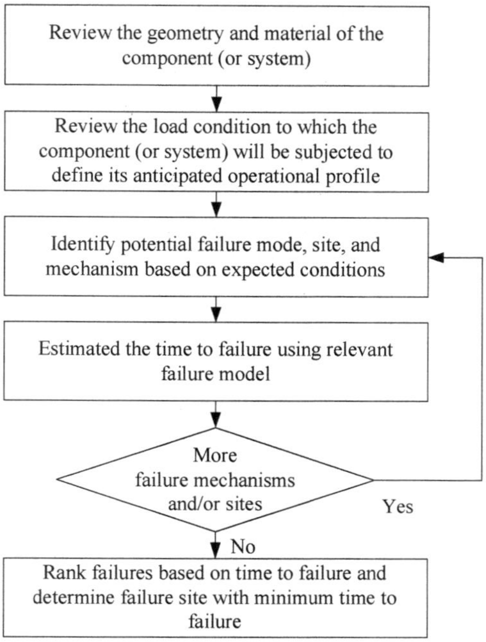

Figure 1.3. Generic process of estimating the reliability through stress and damage models [2]

4. *Availability of failure models*
   Failure models are currently not available for all possible failures and for all categories of electronic components, although many research projects have been performed to identify failure mechanisms and models.

1.3 Intermittent Failure Models of the Electronic Components

Intermittent failure is a temporal malfunction of a component (or circuit) that repeatedly occurs at generally irregular intervals only when the specified operation conditions are satisfied. A good explanation for intermittent failure is found by observing the phenomena that occur on interconnection points of a ball grid array (BGA) under shear torsion stress [31].

Intermittent failure is caused by component defects, such as structural defects, bad electrical contact, loose connections, component aging, and IC chip contamination. Each defect is due to the synergetic effect of several environmental stresses, such as thermal variation, humidity, shock, vibration, electro-magnetic interference, pollution, and power surge [32]. For example, electrical contacts can be open intermittently due to mechanical contact motion, such as contact bounce and contact sliding that can be triggered by shock and vibration [33]. Electrical contacts can also be intermittently open due to mismatch of the thermal expansion coefficient between contact materials that is triggered by thermal variation.

Corrosion by environmental stresses, such as high temperature, high humidity, and atmospheric dust can also cause electrical contacts to be intermittently open.

Intermittent failure occurs not only by one individual stress but by several simultaneous stress conditions. Intermittent failure appears only when stress conditions causing it are active, making it difficult to detect and predict intermittent failures.

Intermittent failures of components (or circuit) have critical effects on digital system reliability and safety [34–44]. An automatic error log and an operator's log of 13 VICE file servers over 22 months was collected and analyzed from February 1986 to January 1988 [34]. The file server hardware was composed of a SUN-2 workstation with a Motorola 68010 processor, a Xylogics 450 disk controller, and Fujitsu Eagle disk drives; 29 permanent failures, 610 intermittent errors, 446 transient errors and 296 system crashes were observed. Ten percent of the system crashes were due to permanent faults and 90 percent of system crashes were caused by a combination of intermittent and transient errors. However, system crashes due to intermittent errors and those due to transient errors were not clearly discriminated. Problems involved with the occurrence of No Fault Found (NFF) events during test and repair of the electronic system have been discussed [42–44]. An electronic system is sent to the repair technician for troubleshooting the failure when the electronics system is recognized as failed during operation. The event is reported as a NFF event if the technician cannot duplicate the failure during the test. Roughly, 50 percent of all failures recognized during operation are classified as NFF events [42]. One reason for NFF events is attributed to system failures by intermittent faults, such as intermittent connectivity in the wiring or connectors and partially defective components. NFF events of electronic boxes in military weapon systems were reviewed and analyzed [44]. Electronic boxes experience harsh environmental stress conditions during the military operations, such as vibration, G loading, thermal extremes, shock, and other stresses that are usually absent at a test facility. Therefore, NFF events occur because intermittency occurs while the boxes are under real stress conditions, not under benign test stress conditions.

There are many research projects that are concerned with intermittent behavior of electronics components. Most are concerned with the method used to efficiently detect and diagnose intermittent faults in the system during operation or repair [45–50] and developing system design that tolerate intermittent faults [51, 52]. Other research projects have examined the system reliability model that considers intermittent faults [53, 54]. There are few studies that have analyzed and modeled intermittent failures of electronic components (or circuits) based on physical failure mechanisms from a perspective of reliability [31, 33–41]. Even these studies mainly focused on intermittent failures of electrical contacts [31, 33–38].

Intermittent failure of a Josephson device in the superconducting state due to thermal noise has been observed and the intermittent rate of a junction switching out of the superconducting state has been predicted [39]. The intermittent error rate due to thermal noise corresponded to the probability per unit time of a state transition in a thermal activation process in the form of $\delta e^{-U/kT}$, where δ is a pre-factor taken to be the attempt frequency, U the barrier height, and k the Boltzmann constant.

Intermittent behavior of electrical contacts has been described by the accelerated reliability test that monitors the resistance of BGA component solder joints under shear torsion stress [31]. Four stages of resistance change have been observed as the crack grows in the solder joint as:

- No total-length crack; resistance presents stable values
- Total length crack with low constriction resistance; resistance fluctuation between 0.005 and 1 ohm
- Transient microsecond open circuits; resistance fluctuation between 0.1 and 100 ohm
- Increase of gap between cracked surfaces; transient events longer than 1 second, or permanent open circuit even under unstressed conditions

Intermittent fluctuation of electrical contact resistance can produce an unintended voltage variation in the circuit, especially if the contact is a signal contact for data transmission in a digital circuit. Even a small fluctuation of contact resistance can deform the data information and, eventually, cause intermittent failure of the circuit. Intermittent failure of the circuit, however, depends on the magnitude of contact resistance fluctuation that the circuits are immune to. That is, there exists a threshold after which the resistance change can cause intermittent failure in the circuit. It is important to find the threshold level of resistance change to determine intermittent failure criteria and estimate the occurrence rate of intermittent electrical contact failure.

A multi-contact reliability model that considers intermittent failure due to fretting corrosion has been proposed [40]. A mathematical reliability model has been proposed, assuming that each single contact failure exhibits Weibull distribution and that all contact interfaces were in the failed state for multi-contact failure.

1.4 Transient Failure Models of the Electronic Components

Transient failures of electronic components are caused by cosmic–galactic particle radiation, electromagnetic interference, lightning, power supply noise, thermal hot spots, and active voltage [55, 56]. Transient failures are a huge concern in advanced semiconductor components, such as static random access memory (SRAM), dynamic random access memory (DRAM), field programmable gate arrays (FPGA), and microprocessors. Transient failures caused by energetic nuclear particles are a main focus since nuclear particles have attracted the greatest concern as the main cause of transient failures in semiconductors [55].

An introduction to the basic mechanism and models of transient failures induced by radiation has been given [56–57]. Terms like “soft error” or “single-event upset (SEU)” are more generally used by researchers for indicating transient failure caused by energetic nuclear particles. The term “soft error” is used since the characteristic of nuclear particle-induced failure is “soft” in that the component is not persistently damaged and operates correctly if erroneous data in the component are corrected by overwriting or reset. The term “single-event upset” is used in that the data state in the electronic component is reversed or flipped by a single energetic nuclear particle strike on the component.

Soft errors are mainly caused by energetic nuclear particles, such as alpha particles, protons, neutrons, and heavier particles. Primary sources of these nuclear particles are from primordial and induced radioactive atoms that exist in all materials and from cosmic rays, and galactic and solar particles.

Alpha particles are emitted from radioactive atoms, particularly from uranium (U-238) and thorium (Th-232), that are present in device material and packaging material of semiconductors [58, 59]. Alpha particles outside the packaged semiconductor are not of concern since they do not have sufficient energy to penetrate the package and substrates of the device. Alpha particles are doubly charged helium nuclei consisting of two protons and two neutrons. Decay of alpha particle emission is common for atomic nuclei heavier than lead (Pb). Alpha particles emitted from the decay of U-238 and Th-232 in semiconductor packaging material deposits all their kinetic energy during passing and ionize directly by generating electron–hole pairs along their trajectory. Electron–hole pairs diffuse and electrons are swept by a high electric field to the depletion region when the alpha particle strikes near the sensitive region, especially reverse-biased p/n junction, in microelectronic circuits of semiconductors. The digital value stored in this region can be changed from a "1" to a "0", or *vice versa* if the number of collected electrons exceeds a critical charge that differentiates a "1" from a "0".

Other causes of soft errors are cosmic rays which constantly hit the earth's surface [60, 61]. Cosmic rays include galactic cosmic rays and solar particles from the sun. Cosmic rays interact with nitrogen and oxygen atoms of the atmosphere and produce a shower of secondary particles, consisting of protons, electrons, neutrons, heavy ions, muons, and pions when they hit the earth's atmosphere. Secondary particles in turn penetrate into lower altitude and generate third-generation particles. Cosmic radiation on the earth's surface is from products of the sixth and seventh reaction of galactic cosmic rays and solar particles with the atmosphere. Ions are dominant causes of soft errors at altitudes higher than 6500 feet. Neutrons are dominant causes of soft error at lower altitudes and at the earth's surface [62-65]. A neutron experiences one of several potential nuclear reactions when it strikes a nucleus in a semiconductor. These reactions include elastic scattering, inelastic scattering, charged-particle-producing reactions, neutron-producing reactions, and fission. Each product from these nuclear reactions deposits its kinetic energy and directly ionizes material. Neutrons can cause the soft errors by indirect ionization as a result of these reactions.

A schematic diagram for soft-error production due to energetic particles is given in Figure 1.4. The ratio of soft-error occurrence of 0.18 μm 8 Mb SRAM by energetic nuclear particles is summarized in Figure 1.5 [66]. The soft-error rate induced by thermal neutrons is approximately three times larger than that induced by high-energy neutrons. Soft errors manifest themselves as various types of mode, including single-bit upset (SBU), multi-bit upset (MBU), single-event interrupt functional interrupt (SEFI), and single-event transient (SET) [67–69]. Energetic particles cause SBU or MBU when they strike a memory cell in SRAM, DRAM, FPGA, and microprocessors. SBU refers to flipping of one bit in a memory element whereas MBU means flipping of two or more bits in a memory element.

The energetic particles can also lead to interruption of normal operation of the affected device, SEFI. Various SEFI phenomena have been described, including inadvertent execution of built-in-test modes in DRAM, unusual data output patterns in EEPROM, halts or idle operations in microprocessors, and halts in analog to digital converters [68]. SET indicates transient transition of voltage or current occurring at a particular node in a combinational logic circuit of SRAM, DRAM, FPGA, and microprocessors when an energetic particle strikes the node. SET propagates through the subsequent circuit along the logic paths to the memory element and causes a soft error under special conditions. The rate at which a component experiences soft errors is called the soft-error rate (SER) and is treated as a constant. SER is generally expressed as number of failures-in-time (FIT). One FIT is equivalent to 1 error per billion hours of component operation.

The SER is estimated by accelerated test, field test, and simulation. An accelerated test is based on exposing the tested component to a specific radiation source whose intensity is much higher than the environmental radiation level, using a high-energy neutron or proton beam [70]. The results obtained from accelerated testing is extrapolated to estimate the actual SER. The field test is a method that tests a large number of components exposed to environmental radiation for enough time to measure actual SER confidently [71]. Another way to estimate SER is numerical computation (simulation) based on mathematical and statistical models of physical mechanisms of soft errors [72]. The field test method requires a long time and many specimens to obtain significant data, although it is the most accurate way to estimate the actual SER of the component. The accelerated test can obtain SER in a short time compared with the field test. This method, however, requires a testing facility to produce a high-energy neutron or proton beam and extrapolated calculation from accelerated test results to estimate the actual SER. SER estimation using a simulation technique is easy because it needs only a computer and a simulation code. But, the accuracy of the SER calculated from a simulation code depends on how well the mathematical model reflects the physical mechanisms of soft errors. This technique also needs input data, such as environmental radiation flux, energy distribution and component structure. The

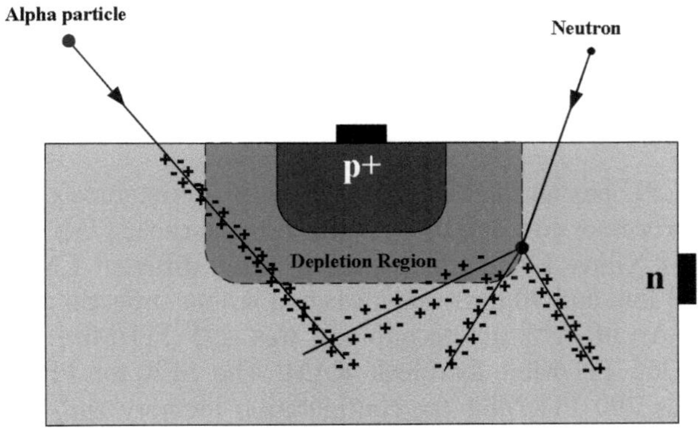

Figure 1.4. Soft-error mechanisms induced by energetic particles

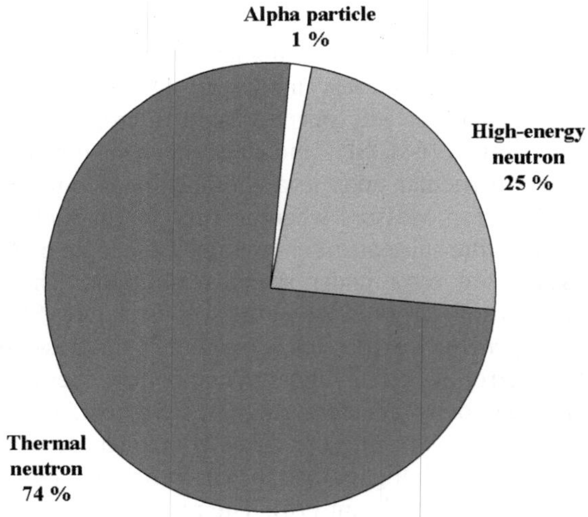

Figure 1.5. Ratio of the SERs of 0.18 μm 8 Mb SRAM induced by various particles [66]

inaccuracy of the mathematical model and input data can produce results that deviate from the actual SER.

The SER of the semiconductor has a wide variation depending on manufacture, technology generation and environmental radiation level. Nine SRAMs sampled from three vendors were tested to examine the neutron-induced upset and latch-up trends in SRAM using the accelerated testing method [73]. The SRAMs were fabricated in three different design technologies, full CMOS 6-transistor cell design, thin-film transistor-loaded 6-transistor cell design, and polysilicon resistor-loaded 4-transistor cell design. The SER of SRAMs at sea level in New York City varied from 10 FIT/Mbit to over 1000 FIT/Mbit.

The SER of each type of DRAM in a terrestrial cosmic ray environment with hadrons (neutrons, protons, pions) from 14 to 800 MeV has been reported [74]. This experiment included 26 different 16 Mb DRAMs from nine vendors. The DRAMs were classified into three different types according to cell technologies for bit storage: stacked capacitors (SC), trenches with internal charge (TIC), and trenches with external charge (TEC). TEC DRAMs had an SER ranging from 1300 FIT/Mbit to 1500 FIT/Mbit and SC DRAMs had an SER ranging from 110 FIT/Mbit to 490 FIT/Mbit. TIC DRAMs had an SER ranging from 0.6 FIT/Mbit to 0.8 FIT/Mbit.

A typical CPU has an SER of 4000 FIT with approximately half of the errors affecting the processor core and the rest affecting the cache [75].

An SER of Xilinx FPGAs fabricated in three different CMOS technologies (0.15 μm, 0.13 μm, and 90 nm) was measured at four different altitudes [71]. The SER for FPGAs of 0.15 μm technology was 295 FIT/Mbit for configuration memory and 265 FIT/Mbit for block RAM. The SER for FPGAs of 0.13 μm technology was 290 FIT/Mbit for configuration memory and 530 FIT/Mbit for block RAM.

Not every soft error in electronic components cause a failure because some types of soft errors are masked and eliminated by system dynamic behavior, such as hardware/software interaction and error handling. Error-handling techniques are generally implemented at any level in the system hierarchy and are realized by hardware, software, or both. Error-handling techniques include error detection codes, error correction codes, self-checking logic, watchdog timers, processor-monitoring techniques, voting, and masking. Typical error detection codes are parity codes, checksums, arithmetic codes, and cyclic codes. Error correction codes are single-error correction (SEC) codes, multiple-error correction codes, burst-error correction codes, and arithmetic-error correction codes.

Error-handling techniques only detect or correct special types of soft errors because of their limited ability. These techniques cannot eliminate all soft errors but reduce the opportunity that soft errors cause system malfunction. For example, SBU is detected and corrected if it occurs on memory that has the capability of SEC code because SEC codes have the ability to detect and correct SBU. However, MBU on memory can escape the protection by SEC code and cause the system malfunction because SEC code cannot detect and correct MBU.

There is a need to quantify the effectiveness of error-handling techniques for estimating how many soft errors in components will cause system failures. The concept of coverage is borrowed and used for this object [76]. The mathematical definition of coverage is the condition probability that an error is processed correctly given that an error has occurred. It is written as:

$$\text{Coverage} = \Pr\{\text{error processed correctly} \mid \text{error existence}\} \tag{1.11}$$

Therefore, the transient failure of the electronic component can be modeled:

$$h_{\text{transient}}(t) = \lambda_{\text{transient}} = (1 - \text{Coverage}) \times SER \tag{1.12}$$

It is difficult to analytically model coverage because it is a complex and laborious task. Fault injection experiments are commonly used for estimating the coverage parameter. These fault injection techniques are placed in three general categories: physical fault injection, software fault injection, and simulated fault injection [77, 78]. Physical fault injection is a method that injects faults directly into physical hardware and monitors the results of fault injection. The mostly frequently used means for generating faults in hardware are to use heavy ion radiation, modify the value of integrated circuit pins, use electromagnetic interferences, or use a laser. Software fault injection reproduces errors at the software level that occurs in either hardware or software. Fault injection is performed by modifying the content of memory or register to emulate hardware or software faults. Simulation-based fault injection is based on the simulation model which represents the architecture or behavior of the component at a variety of levels of abstraction ranging from the transistor level to an instruction level. The simulation model is constructed using hardware description language, such as VHDL, Verilog, system C, or SPICE simulation language.

The physical fault injection technique supplies more realistic results. But, this requires special hardware equipment for instrumentation, and interfaces to the target system. The types of injected faults cannot be controlled. The software-based fault injection technique provides low cost and is easy to control faults. But, this technique concentrates on software errors rather than hardware errors. A simulation technique can easily control the injected fault types and provide early checks in the design process of fault handling techniques, whereas modeling the component and error handling techniques is laborious.

Soft errors are related to technology advances and environmental conditions. The components with higher density, higher complexity, and lower power consumption are being developed as the results of technology advances, making components more vulnerable to soft errors. Additionally, many studies indicate that SER generally exceeds the occurrence rate of other failure modes, including intermittent and permanent failures.

1.5 Concluding Remarks

Electronics reliability has a history of almost six decades from the early 1950s when the reliability of electronic components and systems attracted great interest after recognizing their importance in military application after World War II. There still exist many issues to be resolved.

This chapter reviews current studies and issues related to electronics reliability which are classified into three categories according to the time duration characteristic of failure: permanent, intermittent, and transient. Permanent failure is persistent, consistent, and reproducible and continues to exist indefinitely. Intermittent failure appears for a specified time interval and disappears for a specified time interval repeatedly. Transient failure occurs once and then disappears within a short period of time.

Reliability models for permanent failures have been studied and continue to be controversial compared with intermittent and transient failure modes. Empirical and physics-of-failure models have been used to predict permanent failure. Empirical models are based on historical failure data and tests. But empirical models have been criticized due to intrinsic problems, such as an unrealistic constant failure rate assumption, the lack of accuracy, out of date information, and lack of information about failure modes and mechanisms. Permanent failure models based on the physics-of-failure approach generally predict the time to failure by analyzing root-cause failure mechanisms that are governed by fundamental mechanical, electrical, thermal, and chemical processes. These models also have problems, such as difficulty in obtaining input parameters for the model, difficulty in applying multiple failure models for one component reliability prediction, and limitations and availability of failure models.

Intermittent failure is caused by component defects, such as structural defects, bad electrical contact, loose connections, component aging, and IC chip contamination. Few studies have analyzed and modeled intermittent failures of electronic components (or circuits) from a perspective of reliability because of difficulty in detecting and analyzing the component intermittent failures, although there is sufficient evidence that intermittent failures of components (or circuit) have critical effects on digital system reliability and safety. Studies of intermittent failures have mainly focused on electrical contacts.

Transient failures of electronic components are caused by atmospheric nuclear particles, electromagnetic interference, radiation from lightning, power supply noise, thermal hot spots, and active voltage. Many research projects have focused on transient failures caused by energetic nuclear particles since nuclear particles are the main cause of transient failures in semiconductors. Transient failure is generally estimated by accelerated test, field test, and simulation. The accelerated test is based on exposing components to a specific radiation source whose intensity is much higher than environmental radiation level using a high-energy neutron or proton beam. The field test evaluates a large number of components at environmental radiation levels for sufficient time to confidently measure actual SER. Another way to estimate SER is numerical computation (simulation) which is based on mathematical and statistical models of physical mechanisms of soft errors. Many studies indicate that SER generally exceeds the occurrence rate of other failure modes, including intermittent and permanent failures.

References

[1] Wikipedia, http://en.wikipedia.org/wiki/Electronics

[2] IEEE Std. 1413.1 (2003) IEEE Guide for Selecting and Using Reliability Predictions Based IEEE 1413, February

[3] Siewiorek DP and Swarz RS (1998) Reliable Computer Systems: Design and Evaluation, pub. A K Peters, Ltd.

[4] Modarres M, Kaminskiy M, and Krivtsov V (1999) Reliability Engineering and Risk Analysis: Practical Guide, pub. Marcel Dekker, Inc.

[5] MIL-HDBK-217F N2 (1995) Reliability Prediction of Electronic Equipment, February

[6] Denson W (1998) The History of Reliability Prediction, IEEE Transactions on Reliability, Vol. 47, No. 3-SP, pp. 321–328

[7] RIAC-HDBK-217 Plus (2006) Handbook of 217Plus Reliability Prediction Models

[8] Wong KL (1981) Unified Field (Failure) Theory – Demise of the Bathtub Curve, Proceedings of the Annual Reliability and Maintainability Symposium, pp. 402–407

[9] Wong KL and Linstrom DL (1988) Off the Bathtub onto the Roller-Coaster Curve, Annual Reliability and Maintainability Symposium, pp. 356–363

[10] Jensen F (1989) Component Failures Based on Flow Distributions, Annual Reliability and Maintainability Symposium, pp. 91–95

[11] Klutke GA, Kiessler PC, and Wortman MA (2003) A Critical Look at the Bathtub Curve, IEEE Transactions on Reliability, Vol. 52, No. 1, pp. 125–129

[12] Brown LM (2003) Comparing Reliability Predictions to Field Data for Plastic Parts in a Military, Airborne Environment, Annual Reliability and Maintainability Symposium

[13] Wood AP and Elerath JG (1994) A Comparison of Predicted MTBF to Field and Test Data, Annual Reliability and Maintainability Symposium

[14] Pecht MG, Nash FR (1994) Predicting the Reliability of Electronic Equipment, Proceedings of IEEE, Vol. 82, No. 7, pp. 992–1004

[15] Pecht MG (1996) Why the Traditional Reliability Prediction Models Do Not Work – Is There an Alternative?, Electron. Cooling, Vol. 2, No. 1, pp. 10–12

[16] Evans J, Cushing MJ, and Bauernschub R (1994) A Physics-of-Failure (PoF) Approach to Addressing Device Reliability in Accelerated Testing of MCMS, Multi-Chip Module Conference, pp. 14–25

[17] Moris SF and Reillly JF (1993) MIL-HDBK-217-A Favorite Target, Annual

Reliability and Maintainability Symposium, pp. 503–509
[18] Lall P (1996) Tutorial: Temperature as an Input to Microelectronics Reliability Models, IEEE Transactions on Reliability, Vol. 45, No. 1, pp. 3–9
[19] Hakim EB (1990) Reliability Prediction: Is Arrhenius Erroneous, Solid State Technology
[20] Bowles JB (1992) A Survey of Reliability-Prediction Procedures for Microelectronic Devices, IEEE Transactions on Reliability, Vol. 41, No. 1, pp. 2–12
[21] Jones J and Hayes J (1999) A Comparison of Electronic-Reliability Prediction Models, IEEE Transactions on Reliability, Vol. 48, No. 2, pp. 127–134
[22] Ebel GH (1998) Reliability Physics in Electronics: A Historical View, IEEE Trans on Reliability, Vol. 47, No. 3-SP, pp. sp-379–sp389
[23] Blish R, Durrant N (2000) Semiconductor Device Reliability Failure Models, International SEMATECH, Inc.
[24] Haythornthwaite R (2000) Failure Mechanisms in Semiconductor Memory Circuits, IEEE International Workshop on Memory Technology, Design and Testing, pp. 7–13
[25] Black JR (1983) Physics of Electromigration, IEEE Proceedings of the International Reliability and Physics Symposium, pp. 142–149
[26] Kato M, Niwa H, Yagi H, and Tsuchikawa H (1990) Diffusional Relaxation and Void Growth in an Aluminum Interconnect of Very Large Scale Integration, Journal of Applied Physics, Vol. 68, pp. 334–338
[27] Chen IC, Holland SE, and Hu C (1985) Electrical Breakdown in Thin Gate and Tunneling Oxides, IEEE Transactions on Electron Devices, ED-32, p. 413.
[28] Manson S (1966) Thermal Stress and Low Cycle Fatigue, McGraw-Hill, New York
[29] Peck D (1986) IEEE International Reliability Physics Symposium Proceedings, p. 44
[30] Mortin DE, Krolewski JG, and Cushing MJ (1995) Consideration of Component Failure Mechanisms in the Reliability Assessment of Electronic Equipment – Addressing the Constant Failure Rate Assumption, Annual Reliability and Maintainability Symposium, pp. 54–59
[31] Maia Filho WC, Brizoux M, Fremont H, and Danto Y (2006) Improved Physical Understanding of Intermittent Failure in Continuous Monitoring Method, Microelectronics Reliability, Vol. 46, pp. 1886–1891
[32] Skinner DW (1975) Intermittent Opens in Electrical Contacts Caused by Mechanically Induced Contact Motion, IEEE Transactions on Parts, Hybrids, and Package, Vol. PHP-11, No. 1, pp. 72–76
[33] Kulwanoski G, Gaynes M, Smith A, and Darrow B (1991) Electrical Contact Failure Mechanisms Relevant to Electronic Package, Proceedings of the 37th IEEE Holm Conference on Electrical Contacts, pp. 288–292
[34] Lin TY and Siewiorek DP (1990) Error Log Analysis: Statistical Modeling and Heuristic Trend Analysis, IEEE Transactions on Reliability, Vol. 39, No. 4, pp. 419–432
[35] Abbott W (1984) Time Distribution of Intermittents Versus Contact Resistance for Tin-Tin Connector Interfaces During Low Amplitude Motion, IEEE Transactions on Components, Hybrids, and Manufacturing Technology, Vol. CHMT-7, No. 1, pp. 107–111
[36] Malucci RD (2006) Stability and Contact Resistance Failure Criteria, IEEE Transactions on Component and Packaging Technology, Vol. 29, No. 2, pp. 326–332
[37] Wang N, Wu J, and Daniel S (2005) Failure Analysis of Intermittent Pin-To-Pin Short Caused by Phosphorous Particle in Molding Compound, 43rd Annual International Reliability Physics Symposium, pp. 580–581
[38] Kyser EL (1997) Qualification of Surface Mount Technologies for Random Vibration in Environmental Stress Screens, Annual Reliability and Maintainability Symposium, pp. 237–241

[39] Raver N (1982) Thermal Noise, Intermittent Failures, and Yield in Josephson Circuits, IEEE Journal of Solid-State Circuits, Vol. SC-17, No. 5, pp. 932–937
[40] Swingler J and McBride JW (2002) Fretting Corrosion and the Reliability of Multicontact Connector Terminals, IEEE Transactions on Components and Packaging Technologies, Vol. 25, No. 24, pp. 670–676
[41] Seehase H (1991) A Reliability Model for Connector Contacts, IEEE Transactions on Reliability, Vol. 40, No. 5, pp. 513–523
[42] Soderholm R (2007) Review: A System View of the No Fault Found (NFF) Phenomenon, Reliability Engineering and System Safety, Vol. 92, pp. 1–14
[43] James I, Lumbard K, Willis I, and Globe J (2003) Investigating No Faults Found in the Aerospace Industry, Proceedings of Annual Reliability and Maintainability Symposium, pp. 441–446
[44] Steadman B, Sievert S, Sorensen B, and Berghout F (2005) Attacking "Bad Actor" and "No Fault Found" Electronic Boxes, Autotestcon, pp. 821–824
[45] Contant O, Lafortune S, and Teneketzis D (2004) Diagnosis of Intermittent Faults, Discrete Event Dynamic Systems: Theory and Applications, Vol. 14, pp. 171–202
[46] Bondavlli A, Chiaradonna S, Giandomeenico FD, and Grandoni F (2000) Threshold-Based Mechanisms to Discriminate Transient from Intermittent Faults, IEEE Transactions on Computers, Vol. 49, No. 3, pp. 230–245
[47] Ismaeel A, and Bhatnagar R (1997) Test for Detection & Location of Intermittent Faults in Combinational Circuit, IEEE Transactions on Reliability, Vol. 46, No. 2, pp. 269–274
[48] Chung K (1995) Optimal Test-Times for Intermittent Faults, IEEE Transactions on Reliability, Vol. 44, No. 4, pp. 645–647
[49] Spillman RJ (1981) A Continuous Time Model of Multiple Intermittent Faults in Digital Systems, Computers and Electrical Engineering, Vol. 8, No. 1, pp. 27–40
[50] Savir J (1980) Detection of Single Intermittent Faults in Sequential Circuits, IEEE Transactions on Computers, Vol. C-29, No. 7, pp. 673–678
[51] Roberts MW (1990) A Fault-tolerant Scheme that Copes with Intermittent and Transient Faults in Sequential Circuits, Proceedings on the 32nd Midwest Symposium on Circuits and Systems, pp. 36–39
[52] Hamilton SN, and Orailoglu A (1998) Transient and Intermittent Fault Recovery Without Rollback, Proceedings of Defect and Fault Tolerance in VLSI Systems, pp. 252–260
[53] Varshney PK (1979) On Analytical Modeling of Intermittent Faults in Digital Systems, IEEE Transactions on Computers, Vol. C-28, pp. 786–791
[54] Prasad VB (1992) Digital Systems with Intermittent Faults and Markovian Models, Proceedings of the 35th Midwest Symposium on Circuits and Systems, pp. 195–198
[55] Vijaykrishnan N (2005) Soft Errors: Is the concern for soft-errors overblown?, IEEE International Test Conference, pp. 1–2
[56] Baunman RC (2005) Radiation-Induced Soft Errors in Advanced Semiconductor Technologies, IEEE Transactions on Device and Material Reliability, Vol. 5, No. 3, pp. 305–316
[57] Dodd PE and Massengill LW (2003) Basic Mechanisms and Modeling of Single-Event Upset in Digital Microelectronics, IEEE Transactions on Nuclear Science, Vol. 50, No. 3, pp. 583–602
[58] May TC and Woods MH (1978) A New Physical Mechanism for Soft Error in Dynamic Memories, 16th International Reliability Physics Symposium, pp. 34–40
[59] Kantz II L (1996) Tutorial: Soft Errors Induced by Alpha Particles, IEEE Transactions on Reliability, Vol. 45, No. 2, pp. 174–178
[60] Ziegler JF (1996) Terrestrial Cosmic Rays, IBM Journal of Research and Development, Vol. 40, No. 1, pp. 19–39

[61] Ziegler JF and Lanford WA (1981) The Effect of Sea Level Cosmic Rays on Electronic Devices, Journal of Applied Physics, Vol. 52, No. 6, pp. 4305–4312
[62] Barth JL, Dyer CS, and Stassinopoulos EG (2003) Space, Atmospheric, and Terrestrial Radiation Environments, IEEE Transactions on Nuclear Science, Vol. 50, No. 3
[63] Siblerberg R, Tsao CH, and Letaw JR (1984) Neutron Generated Single Event Upsets, IEEE Transactions on Nuclear Science, Vol. 31, pp. 1066–1068
[64] Gelderloos CJ, Peterson RJ, Nelson ME, and Ziegler JF (1997) Pion-Induced Soft Upsets in 16 Mbit DRAM Chips, IEEE Transactions on Nuclear Science, Vol. 44, No. 6, pp. 2237–2242
[65] Petersen EL (1996) Approaches to Proton Single-Event Rate Calculations, IEEE Transactions on Nuclear Science, Vol. 43, pp. 496–504
[66] Kobayashi H, et al. (2002) Soft Errors in SRAM Devices Induced by High Energy Neutrons, Thermal Neutrons and Alpha Particles, International Electron Devices Meeting, pp. 337–340
[67] Quinn H, Graham P, Krone J, Caffrey M, and Rezgui S (2005) Radiation-Induced Multi-Bit Upsets in SRAM based FPGAs, IEEE Transactions on Nuclear Science, Vol. 52, No. 6, pp. 2455–2461
[68] Koga R, Penzin SH, Crawford KB, and Crain WR (1997) Single Event Functional Interrupt (SEFI) Sensitivity in Microcircuits, Proceedings 4th Radiation and Effects Components and Systems, pp. 311–318
[69] Dodd PE, Shaneyfelt MR, Felix JA, and Schwank JR (2004) Production and Propagation of Single Event Transient in High Speed Digital Logic ICs, IEEE Transactions on Nuclear Science, Vol. 51, No. 6, pp. 3278–3284
[70] JEDEC STANDARD (2006) Measurement and Reporting of Alpha Particle and Terrestrial Cosmic Ray-Induced Soft Errors in Semiconductor Devices, JESD89A.
[71] Lesea A, Drimer S, Fabula JJ, Carmichael C, and Alfke P (2005) The Rosetta Experiment: Atmospheric Soft Error Rate Testing in Differing Technology FPGAs, IEEE Transactions on Device and Materials Reliability, Vol. 5, No. 3, pp. 317–328
[72] Yosaka Y, Kanata H, Itakura T, and Satoh S (1999) Simulation Technologies for Cosmic Ray Neutron-Induced Soft Errors: Models and Simulation Systems, IEEE Transactions on Nuclear Science, Vol. 46, No. 3, pp. 774–780
[73] Dodd PE, Shaneyfelt MR, Schwank JR, and Hash GL (2002) Neutron-Induced Soft Errors, Latchup, and Comparison of SER Test Methods for SRAM Technologies, International Electron Devices Meeting, pp. 333–336
[74] Ziegler JF, Nelson ME, Shell JD, Peterson RJ, Gelderloos CJ, Muhlfeld HP, and Nontrose CJ (1998) Cosmic Ray Soft Error Rates of 16-Mb DRAM Memory Chips, IEEE Journal of Solid State Circuits, Vo. 33, No. 2, pp. 246–252
[75] Messer A et al. (2001) Susceptibility of Modern Systems and Software to Soft Errors, In Hp Labs Technical Report HPL-2001-43, 2001
[76] Kaufman JM and Johnson BW (2001) Embedded Digital System Reliability and Safety Analysis, NUREG/GR-0020.
[77] Kim SJ, Seong PH, Lee JS, et al (2006) A Method for Evaluating Fault Coverage Using Simulated Fault Injection for Digitized Systems in Nuclear Power Plants, Reliability Engineering and System Safety, Vol. 91, pp. 614–623
[78] Alert J, Crouzet Y, Karlsson J, Folkesson P, Fuchs E, Leber GH (2003) Comparison of Physical and Software Implemented Fault Injection Techniques, IEEE Transactions on Computers, Vol. 52, No. 9, pp. 1115–1133

2

Issues in System Reliability and Risk Model

Hyun Gook Kang

Integrated Safety Assessment Division
Korea Atomic Energy Research Institute
1045 Daedeok-daero, Yuseong-gu, Daejeon, 305-353, Korea
hgkang@kaeri.re.kr

The application of large-scale digital or computer systems involves many components, elements, and modules. System reliability and safety need to be calculated no matter how complicated is the structure. Estimation of system reliability/safety provides useful information for system design and verification. Risk allocation to the designed system in a balanced manner is an application example.

The most conservative method for estimating system failure probability is summing up failure probabilities of components. The result of this conservative calculation equals system failure probability for a series of independent components. Reliability of a series of components is the lower boundary of system reliability. Redundant components and standby components are considered in order to estimate realistic reliability. A module consists of many components and a system consists of several kinds of modules. An analytic calculation of the reliability or risk based on the information of module structure and component reliabilities is relatively simple. An analytic model for the reliability or risk estimates at the system level is complex and sometimes difficult to develop.

Conventional methodologies are used with rough assumptions if the reliability modeling is for a decision which does not require high accuracy, such as the determination of the number of spare modules for a non-safety-critical system. Even the use of the lower boundary value of system reliability is possible for simplicity.

More complicated relationships among system functions should be modeled for the accurate and realistic estimation of risk from safety-critical systems. The faults in an advanced digital system are monitored by a self-monitoring algorithm and recovered before the fault causes system failure. Protecting a system from catastrophic damage is possible even though it is not possible to recover the fault. Multiple-channel processing systems might have cross-monitoring functions and independent heartbeat-monitoring equipment can also be installed in the systems. Intelligence and flexibility from microprocessors and software successfully accommodates these sophisticated reliability-enhancing mechanisms.

Clarification and definition of reliability or risk-modeling objectives are very important because the hazard state varies along this definition. Failure of a status indication lamp or trouble in a cabinet lock button is considered as part of system failure for maintenance purposes. Faults which do not disturb shutdown signal generation will not be considered for risk estimation of safety-critical systems, such as an automatic shutdown system in a nuclear power plant.

Modern digital technologies are expected to significantly improve both economical efficiency and safety of plants owing to the general progress of instrumentation and control (I&C) technologies for process engineering, such as computer technology, control engineering, data processing and transfer technology, and software technology. The assessment of digital system safety becomes a more sensitive issue rather than maintenance reliability, when a digital technology is applied to a safety-critical function.

Economical efficiency improvement due to digital applications seems clear, while safety improvement is not well accepted. There are still many arguable safety issues, even though the use of digital equipment for safety-critical functions provides many advantageous features.

Digital signal-processing system unavailability severely affects total plant safety because many safety signals are generated by the same digital system [1]. A sensitivity study showed that the protection and monitoring system is the most important system for the safety of a Westinghouse AP1000 nuclear power plant [2]. The potential impact of digital system malfunctions on core damage frequency in the advanced boiling water reactor is high [3].

Safety assessment, based on unavailability estimation, can quantitatively show improvement because it demonstrates that a balanced design has been achieved, by showing that no particular class of accident of the system makes a disproportionate contribution to overall risk. The importance of probabilistic risk assessment (PRA) for safe digital applications is pointed out in the HSE guide [4]. The PRA of digital safety-critical systems plays the role of a decision-making tool and has sufficient accuracy.

Characteristics of digital systems from a safety assessment viewpoint are:

- The utilization of hardware is determined by software and inputs.
- The system is multi-purpose.
- The failure modes are not well defined.
- Software might hide the transient faults of hardware.
- Software fails whenever it executes the faulty part of the code.
- Greater effort in the management of software quality can cause a lower expectation for software failure in the operation phase, while quantification is still challenging.
- Various monitoring and recovery mechanisms are adopted, but their coverage is not well defined.
- Apparently different components might cause common-cause failure (CCF) because electronic components consist of a lot of small modules, which are manufactured in a globally standardized environment.
- Digital systems are more sensitive to environmental conditions, such as ambient temperature, than conventional analog systems.
- There might be no warning to operators when a system fails.

- The system failure might cause the blockage of safety-critical information from field to operators.
- New initiating events may be induced by digital system failure.

Many assumptions are used for quantitative analysis of a digital I&C system. Some are intentionally used for analysis simplicity, while others are caused by failing to show enough caution. Unreasonable assumptions result in unreasonable safety evaluation. Fault-free software and perfect coverage of fault tolerance mechanisms are typical examples of unreasonable assumptions.

The characteristics of digital applications are different from those of conventional analog I&C systems, because their basic elements are microprocessors and software, which make the system more complex to analyze. There are important issues in digital system safety analysis which are complicated and correlated [5]. Several system reliability and risk estimation methodologies and important issues related to the reliability and safety modeling of large digital systems are described in this chapter. A brief introduction to these methodologies for reliability calculation or hazard identification is given in Section 2.1. The related issues are categorized into six groups from the viewpoint of PRA: the modeling of the multi-tasking of digital systems, the estimation of software failure probability, the evaluation of fault tolerance mechanisms, the assessment of network safety, the assessment of human failure probability, and the assessment of CCF (Sections 2.2 to 2.7).

2.1 System Reliability Models

Failure of either component will result in system failure for a system composed of two independent modules, in the same way as for a component reliability model. The system is represented by a series of blocks, as shown in a reliability block diagram (RBD) (Figure 2.1(a)). If λ_1 and λ_2 are the hazard rates of the two modules, the system hazard rate will be $\lambda_1 + \lambda_2$. The reliability of the system is the combined probability of no failure of either modules: $R_1R_2=\exp[-(\lambda_1 + \lambda_2)t]$. In the case of *s*-independent modules, for a series of *n* modules:

$$R = \prod_{i=1}^{n} R_i \tag{2.1}$$

This is the simplest basic model. Parts count reliability prediction is based on this model. The failure logic model of the system in real applications is more complex, if there are redundant subsystems or components.

The Markov model is a popular method for analyzing system status. The Markov model provides a systematic method for analysis of a system which consists of many modules and adopts a complex monitoring mechanism. The Markov model is especially useful for a more complicated transition among the system states or the repair of the system. A set of states and probabilities that a system will move from one state to another must be specified to build a Markov model. Markov states represent all possible conditions the system can exist in. The system can only be in one state at a time. A Markov model of the series system is shown in Figure 2.1(b). State *S0* is an initial state. States *S1* and *S2* represent the state of module 1 failure and module 2 failure, respectively. Both are defined as hazard states.

Fault tree modeling is the most familiar tool for analysis staff, whose logical structure makes it easy for system design engineers to understand models. A fault tree is a top-down symbolic logic model generated in the failure domain. That is, a fault tree represents the pathways of system failure. A fault tree analysis is also a p owerful diagnostic tool for analysis of complex systems and is used as an aid for de sign improvement.

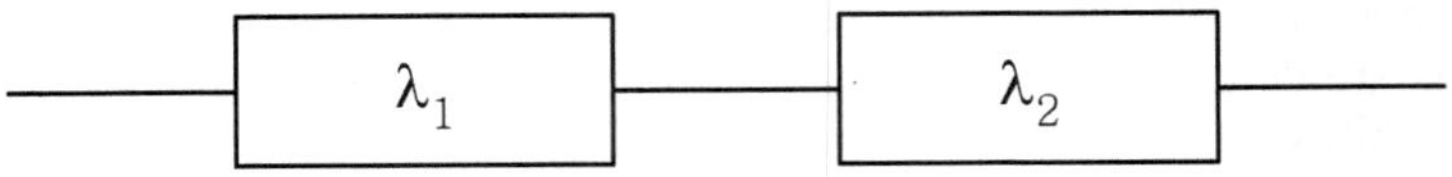

(a) Reliability block diagram

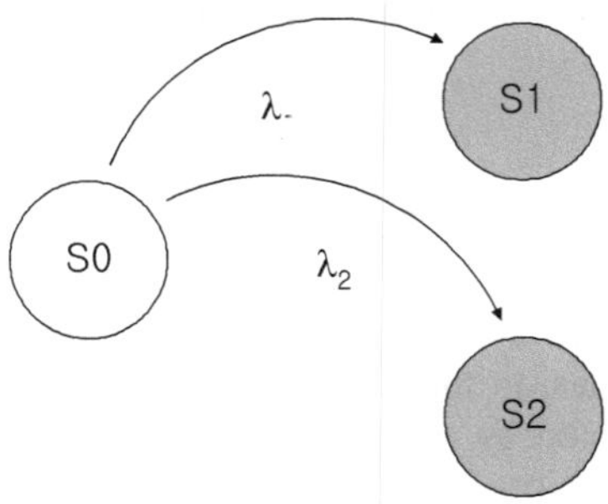

(b) Markov model

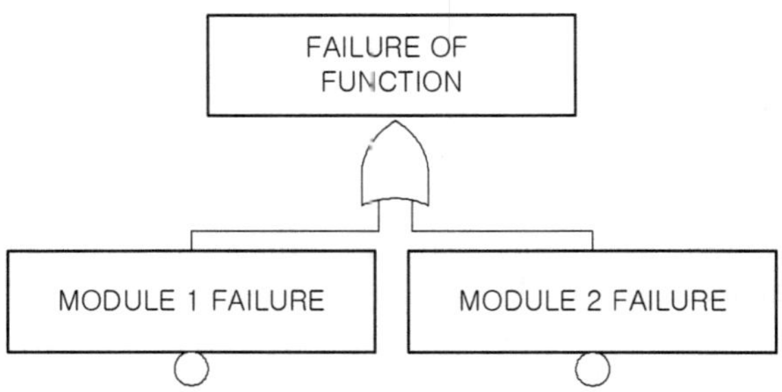

(c) Fault tree model

Figure 2.1. Series system

The analyst repeatedly asks, "What will cause a given failure to occur?" in using backwards logic to build a faulttree model. The analyst views the system from a top-down perspective. This means he starts by looking at a high-level system failure and proceeds down into the system to trace failure paths. Fault trees are generated in the failure domain, while reliability diagrams are generated in the success domain. Probabilities are propagated through the logic models to determine the probability that a system will fail or the probability the system will operate successfully (*i.e.*, the reliability). Probability data may be derived from available empirical data or found in handbooks.

Fault tree analysis (FTA) is applicable both to hardware and non-hardware systems and allows probabilistic assessment of system risk as well as prioritization of the effort based upon root cause evaluation. An FTA provides the following advantages [6]:

1. Enables assessment of probabilities of combined faults/failures within a complex system.
2. Single-point and common-cause failures can be identified and assessed.
3. System vulnerability and low-payoff countermeasures are identified, thereby guiding deployment of resources for improved control of risk.
4. This tool can be used to reconfigure a system to reduce vulnerability.
5. Path sets can be used in trade studies to compare reduced failure probabilities with increases in cost to implement countermeasures.

2.1.1 Simple System Structure

The probability of failure (P) for a given event is defined as the number of failures per number of attempts, which is the probability of a basic event in a fault tree. The sum of reliability and failure probability equals unity. This relationship for a series system can be expressed as:

$$\begin{aligned} P &= P_1 + P_2 - P_1 P_2 \\ &= (1 - R_1) + (1 - R_2) - (1 - R_1)(1 - R_2) \\ &= 1 - R_1 R_2 \\ &= 1 - R \end{aligned} \tag{2.2}$$

The reliability model for a dual redundant system is expressed in Figure 2.2. Two s-independent redundant modules with reliability of R_1 and R_2 will successfully perform a system function if one out of two modules is working successfully. The reliability of the dual redundant system, which equals the probability that one of modules 1 or 2 survives, is expressed as:

$$\begin{aligned} R &= R_1 + R_2 - R_1 R_2 \\ &= e^{-\lambda_1 t} + e^{-\lambda_2 t} - e^{-(\lambda_1 + \lambda_2)t} \end{aligned} \tag{2.3}$$

This is often written as:

$$R = 1 - (1 - R_1)(1 - R_2) \tag{2.4}$$

$$1 - R = (1 - R_1)(1 - R_2) \tag{2.5}$$

In the case of *s*-independent modules, for *n* redundant modules, the reliability of a system is generally expressed as:

$$R = 1 - \prod_{i=1}^{n} (1 - R_i) \tag{2.6}$$

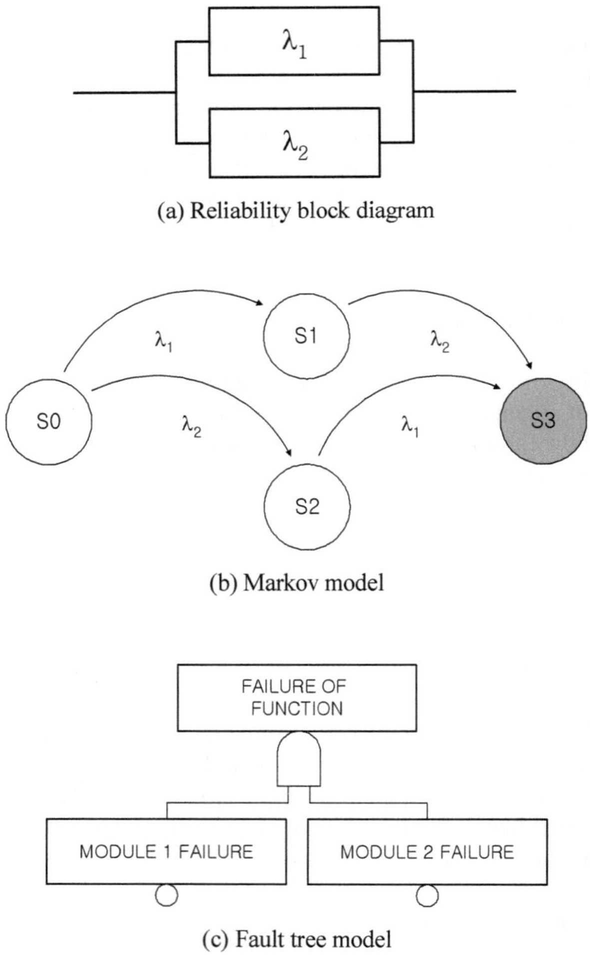

Figure 2.2. Dual redundant system

2.1.2 Complicated System Structure

Not all systems can be modeled with simple RBDs. Some complex systems cannot be modeled with true series and parallel branches. Module 2 monitors status information from module 1 and module 2 automatically takes over the system function when an erroneous status of module 1 is detected in a more complicated system. The system is conceptually illustrated in Figure 2.3.

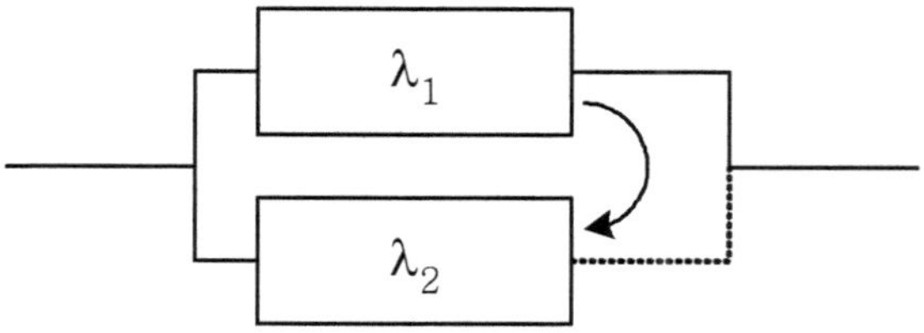

Figure 2.3. Standby and automatic takeover system

In this case, using a successful takeover probability of the module 2, μ, the reliability of the system is generally expressed as:

$$\begin{aligned} 1-R &= (1-R_1)\{(1-R_2)+(1-\mu)-(1-R_2)(1-\mu)\} \\ &= (1-R_1)\{(1-R_2)\mu+(1-\mu)\} \end{aligned} \tag{2.7}$$

The Markov model is shown in Figure 2.4. A fault tree is shown in Figure 2.5.

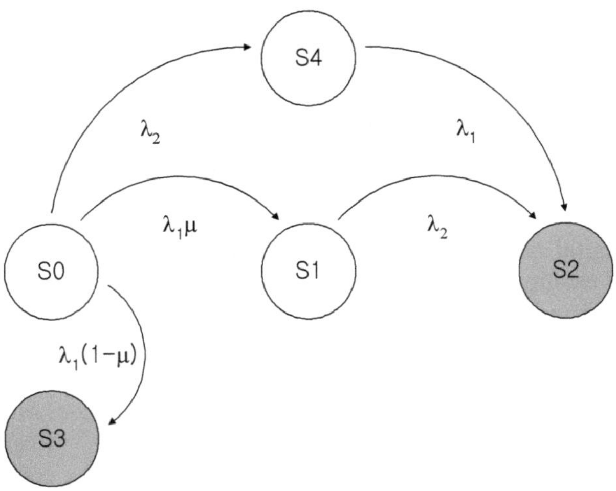

Figure 2.4. Markov model for standby and automatic takeover system

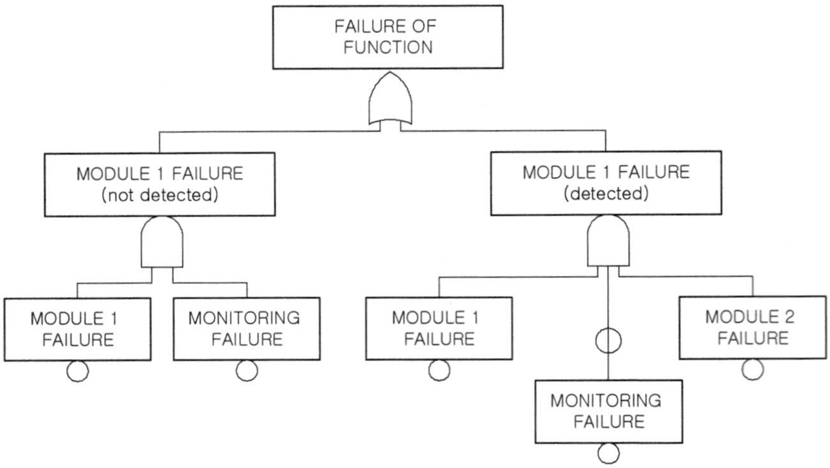

Figure 2.5. Fault tree for standby and automatic takeover system

2.2 Modeling of the Multi-tasking of Digital Systems

2.2.1 Risk Concentration

Microprocessors and software technologies make the digital system multi-functional because a system performs several sequential or conditional functions. This multi-tasking feature is represented in safety assessment because it will cause risk concentration and deteriorate the reliability of the system.

The use of a single microprocessor module for multiple safety-critical functions will cause severe concentration of risk in the single microprocessor. Safety-critical applications have adopted a conservative design strategy, based on functional redundancies. However, software programs of these functions are executed by one microprocessor in the case of digital systems. The effects of multi-tasking on safety should be carefully modeled and evaluated in order to compare the developed digital system with the conventional analog system.

A typical example for finding two ways of handling diverse process parameters and functional redundancy is shown in Figure 2.6, when considering the main steam line break accident in a nuclear power plant. Several parameters affected by this accident will move to an abnormal region. First, the "Low steam generator pressure" parameter triggers the output signal A. As time goes on, the parameters of "Low pressurizer pressure," "Low steam generator level," "Reactor overpower" will trigger the output signals B, C, and D, respectively. In a conventional analog circuit system (Figure 2.6(a)), the first triggered signal A makes trip circuit breakers open and initiates reactor shutdown. Signals B, C, and D are sequentially

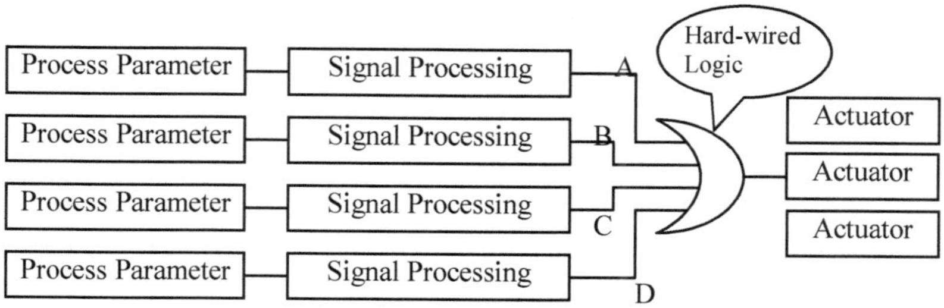

(a) Typical process of signal processing using conventional analog

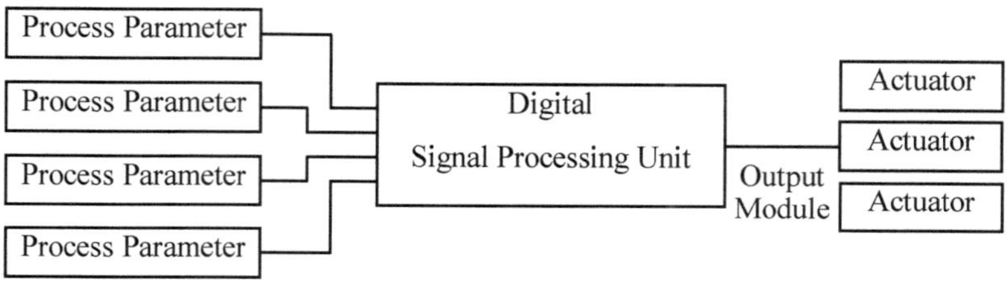

(b) Typical process of signal processing using digital units

Figure 2.6. Schematic diagram of signal processing using analog circuit and digital processor unit

generated if the signal processing circuits for parameter A fail. However, parameters A, B, C, and D use the same equipment for signal-processing in the case of digital system (Figure 2.6 (b)). There is no functional backup if the digital signal-processing unit fails.

The risk concentration on a processing unit is demonstrated in Figure 2.7 by fault trees for the systems in Figure 2.6. Component reliabilities should be carefully analyzed. Self-monitoring and fault-tolerant mechanisms for these components should be strengthened in the design phase to improve system reliability.

There are two or more duplicated trip channels in safety-critical applications that are not functional backups and are vulnerable to the CCF. The dominant contributor to system unavailability is the CCF of digital modules in 2-out-of-3 trains voting logic (Figure 2.8). The importance of precise estimation of digital equipment CCF should be emphasized. Products from different vendors do not guarantee the independence of faults, since global standardization and the large manufacturer in the electronic parts market lead to similar digital hardware products.

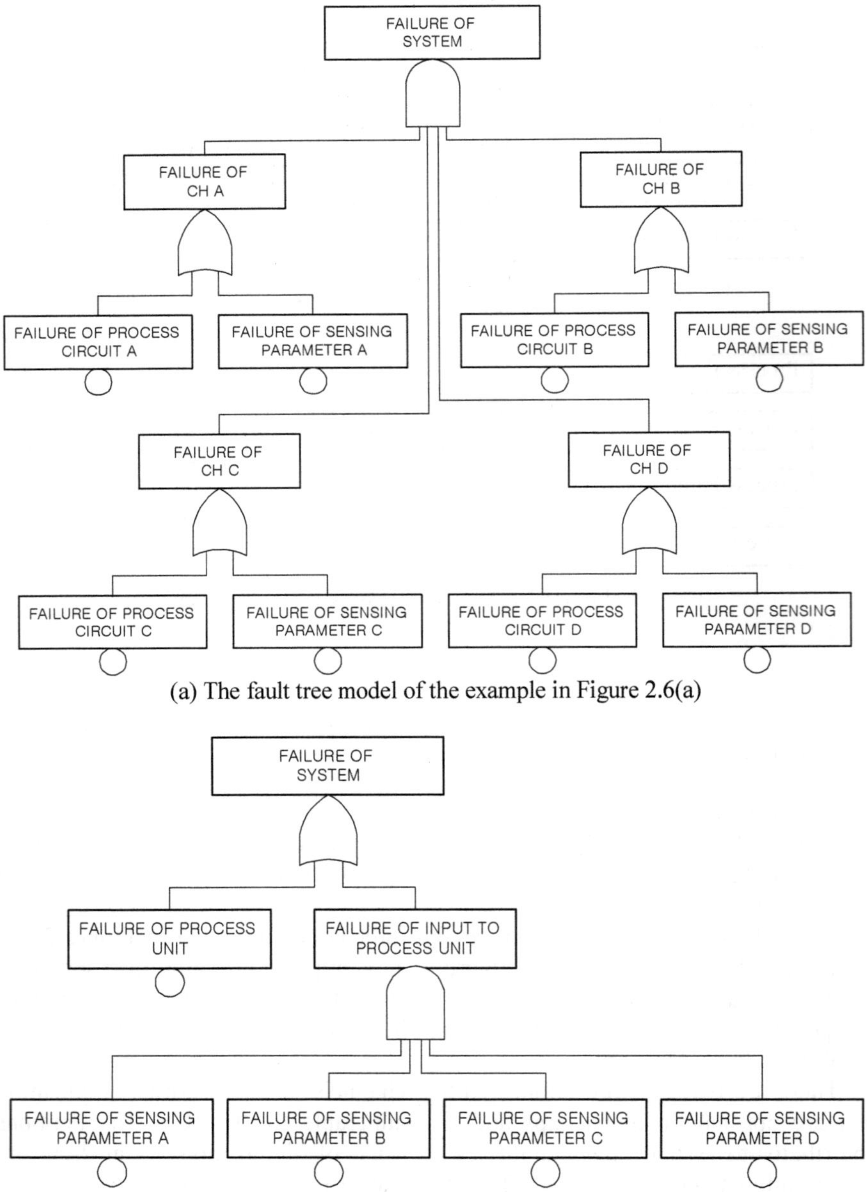

(a) The fault tree model of the example in Figure 2.6(a)

(b) The fault tree model of the example in Figure 2.6(b)

Figure 2.7. The fault trees for the systems shown in Figure 2.6

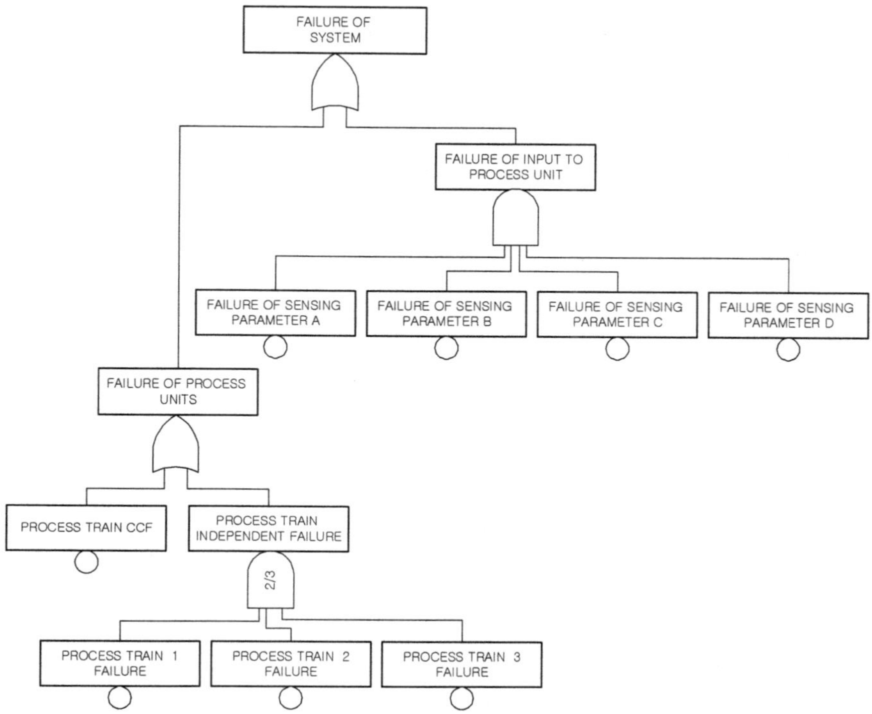

Figure 2.8. The fault tree model of a three-train signal-processing system which performs 2-out-of-3 auctioneering

2.2.2 Dynamic Nature

Static modeling techniques, such as a classical event tree and a fault tree, do not simulate the real world without considerable assumptions, since the real world is dynamic. Dynamic modeling techniques, such as a dynamic fault tree model, accommodate multi-tasking of digital systems [7], but are not very familiar to designers.

Estimating "how many parameters will trigger the output signals within the specific time limit for specific kind of accident" is very important, in order to build a sophisticated model with the classical static modeling techniques. Several assumptions, such as the time limit and the severity of standard accidents are required. Parameters for several important standard cases should be defined. For example, a reactor protection system should complete its actuation within 2 hours and the accident be detected through changes in several parameters, such as "Low steam generator pressure," "Low pressurizer pressure," and "Low steam generator level" in the case of a steam line break accident in nuclear power units. The digital system also provides signals for human operators. The processor module in some cases generates signals for both the automated system and human operator. The effect of digital system failure on human operator action is addressed in Section 2.6.

2.3 Estimation of Software Failure Probability

Software is a basis for many of the important safety issues in digital system safety assessment. This section discusses the effect of safety software on safety modeling of a digital system. Software-related issues are dealt with in Chapters 4 and 5 in a more detailed manner.

2.3.1 Quantification of Software Reliability

There is much discussion among software engineering researchers about whether software failure can be treated in a probabilistic manner [8]. Software faults are design faults by definition. That is, software is deterministic and its failure cannot be represented by "failure probability." However, software could be treated based on a probabilistic method because of the randomness of the input sequences, if the software of a specific application is concerned. This is the concept of "error crystals in software," which is the most common justification for the apparent random nature of software failure. Error crystals are the regions of the input space that cause software to produce errors. A software failure occurs when the input trajectory enters an error crystal.

Prediction of software reliability using a conventional model is much harder than for hardware reliability. Microprocessor applications fail frequently when first installed and then become reliable after a long sequence of revisions. The software reliability growth model is the most mature technique for software dependability assessment, which estimates the increment of reliability as a result of fault removal. The repeated occurrence of failure-free working is inputted into probabilistic reliability growth models, which use these data to estimate the current reliability of the program, and to predict how the reliability will change in the future. However, this approach is known to be inappropriate in safety-critical systems since the fixes cannot be assumed effective and the last fix may have introduced new faults [9].

The lower limit of software-failure probability estimated conservatively by testing can be an alternative. The feasibility of reliability quantification of safety-critical software using statistical methods is not accepted by some researchers because exorbitant amounts of testing when applied to safety-critical software are required [10]. However, the developed software must undergo a test phase to show integrity, even if it is not for calculating reliability. Carefully designed random tests and advanced test methodologies provide an estimate of the lower bound of the reliability that is experienced in actual use.

The number of observed failures of highly reliable software during the test is expected to be zero because found errors will be debugged in the corresponding code and the test will be performed again. The concept of software failure probability implies the degree of fault expectation due to software that showed no error in the testing phase. The conventional method to calculate the required number of tests is easily derived. The confidence level C is expressed using the random variable T as the number of tests before the first failure and U as the required number of tests as:

$$C = \Pr(T \le U) = \sum_{t=1}^{U} p(1-p)^{t-1} = p\left[\frac{1-(1-p)^U}{1-(1-p)}\right] \tag{2.8}$$

The failure probability is denoted p. This equation can be solved for U as:

$$U = \frac{\ln(1-C)}{\ln(1-p)} \tag{2.9}$$

An impractical number of test cases may be required for some ultra-high reliable systems. A failure probability that is lower than 10^{-6} with 90% confidence level implies the need to test the software for more than 2.3×10^6 cases without failure. Test automation and parallel testing in some cases is able to reduce the test burden, such as sequential processing software which has no feedback interaction with users or other systems. The validity of test-based evaluation depends on the coverage of test cases. Test cases represent the inputs which are encountered in actual use. This issue is addressed by the concept of reliability allocation [11]. The required software reliability is calculated with target reliability of the total system. The cases of no failure observed during test are covered by Equations 2.8 and 2.9. Test stopping rules are also available for the cases of testing restart after error fixing [11]. The number of needed test cases for each next testing is discussed in a more detailed manner in Chapter 4.

2.3.2 Assessment of Software Development Process

The development process of software is considered in order to assess the expected software failure rate. The application of formal methods to the software development process and usage of mathematical verification of software specifications reduces the possibility of failures due to design faults [12]. The number of remaining potential faults in software is reduced by using software verification and validation (V&V) methodologies. This effect is reflected on the probability estimation of basic events. Thus, the quantification of rigidity of software V&V is performed through the PRA process.

Formal methods, including the formal specification technique, are examples of software V&V processes. Formal methods use ideas and techniques from mathematical or formal logic to specify and reason about computational systems [13, 14]. Formal methods are one of the strongest aids for developing highly reliable software, even though the extent of this kind of proofs is limited. These methods had been widely shown to be feasible in other industries [15]. There are many kinds of approaches for improving the quality of software production besides these formal methods.

The Bayesian belief network (BBN) can be used for estimating the effectiveness of these quality-improving efforts in a more systematic manner [16, 17]. Applying the BBN methodology to the PRA of digital equipment is helpful to integrate many aspects of software engineering and quality assurance. This estimation is performed in consideration of various kinds of activities from each stage of software lifecycle. Difficulties in establishing the BBN include topology and data gathering.

2.3.3 Other Issues

Issues in software reliability are diversity in software codes and hardware–software interaction, in addition to quantification and lifecycle management. Diversity of software plays an important role in fault tolerance of digital systems. Diversity is implemented without modification of hardware components by installing two or more versions of software which are developed by different teams. Faults are expected to be different. As a result, failures can be masked by a suitable voting mechanism. Proving the high reliability of software contains many difficulties. Diversity is helpful in reducing the degree of proof.

Design diversity brings an increase in reliability compared with single versions. This increase is much less than what completely independent failure behavior would imply. The assumption of independence is often unreasonable in practice [18]. Therefore, the degree of dependence must be estimated for each particular case.

Estimation of digital system reliability by calculating the reliability of hardware and software separately [19] does not reflect the effect of hardware–software interactions. An obvious effect of hardware fault masking by software has been reported [20]. A substantial number of faults do not affect program results. Hardware–software interaction may be a very important factor in estimating the dependability of systems. Therefore, the effect of such interactions should be considered.

The interaction problem becomes more complex when aging of hardware is considered. The aging effect induces slight changes in hardware. Different software may cause faulty output. The correlated effect of hardware design and software faults and the correlation between diverse hardware and software should also be considered. These considerations result in very complex and impractical models. The realistic modeling of interactions between hardware and software requires extensive investigation.

Software safety is an important issue in safety assessment of a large digital system. Further discussions regarding these issues are found in Chapters 4 and 5.

2.4 Evaluation of Fault Tolerance Features

Microprocessor and software technologies are used to implement fault-tolerant mechanisms and network communication, to improve efficiency and safety. Fault-tolerant mechanisms are implemented to check the integrity of system components.

Greater attention to watchdog timers and duplication techniques are needed. These are popular and simple ways to establish a fault-tolerant system in industry. Fault-tolerant mechanisms effectively enhance system availability, although their coverage is limited. Digital systems have various faults. Fault-tolerant mechanisms are unable to cover all the faults. The limitation of a fault-tolerant mechanism is expressed using the concept of coverage factor, which must be considered in developing a fault tree. The coverage factor plays a critical role in assessing the safety of digital systems, if a safety-critical system adopts the "fail-safe" concept.

Watchdog devices have been widely adopted as a fault-tolerance feature for safety systems to generate a protective signal at failure of microprocessor-based devices. A simple example of a watchdog timer is illustrated in Figure 2.9. The power for signal generation will be isolated when the watchdog timer detects the failure of a processor. A fault tree of a watchdog timer application (Figure 2.9) is shown in Figure 2.10. Watchdog timer failures are categorized into two groups: failure of the watchdog timer switch (recovery failure); and failure of the watchdog timer to detect microprocessor failure (functional failure). Assume the values of p and w as 10^{-3} failure/demand and 10^{-7} failure/demand, respectively. They are reasonable failure probabilities for typical programmable logic processors and contact relays. System unavailability (Figure 2.10) equals 10^{-20} with a perfect watchdog mechanism (c = 1). System unavailability equals 10^{-6} if the coverage equals zero (c = 0). The effect of watchdog timer coverage estimation on the system unavailability is shown in Figure 2.11.

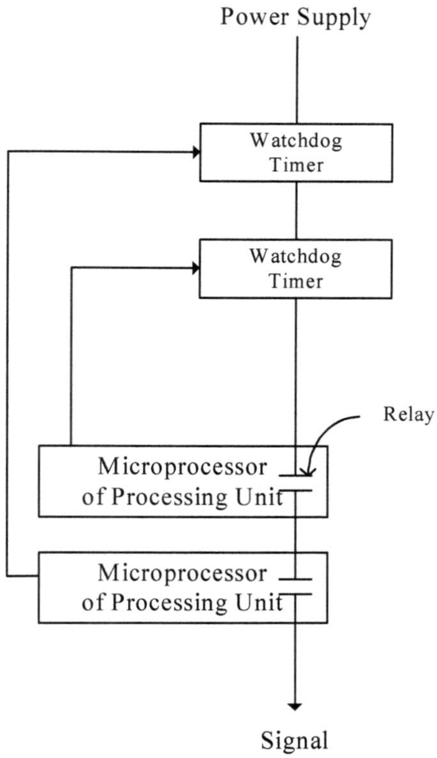

Figure 2.9. Schematic diagram of a typical watchdog timer application

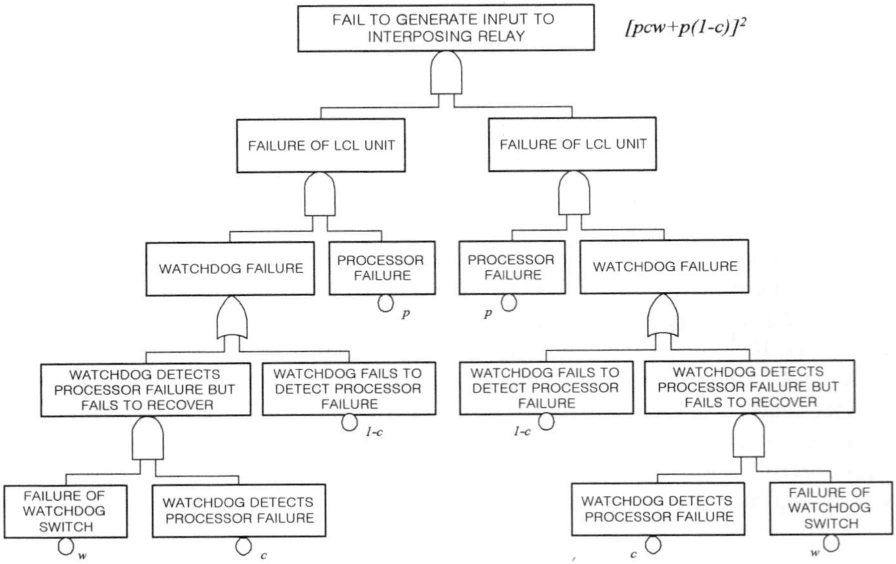

Figure 2.10. Fault tree model of the watchdog timer application in Figure 2.9 (p: the probability of processor failure, c: the coverage factor, w: the probability of watchdog timer switch failure)

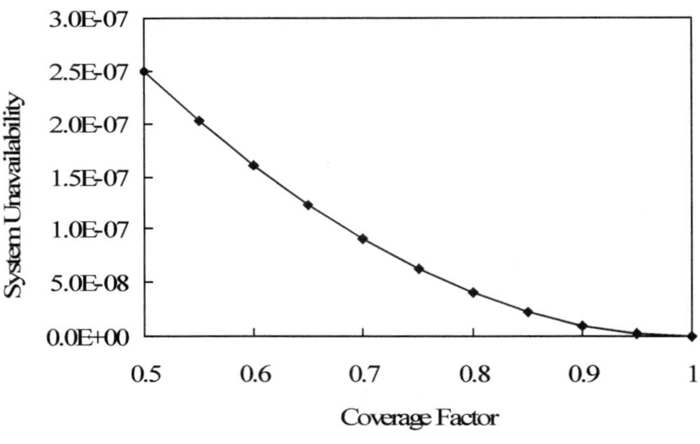

Figure 2.11. System unavailability along the coverage factor of watchdog timer in Figure 2.9

Coverage of the watchdog timer depends on the completeness of the integrity-checking algorithm. Fault coverage of the processor-based monitoring systems or fully duplicated backups are higher than those of watchdog timers because the former has a higher computing power, wider monitoring range, and more sophisticated algorithm.

Quantification of the coverage factor is very important. There is no widely accepted method except experiment for each specific system. Simulation using fault injection is one of the promising methods for estimating coverage factor. The expert knowledge might be used to estimate the rough bounds of the coverage.

2.5 Evaluation of Network Communication Safety

Application of the network communication technique is useful in reducing the cabling number when a system consists of many components and processor modules. The use of signal transmission components, such as fiber-optic modems and opto-couplers, is reduced by using network communication.

Distributed real-time systems have found widespread use in most major industries. Protocols and networking are at the heart of systems. Reliable data acquisition and distribution are essential. Safety-critical networks include information networks in nuclear plants, distributed battle management, intelligent transportation systems, distributed health care, and aviation traffic monitoring systems [21].

Network equipment and functions are closely monitored and controlled to ensure safe operation and prevent costly consequences. The probability of system failure increases as networks become more complex. Failure of any network element can cause an entire network break-down, and in safety-critical settings, the consequences can be severe. A well-known example of such failure is the 1990 nationwide AT&T network failure.

Metrics, such as the processing delay at each node, the capacity of each link, round trip propagation delay, the average queue length of messages awaiting service, utilization, throughput, node delay, and end-to-end delay metrics, have been used as performance criteria. Redundancy is also a metric that is often considered in network evaluations. Redundancy is considered a key feature of a safety-critical network, which drastically improves safety. However, redundancy may increase network complexity and increase network usage, especially in applications where network survivability is crucial [21].

Ethernet is the most widely deployed computer networking technology in the world. Applicability of common networks to safety-critical systems is impractical due to non-determinism. Ethernet cannot establish bounds on time required for a packet to reach its destination. This behavior is not acceptable in safety-critical systems, where a timely response is considered vital.

Safety network communication in the safety-critical system must be evaluated and proved, even though the technique provides many advantages for system development. Proving safety based on the "fail safe" concept is possible in some applications. The system is designed to perform safety actions when the network fails to transfer the information. This is intrinsic safety. Increased spurious transients and expense are also noted.

The probability that the system becomes unsafe due to network failure is evaluated to quantify the risk. Hazard analysis and the identification of paths which might lead the system to an unsafe state are performed, and the probabilistic quantification of each path is also required. Network failure is caused by defects in hardware of network modules or a fault in network protocol, which is the basis of network software.

The main issues in network safety quantification are grouped into two categories: software and hardware. Network configuration and hazard states should be reflected and carefully modeled in a safety assessment model.

2.6 Assessment of Human Failure Probability

The PRA provides a unifying means of assessing the system safety, including the activities of human operators. Human factors and human failure probability are described in Chapters 7 and 8. Issues caused by the interaction of human operators and the digital information system are described in this section.

Two aspects are considered for human failure: the human operator as a generator of manual signals for mitigation when an accident happens, and the human operator as an initiator of spurious plant transients. Both are related to the digital system because its purpose is not only the generation of an automatic signal but also the provision of essential information, such as pre-trip and trip alarms to the operator. These are treated in a different manner from a PRA viewpoint, because the former is related to the accident mitigation, while the latter is related to accident initiation.

The multi-tasking feature of digital systems enables safety-critical signal generation systems to supply the alarms and key information to the human operator. Several functions, such as alarm generation, trip signal generation, and a safety-function-actuation signal generation are simultaneously performed for all the parameters by the digital processing system. An operator will not receive adequate data regarding plant status in the event of system failure.

Reasons for a specific safety function failure are expressed by these relationships (Figure 2.12). A signal generation failure implies human operator interception of an automatically generated signal or the concurrent occurrence of an automatic signal-generation failure and a manual signal generation failure, since a human operator or an automatic system generates safety-actuation signals. A human operator does not generate the signal if an automatic system successfully generates the safety signal. Human error probability (HEP) of a manual signal generation is a conditional probability, given that the automatic signal generation fails. This is an error of omission (EOO). The reason for automatic generation failure is the failure of processing systems or of instrumentation sensors. A processing system failure deteriorates the performance of a human operator, since it implies that the alarms from the processing system will not be provided to the operator. Concurrent failure of multiple redundant sensors also deteriorates human performance, since it causes the loss of corresponding sensor indications and failure of the automated signal generation system, causing the loss of corresponding alarms.

An operator may also wrongly turn off the automatically generated signal (Figure 2.12). This is an error of commission (EOC). The probability of an EOC is a conditional probability if the automatic system successfully generates a proper signal using the sound sensors.

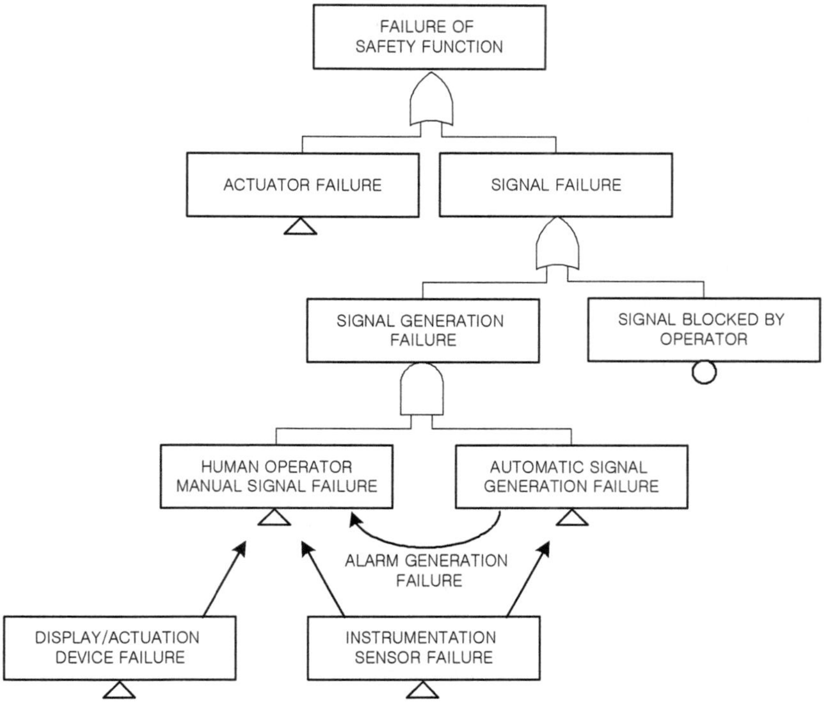

Figure 2.12. The schematic of the concept of the safety function failure mechanism [22]

The failure of a human operator to generate a safety-action signal (EOO) is modeled in a typical PRA. A human operator is usually treated as the backup for an automated digital system. The event of an EOO is followed by the failure of automatic signal generation. The probability of an EOO is evaluated, based on assumptions which reflect the reasons for automatic generation failure. Situations after digital processing system failure are different from that of a conventional analog system failure (trip and pre-trip alarms will be provided to the operator in a more confusing manner in the case of digital system). The probability of an EOO will increase [23–25]. The increased EOO probability results in higher plant risk.

The initiation of a spurious transient by a human operator is treated as an EOC. The loss or the faulty provision of essential information results in an increase in EOCs. EOCs have a greater potential for being significant contributors to plant risk [26].

2.7 Assessment of Common-cause Failure

Safety-critical systems in nuclear power plants adopt multiple-redundancy design in order to reduce the risk from single component failure. The digitalized safety-signal generation system is based on a multiple-redundancy strategy that consists of redundant components. The level of redundant design of digital systems is usually higher than those of conventional mechanical systems. This higher redundancy will clearly reduce the risk from a single component failure, and raise the importance of CCF analysis. CCF stands for failure of multiple items occurring from a single cause that is common to all.

Environmental causes for digital system failure are smoke, high temperature, manufacturing fault, and design fault. There are several definitions of CCF events. A common-cause event is defined as "A dependent failure in which two or more component fault states exist simultaneously, or within a short time interval, and are a direct result of a shared cause," according to NUREG/CR-6268.

Two kinds of dependent events have been identified by OEDC/NEA when modeling common-cause failures in systems consisting of redundant components [27, 28]:

- *Unavailability of a specific set of components of the system, due to a common dependency, for example on a support function. If such dependencies are known, they can be explicitly modeled in a PRA.*
- *Unavailability of a specific set of components of the system due to shared causes that are not explicitly represented in the system logic model. Such events are also called "residual" CCFs, and are incorporated in PRA analyses by parametric models.*

Arguments in the analysis of CCF events have been raised concerning the applicability of multiple failure data. Acquiring data for CCF analysis is difficult since the components and modules in a newly designed digital system are different from those in old ones. CCF events tend to involve very plant-specific features. Whether events occurring at a specific system in one plant are directly applicable in the analysis of another system in a different plant is not clear.

A higher level of redundancy increases the difficulty of a CCF analysis, since an impractically large number of CCF events need to be modeled in the fault tree, if conventional CCF modeling methods are applied. For example, in some nuclear power plants, there are four signal-processing channels for the safety parameters, and each channel consists of two or four microprocessor modules for the same function. If the number of redundancy for safety signal-processing modules is 16, the system model will have 65,519 CCF events ($_{16}C_2 + {}_{16}C_3 + {}_{16}C_4 + \ldots + {}_{16}C_{15} + {}_{16}C_{16} = 2^{16} - 16 - 1 = 65{,}519$). The number of CCF events in a model will increase to 131,054, 262,125, and 524,268, if a system has redundancies of 17, 18, and 19, respectively. These large numbers of CCF events are not practical for treatment in a PRA.

CCFs are a major cause of system failure for highly redundant systems. The occurrence of a CCF event will also affect operator performance, if the system provides important operational information. CCF events in a digital system model are carefully treated with consideration of:

- Proper CCF data are collected and analyzed to estimate the probability of each CCF event.
- Large number of CCF events is reduced in an acceptable manner for developing a practical PRA model.

- Information for operator performance estimation is available after the reduction in number of events.

2.8 Concluding Remarks

The factors which are carefully considered in modeling the safety of digital systems are listed:

- CCF estimation
- Modeling for dynamic system
- Software testing for failure probability estimation
- Evaluation of software verification and validation
- Dependency between diverse software programs
- Effect of hardware and software interaction
- The fault coverage of fault-tolerant mechanisms
- The safety of network communication
- The probability of human error of omission
- The probability of human error of commission

The proper consideration of these factors makes the safety assessment results more realistic. The active design feedback of insight from the risk assessment will improve the large safety-critical system reliability in an effective manner. Fault monitoring of input/output modules in addition to the processor module is an example of extended design feedback of risk information. Properly designed on-line testing and monitoring mechanisms will improve system integrity by reducing inspection intervals.

References

[1] Kang HG, Jang SC, Ha JJ (2002) Evaluation of the impact of the digital safety-critical I&C systems, ISOFIC2002, Seoul, Korea, November 2002

[2] Sancaktar S, Schulz T (2003) Development of the PRA for the AP1000, ICAPP '03, Cordoba, Spain, May 2003

[3] Hisamochi K, Suzuki H, Oda S (2002) Importance evaluation for digital control systems of ABWR Plant, The 7th Korea-Japan PSA Workshop, Jeju, Korea, May 2002

[4] HSE (1998) The use of computers in safety-critical applications, London, HSE books

[5] Kang HG, et al. (2003) Survey of the advanced designs of safety-critical digital systems from the PSA viewpoint, Korea Atomic Energy Research Institute, KAERI/AR-00669/2003

[6] Goldberg BE, Everhart K, Stevens R, Babbitt N III, Clemens P, Stout L (1994) System engineering “Toolbox” for design-oriented engineers, NASA Reference Publication 1358

[7] Meshkat L, Dugan JB, Andrews JD (2000) Analysis of safety systems with on-demand and dynamic failure modes, Proceedings of 2000 RM

[8] White RM, Boettcher DB (1994) Putting Sizewell B digital protection in context, Nuclear Engineering International, pp. 41–43

[9] Parnas DL, Asmis GJK, Madey J (1991) Assessment of safety-critical software in nuclear power plants, Nuclear Safety, Vol. 32, No. 2
[10] Butler RW, Finelli GB (1993) The infeasibility of quantifying the reliability of life-critical real-time software, IEEE Transactions on Software Engineering, Vol. 19, No. 1
[11] Kang HG, Sung T, et al (2000) Determination of the Number of Software Tests Using Probabilistic Safety Assessment KNS conference, Proceeding of Korean Nuclear Society, Taejon, Korea
[12] Littlewood B, Wright D (1997) Some conservative stopping rules for the operational testing of safety-critical software, IEEE Trans. Software Engineering, Vol. 23, No. 11, pp. 673–685
[13] Saiedian H (1996) An Invitation to formal methods, Computer
[14] Rushby J (1993) Formal methods and the certification of critical systems, SRI-CSL-93-07, Computer Science Laboratory, SRI International, Menlo Park
[15] Welbourne D (1997) Safety critical software in nuclear power, The GEC Journal of Technology, Vol. 14, No. 1
[16] Dahll G (1998) The use of Bayesian belief nets in safety assessment of software based system, HWP-527, Halden Project
[17] Eom HS, et al. (2001) Survey of Bayesian belief nets for quantitative reliability assessment of safety critical software used in nuclear power plants, Korea Atomic Energy Research Institute, KAERI/AR-594-2001, 2001
[18] Littlewood B, Popov P, Strigini L (1999) A note on estimation of functionally diverse system, Reliability Engineering and System Safety, Vol. 66, No. 1, pp. 93-95
[19] Bastl W, Bock HW (1998) German qualification and assessment of digital I&C systems important to safety, Reliability Engineering and System Safety, Vol. 59, pp. 163-170
[20] Choi JG, Seong PH (2001) Dependability estimation of a digital system with consideration of software masking effects on hardware faults, Reliability Engineering and System Safety, Vol. 71, pp. 45-55
[21] Bayrak T, Grabowski MR (2002) Safety-critical wide area network performance evaluation, ECIS 2002, June 6–8, Gdańsk, Poland
[22] Kang HG, Jang SC (2006) Application of condition-based HRA method for a manual actuation of the safety features in a nuclear power plant, Reliability Engineering & System Safety, Vol. 91
[23] Kauffmann JV, Lanik GT, Spence RA, Trager EA (1992) Operating experience feedback report – human performance in operating events, USNRC, NUREG-1257, Vol. 8, Washington DC
[24] Decortis F (1993) Operator strategies in a dynamic environment in relation to an operator model, Ergonomics, Vol. 36, No. 11
[25] Park J, Jung W (2003) The requisite characteristics for diagnosis procedures based on the empirical findings of the operators' behavior under emergency situations, Reliability Engineering & System Safety, Volume 81, Issue 2
[26] Julius JA, Jorgenson EJ, Parry GW, Mosleh AM (1996) Procedure for the analysis of errors of commission during non-power mode of nuclear power plant operation, Reliability Engineering & System Safety, Vol. 53
[27] OECD/NEA Committee on the safety of nuclear installations, 1999, ICDE project report on collection and analysis of common-cause failures of centrifugal pumps, NEA/CSNI/R(99)2
[28] OECD/NEA Committee on the safety of nuclear installations, 2003, ICDE project report: Collection and analysis of common-cause failures of check valves, NEA/CSNI/R(2003)15

3

Case Studies for System Reliability and Risk Assessment

Jong Gyun Choi[1], Hyun Gook Kang[2] and Poong Hyun Seong[3]

[1]I&C/Human Factors Division
Korea Atomic Energy Research Institute
1045 Daedeok-daero, Yuseong-gu, Daejeon, 305-353, Korea
choijg@kaeri.re.kr

[2]Integrated Safety Assessment Division
Korea Atomic Energy Research Institute
1045 Daedeok-daero, Yuseong-gu, Daejeon, 305-353, Korea
hgkang@kaeri.re.kr

[3]Department of Nuclear and Quantum Engineering
Korea Advanced Institute of Science and Technology
373-1 Guseong-dong, Yuseong-gu, Daejeon, 305-701, Korea
phseong@kaist.ac.kr

Case studies of countermeasures mentioned in Chapters 1 and 2 are presented. The safety of digital applications in NPPs has been discussed by the National Research Council [1]. Appropriate methods for assessing safety and reliability are keys to establishing the acceptability of digital instrumentation and control systems in NPPs.

PRA since 1975 has been widely used in the nuclear industry for licensing and identifying plant safety vulnerabilities. PRA techniques are used in the nuclear industry to assess the relative effects of contributing events on plant risk and system reliability. PRA involves basic tasks, such as the definition of accident sequences, an analysis of plant systems and their operation, the collection of component data, and an assessment of accident-sequence frequencies [2]. PRA is used for operation, regulation, and design. PRA is also used to assess the relative effects of the contributing events on system-level safety and to provide a unifying means of assessing physical faults, recovery processes, contributing effects, human actions, and other events that have a high degree of uncertainty.

Models of digitalized systems are part of plant risk assessment. For example, the unavailability of a digitized reactor trip signal generation system is one of the main reasons for “transients without safe reactor scram.” Failures of safety components during an abnormal condition also deteriorate the mitigation of an accident.

Digital systems consist of input, processor, output, and network modules. A methodology for reliability calculation of a module is introduced in Section 3.1. A method for reliability assessment of an embedded digital system using a multi-state function is described in Section 3.2. An analysis framework consisting of the process of system model development and sensitivity studies is introduced in Section 3.3.

3.1 Case Study 1: Reliability Assessment of Digital Hardware Modules

The components of a typical digital hardware module are categorized into four sub-function groups according to their functions (Figure 3.1):

> The components in group ***a*** receive input signals and transform them, and transfer the transformed signal to group ***b***. Group ***a*** compares the transformed signal with the feedback signal. The comparison between these two signals is used for the loop-back test, and generates an error signal to the external module and the operator through group ***d*** whenever a deviation happens between these two signals. The transmitted signal from group ***a*** is processed in group ***b***. The components in this group provide the final output to the external module and provide the feedback signal to group ***c***. The components in group ***c*** transform the final output for the loop-back test. The transformed final output is given to group ***a*** for a comparison. The components in group ***d*** transport the error signal to the external module or operator to alert them that failures happened in the module.

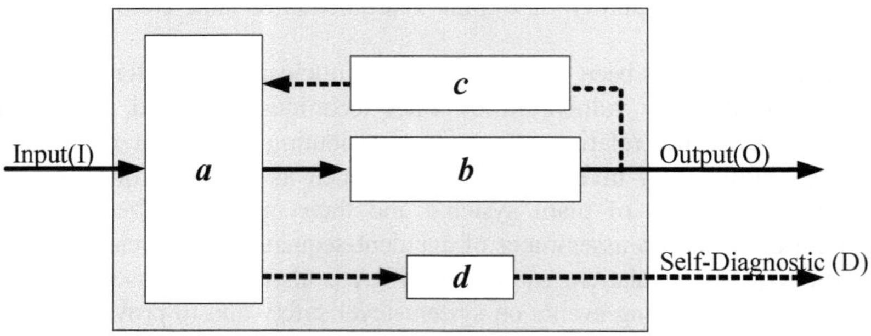

Figure 3.1. Functional block diagram of a typical digital hardware module

Table 3.1. Failure status of a typical digital hardware module

Failure combination (*abcd*)	Output status	Diagnostic status	Module failure
1111	1	1	S
0111	0	0	USF
1011	0	1	SF
1101	1	0	S
1110	1	0	S
0011	0	0	USF
0101	0	0	USF
0110	0	0	USF
1001	0	0	USF
1010	0	0	USF
1100	1	0	S
0001	0	0	USF
0010	0	0	USF
0100	0	0	USF
1000	0	0	USF
0000	0	0	USF

All the groups correctly perform their allotted functions if there is no failure in the module. The programmable logic controller (PLC) module performs its mission successfully when in the success state. The module does not make the final output to the external module and the module comes to a failure state if group ***b*** has failed and the other function groups operate properly. The module immediately generates an error alarm signal to the external module because the self-diagnostic function correctly operates by a loop-back test in group ***a***. The failure of group ***b*** is called safe-failure since the operator makes the system safe and starts maintenance activities immediately after an error alarm signal. The module does not make the transformed signal for group ***b*** if group ***a*** fails. Then the module does not conduct the loop-back test. As a result, the module comes to a failure state. The failure case of group ***a*** is in an unsafe failure state. The module is in an unsafe failure state if all the groups have failed.

The failure status of a typical digital hardware module is shown (Table 3.1). The first column represents the failure combination for each function group. "0" indicates the failure status of the allotted function group and "1" indicates the successful operation status of the given function group. The second and third columns indicate the output and diagnostic status, respectively. The fourth column represents the failure status of the module according to the combination of each function group failure. S, USF, and SF represent the success, unsafe failure, and safe failure state, respectively. Only the unsafe failure state directly affects the RPS safety directly.

Unsafe failure of the module (Table 3.1) is expressed as:

$$\begin{aligned}\text{USF of the module} &= \bar{a}bcd + \bar{a}\bar{b}cd + \bar{a}b\bar{c}d + \bar{a}bc\bar{d} + a\bar{b}\bar{c}d \\ &\quad + a\bar{b}c\bar{d} + \bar{a}\bar{b}\bar{c}d + \bar{a}\bar{b}c\bar{d} + \bar{a}b\bar{c}\bar{d} + a\bar{b}\bar{c}\bar{d} + \bar{a}\bar{b}\bar{c}\bar{d} \\ &= \bar{a}d + \bar{a}\bar{d} + a\bar{b}\bar{c} + a\bar{b}c\bar{d} \\ &= \bar{a} + a\bar{b}(\bar{c} + \bar{d})\end{aligned} \tag{3.1}$$

With rare event approximation and the assumption of independence among *a*, *b*, *c*, and *d*, the unsafe failure probability of the module is expressed as:

$$\begin{aligned}P\{USF \text{ of the module}\} &= P\{\bar{a} + a\bar{b}(\bar{c} + \bar{d})\} \\ &= P(\bar{a}) + P(a)P(\bar{b})P(\bar{c}) + P(a)P(\bar{b})P(\bar{d})\end{aligned} \tag{3.2}$$

Generally, the module in unsafe failure state can be repaired or replaced by periodic surveillance tests at intervals of *T*. The failure of each group is distributed exponentially when the repair restores the module to a state as good as new. The unavailability of the module due to USF is expressed as:

$$\begin{aligned}q_{USF} &= \frac{\lambda_a}{2}T + \left(1 - \frac{\lambda_a}{2}T\right)\frac{\lambda_b}{2}T \cdot \frac{\lambda_c}{2}T + \left(1 - \frac{\lambda_a}{2}T\right)\frac{\lambda_b}{2}T \cdot \frac{\lambda_d}{2}T \\ &= \frac{\lambda_a}{2}T + \frac{\lambda_b\lambda_c}{4}T^2 + \frac{\lambda_b\lambda_d}{4}T^2 - \frac{\lambda_a\lambda_b\lambda_c}{8}T^3 - \frac{\lambda_a\lambda_b\lambda_d}{8}T^3 \\ &\cong \frac{\lambda_a}{2}T, \text{ when } \lambda_a, \lambda_b, \lambda_c << T\end{aligned} \tag{3.3}$$

where:
q_{USF}: The module unavailability due to a unsafe failure
λ_a: The failure rate of function group ***a***
λ_b: The failure rate of function group ***b***
λ_c: The failure rate of function group ***c***
λ_d: The failure rate of function group ***d***
T: The periodic test interval in hours

Unavailability due to unsafe failure rate is approximated by the unavailability of the ***a*** function group. The failure rate of group ***a*** is calculated by summation of failure rates of all components in group ***a***:

$$\lambda_a = \sum_i \lambda_i \tag{3.4}$$

where:
λ_i: The failure rate of each component in a group

Table 3.2. Failure rates of the typical PLC modules

Module name	Failure rates ($\times 10^{-6}$/h)
CPU	21.43
DC 24 V digital input module	1.17
Analog input module	5.36
Analog output module	15.41
DC 24 V digital output module	4.22
AC 250 V relay output module	5.17
Communication module	7.18

The part stress method in MIL-HDBK-217F is employed for the prediction of each component failure rate. For example, the following equation from MIL-HDBK-217F is used to estimate the failure rate of integrated microcircuits (digital gate/logic arrays) in the module [3, 4]:

$$\lambda_p \approx (C_1\pi_T + C_2\pi_E)\pi_Q\pi_L \text{ failures per } 10^6 \text{ hour} \qquad (3.5)$$

where:
C_1 : Die complexity failure rate
C_2 : Packaging failure rate
π_T : Temperature factor
π_E : Environment factor
π_Q : Quality factor
π_L : Learning factor

Values for the above factors are based on applicable plant conditions and configuration details of microcircuits. Suitable values of these parameters are chosen for perceived device specifications and control room conditions. The failure rates of the typical PLC modules, using the proposed failure model, are shown in Table 3.2 assuming 30°C ambient temperature and ground benign environment.

3.2 Case Study 2: Reliability Assessment of Embedded Digital System Using Multi-state Function

The use of digital systems in a nuclear instrument and control system (I&C) is prevalent because of increased capability and superior performance compared to analog systems. However, it is difficult to evaluate the reliability of digital systems because they include fault-handling techniques at various levels of systems, such as Hamming code, watchdog timer and N-version programming [3, 5–7]. Software is another issue in reliability assessment of systems that require ultra-high reliability [8, 9]. In addition, the reliability of digital systems has to be assessed with respect to software, hardware, and software/hardware interactions. Software considerations cannot be fully understood apart from hardware considerations and *vice versa* [4, 10–13].

Software is designed to accomplish functions that the digital system is required to perform at level 0 in the hierarchical functional view of a digital system (Figure 3.2). Software is composed of software modules. Software modules of level 1 perform their allotted tasks through a combination of instruction sets provided by the microprocessor. Parts of hardware components at level 3, such as microprocessors and memories, are used for processing of one instruction at level 2. That is, in order for the digital system to complete its required function, the software determines the correct sequence in which the hardware resources should be used. System failure occurs when software cannot correctly arrange the sequence of use of hardware resources or when one or more hardware resources have faults. A combinatorial model for estimating the reliability of an embedded digital system by means of multi-state function is described below. This model considers not only fault-handling techniques implemented in digital systems but also the interaction between hardware and software with consideration of a software operational profile. The software operational profile determines the use frequency of each software module which control the use frequency of hardware components. The software operational profile is modeled through the adaptation of software control flow into a multi-state function. In this study, the concept of coverage model [14] is extended for modeling fault-handling techniques, implemented hierarchically in digital systems. The discrete function theory [15] provides a complete analysis of a multi-state digital system since fault-handling techniques make it difficult for many types of components in the system to be treated in the binary state [16–18]. Software should not be separately considered from hardware when system reliability is estimated. The effects of the software operational profile on system reliability are also considered. This model was applied for a one-board controller in a digital system.

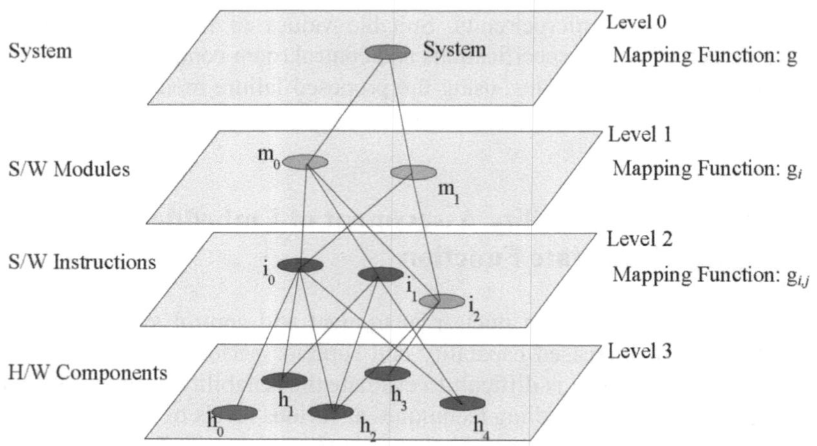

Figure 3.2. Hierarchical functional architecture of digital system at board level

The simplification of this model becomes a conventional model that treats the system in a binary state. This modeling method is particularly attractive for embedded systems in which small-sized application software is implemented, since it will require laborious work for this method to be applied to systems with large software.

3.2.1 Model

The definition of fault coverage, C, is the condition probability that a system recovers, given that a fault has occurred [19]. Fault coverage is mathematically written as:

$$C = \Pr(\text{fault processed correctly} \mid \text{fault existence}) \tag{3.6}$$

A more detailed description of fault coverage types is: fault detection coverage is the system's ability to detect a fault; fault location coverage measures a system's ability to locate the cause of faults; fault containment coverage measures a system's ability to contain faults within a predefined boundary; and fault recovery coverage measures the system's ability to automatically recover from faults and to maintain correct operation [19].

Digital systems are composed of a hierarchy of levels (Figure 3.2). Faults and errors may be generated at any of the levels in the hierarchy. The various techniques for handling a fault, such as fault confinement, fault detection, fault masking, retry, diagnosis, reconfiguration, recovery, restart, repair, and reintegration, are implemented at each level [20]. The detection of the error is left to higher levels if an error is not detected at the level in which it originated. Appropriate information about the detected error must be passed onto a higher level if the current level lacks the capacity to recover from a particular detected error.

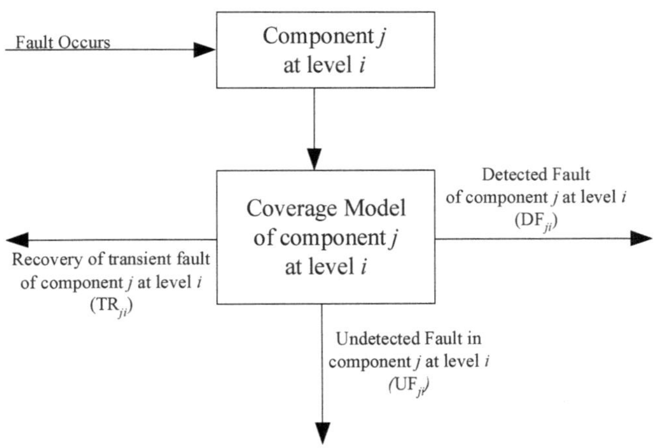

Figure 3.3. Coverage model of a component at level i

The coverage model of a fault-tolerant system was developed to incorporate fault coverage into combinatorial models [14, 21]. The coverage model is generalized with a recovery process initiated when a fault occurs (Figure 3.3). The entry point to the model signifies the occurrence of the fault and the three exits signify three possible outcomes. The transient recovery exit (labeled TR) represents the correct recognition of, and recovery from, a transient fault. Successful recovery from a transient fault restores the system to a consistent state without discarding any components. The detected fault exit (labeled DF) denotes the determination of the permanent nature of the fault. Recovery from the detected fault is passed onto a higher level, if the current level lacks the capacity to recover from the detected fault, although the fault is detected. The undetected fault exit (labeled UF) is reached when a fault is not detected. This type of fault causes the higher-level components to operate erroneously. Each exit in the coverage model has an exit probability associated with it that can be determined by solving the appropriate coverage model. [TR_{ji}, UF_{ji}, DF_{ji}] is defined as the probability for component j of level i to take the state [Transient Recovery, Undetected Fault, Detected Fault]. The three possible cases are mutually exclusive and the sum of their probabilities is 1.

When a logic function is defined as a mapping:
$f:\{0,1,...,r-1\}^n \rightarrow \{0,1,...,r-1\}$, with r the cardinality of the sets and n the number of domain sets. When r is 3, the logic function is as follows:

$$f:\{0,1,2\}^n \rightarrow \{0,1,2\} \tag{3.7}$$

Three logic gates, AND, OR, and DEPEND are defined with a graphical form (Figure 3.4). Each input of logic gates has integer value, 0, 1, 2. The output of the AND gate is defined as the minimum of all the inputs and denoted by conjunction of its inputs. The output of the OR gate is the maximum of all the inputs and denoted by disjunction of its inputs. The output of the DEPEND gate is 0 when all x_i's are 0's. Otherwise, the output of the DEPEND gate is y.

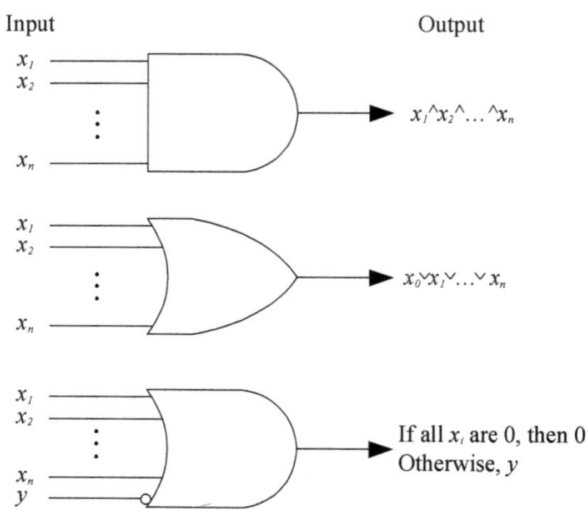

Figure 3.4. Logic gates

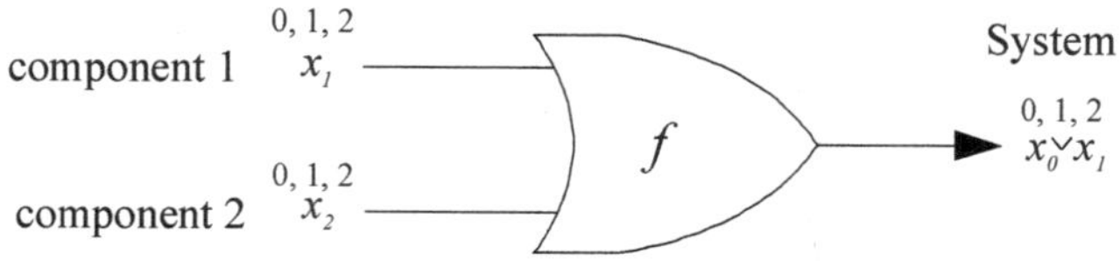

Figure 3.5. Modeling of a series system composed of two components

A logic network is then defined as a circuit composed of these gates. Logic gates can be used to describe the coverage model of a faulty component. For example, the system can be modeled by an OR gate when a system is composed of two components and performs a successful operation when the both components have no faults. The system has a graphical representation (Figure 3.5) in this case and the mapping function of this system has a tabular form (Table 3.3).

When the p_i and q_i are probabilities that the component 1 and 2 are in state i, respectively, the state probability of the system is:

$$\Pr\{\text{System is in state } 0\} = p_0 q_0 \tag{3.8}$$

$$\Pr\{\text{System is in state } 1\} = p_0 q_1 + p_1 q_0 + p_1 q_1 \tag{3.9}$$

$$\Pr\{\text{System is in state } 2\} = p_0 q_2 + p_1 q_2 + p_2 q_0 + p_2 q_1 + p_2 q_2 \tag{3.10}$$

The states of a hardware component have physical meaning as follows:

0: the component operates correctly
1: the component operates erroneously, but is not detected
2: the component operates erroneously, and is detected

The cumulative fault probability is $F_i(t) = 1 - e^{-\lambda_i t}$ if the fault of hardware component i is distributed exponentially and the fault rate of the hardware component i is λ_i. When the state variable of a hardware component i is x_i, the state probability is given by:

$$\begin{aligned}\Pr\{x_i = 0\} &= \Pr\{\text{no fault occurs in component } i\} \\ &\quad + \mathrm{TR}_{j3} \cdot \Pr\{\text{fault occurs in component } i\} \\ &= e^{-\lambda_i t} + \mathrm{TR}_{i3} \cdot (1 - e^{-\lambda_i t}) = \mathrm{TR}_{i3} + (1 - \mathrm{TR}_{i3}) e^{-\lambda_i t}\end{aligned} \tag{3.11}$$

Table 3.3. Function table of the series system

	Component 2 (x_1)		
Component 1 (x_2)	0(p_0)	1(p_1)	2(p_2)
0(q_0)	0(p_0q_0)	1(p_1q_0)	2(p_2q_0)
1(q_1)	1(p_0q_1)	1(p_1q_1)	2(p_2q_1)
2(q_2)	2(p_0q_2)	2(p_1q_2)	2(p_2q_2)

$$\begin{aligned}\Pr\{x_i = 1\} &= \mathrm{UF}_{i3} \cdot \Pr\{\text{fault occurs in component } i\} \\ &= \mathrm{UF}_{i3} \cdot (1 - e^{-\lambda_i t}) = \mathrm{UF}_{i3} - \mathrm{UF}_{i3} e^{-\lambda_i t}\end{aligned} \tag{3.12}$$

$$\begin{aligned}\Pr\{x_i = 2\} &= \mathrm{DF}_{i3} \cdot \Pr\{\text{fault occurs in component } i\} \\ &= \mathrm{DF}_{i3} \cdot (1 - e^{-\lambda_i t}) = \mathrm{DF}_{i3} - \mathrm{DF}_{i3} e^{-\lambda_i t}\end{aligned} \tag{3.13}$$

where $\Pr\{x_i = 0\} + \Pr\{x_i = 1\} + \Pr\{x_i = 2\} = 1$ and $\mathrm{TR}_{i3} + \mathrm{UF}_{i3} + \mathrm{DF}_{i3} = 1$.

The state of an instruction execution depends not only on the states of all the hardware resources required for the instruction execution but also on the instruction itself. All of the hardware resources required for the instruction execution must be in correct operational states in order for one instruction to be executed successfully and the instruction itself must have no faults that are implemented in the instruction by coding errors. The instruction operates correctly, although hardware resources or instruction itself is in fault state, if the fault-handling technique at instruction level recovers the faults. The state of the instruction execution is defined by:

0: the instruction is executed correctly
1: the instruction is executed erroneously, but is not detected
2: the instruction is executed erroneously, and is detected

The state of instruction execution is modeled with the DEPEND gate and the OR gate. For example, an assembly code, ANI 01(Assembly 8085), uses hardware resources: microprocessor and ROM. This instruction execution is modeled by a logic OR gate and a DEPEND gate (Figure 3.6). The input IC variable that represents the fault-handling techniques at instruction level has three states, 0, 1, and 2. The state 0 represents the transient recovery of the fault propagated at sub-level. The state 1 represents the detection of the fault but the lack of capacity to recover the fault. Finally, state 2 represents the inability to detect the fault.

Each of the software modules, $g_0, g_1, \ldots, g_i$ is composed of instruction sets. The state of software module operation is dependent on the execution states of its instruction sets. All of the instructions executed in that software module must be executed successfully in order for a software module to execute its intended function successfully.

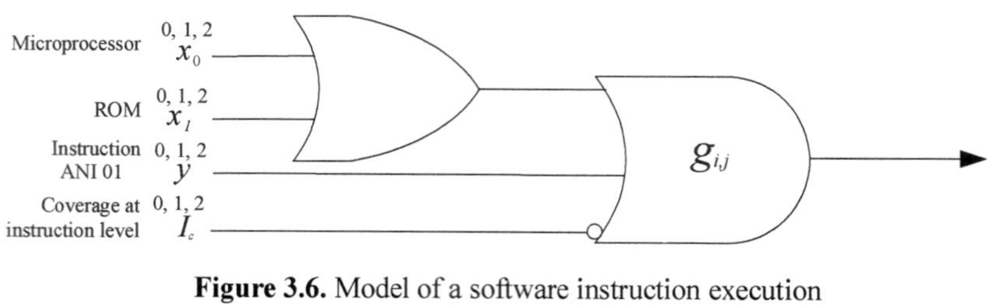

Figure 3.6. Model of a software instruction execution

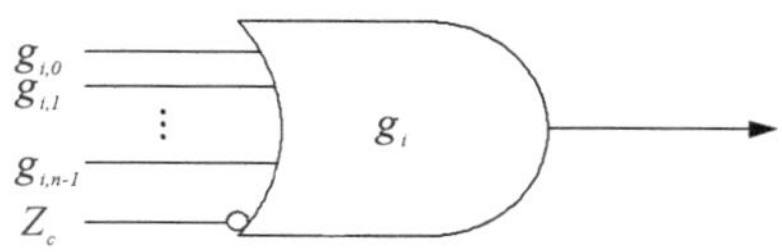

Figure 3.7. Model of a software module operation

The state of the module operation is defined by:

0: the module operates correctly
1: the module operates erroneously, but is not detected at the module level
2: the module operates erroneously, and is detected at the module level

The model of software module operation is shown in Figure 3.7, when the variable Z_c represents the fault coverage at module level. The control flow graph is a graphical representation of a program control structure [22], that uses elements, such as process blocks, decisions, and junctions. A process block is a sequence of program statements uninterrupted by either decisions or junctions. A decision is a program point at which the control flow can diverge. A conditional branch is an example of decisions. A junction is a point in the program where the control flow can merge.

Examples of junctions are the target of a jump or skip instruction in assembly language. A graph using arcs and nodes is obtained in the program control flow graph construction. Nodes are used for the decision and the junction points of the program. Arcs are used for presenting the next points of program execution.

The operational profile of the embedded system determines the control flow of the software. If $\boldsymbol{I}$ is an input domain set of the software, then it can be partitioned into an indexed family, $\boldsymbol{I}_i$, with the following properties:

(a) $\boldsymbol{I} = \bigcup_{i=0}^{n-1} \boldsymbol{I}_i$
(b) $i \neq j \Rightarrow \boldsymbol{I}_i \cap \boldsymbol{I}_j = \phi$

The input domain set $\boldsymbol{I}$ is partitioned by software control flow. The control flow of software by input domain can be modeled with a set $\boldsymbol{S}_i$ called selection set as $\boldsymbol{S}_i = \{s_{0,i}, s_{1,i}, \ldots, s_{n-1,i}\}$, where n is the number of input domains and the element $s_{k,i}$ of $\boldsymbol{S}_i$ is a binary number defined by

$$s_{k,i} = \begin{cases} 2, \text{if module } g_i \text{ is in sequence of software execution by input domain } \boldsymbol{I}_\text{k} \\ 0, \text{otherwise} \end{cases}$$

Additionally, the selection function set is defined as a mapping as follows:

$$\boldsymbol{sf} : \boldsymbol{I} \rightarrow \prod_{i=0}^{p-1} \boldsymbol{S}_i \tag{3.14}$$

where p is the number of software modules and the set $\boldsymbol{sf}$ is $\{sf_0, sf_1, \ldots, sf_{p-1}\}$. The sf_i is a function defined as $sf_i(\boldsymbol{I}_k) = k + 1$ th element of $\boldsymbol{S}_i$.

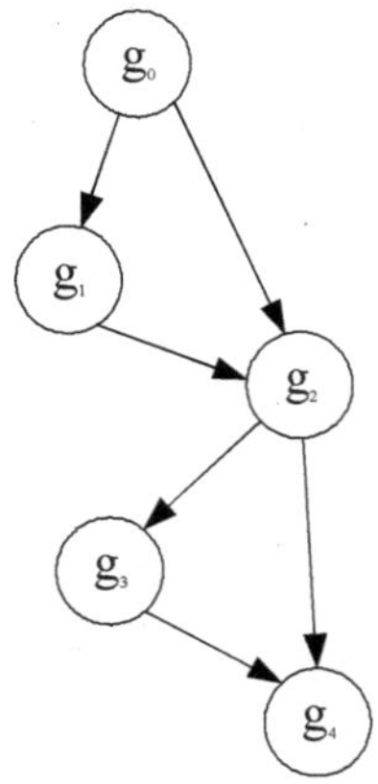

Figure 3.8. Control flow of example software

For example, a control flow of example software has four paths (Figure 3.8). It is assumed that each software module is in one state of three states, {0, 1, 2}. The software input domain can be partitioned into $\{\boldsymbol{I}_0, \boldsymbol{I}_1, \boldsymbol{I}_2, \boldsymbol{I}_3\}$.

The example software is executed in the sequence of software modules, g_0, g_1, g_2, g_3, and g_4 as follows:

If $i \in \boldsymbol{I}_0$, then, $g_0 \rightarrow g_1 \rightarrow g_2 \rightarrow g_3 \rightarrow g_4$
If $i \in \boldsymbol{I}_1$, then, $g_0 \rightarrow g_1 \rightarrow g_2 \rightarrow g_4$
If $i \in \boldsymbol{I}_2$, then, $g_0 \rightarrow g_2 \rightarrow g_3 \rightarrow g_4$
If $i \in \boldsymbol{I}_3$, then, $g_0 \rightarrow g_2 \rightarrow g_4$

When an input value of the example software is an element of input domain set $\boldsymbol{I}_1$, the selection function set is as:

$$\begin{aligned} \boldsymbol{sf}(\boldsymbol{I}_1) &= \{\boldsymbol{sf}_0(\boldsymbol{I}_1), \boldsymbol{sf}_1(\boldsymbol{I}_1), \boldsymbol{sf}_2(\boldsymbol{I}_1), \boldsymbol{sf}_3(\boldsymbol{I}_1), \boldsymbol{sf}_4(\boldsymbol{I}_1)\} \\ &= \{s_{1,0}, s_{1,1}, s_{2,1}, s_{3,1}, s_{4,1}\} = \{2, 2, 2, 0, 2\} \end{aligned}$$

The distribution probability of each input domain set is obtained and use frequency probability of each module in the software is determined if the operational profile of the software is known. The example software has the tabular form of the selection function set (Table 3.3). The logic network of the example software is shown in Figure 3.9 when w_i is the state variable of the software module g_i.

Table 3.4. Selection function set table of the example software

Input domain set	Selection function set
$\boldsymbol{I}_0$	{2, 2, 2, 2, 2}
$\boldsymbol{I}_1$	{2, 2, 2, 0, 2}
$\boldsymbol{I}_2$	{2, 0, 2, 2, 2}
$\boldsymbol{I}_3$	{2, 0, 2, 0, 2}

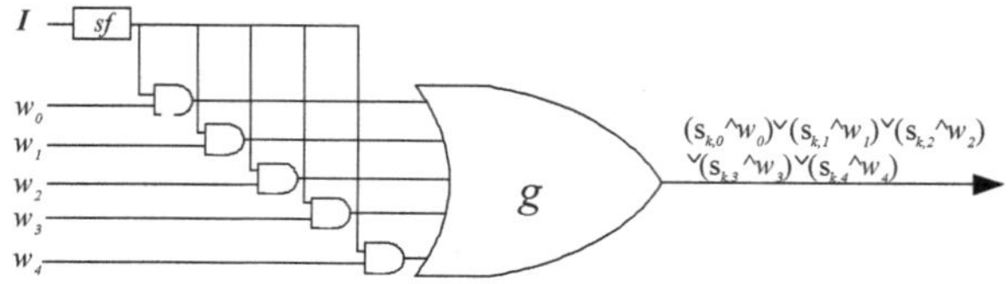

Figure 3.9. Logic gate of example software

3.2.2 A Model Application to NPP Component Control System

The 8085 processor is used in the control system (Table 3.5) which is constructed in the Intel 8085 assembly language using top-down modular design techniques. Onboard memory capabilities include two 1K × 1 read/write memories for single-bit data storage and a 1K × 8 read/write memory for 8-bit data storage. Read-only-memory capability ranges from 1K to a maximum 48 Kbytes. The memory in the system is the Erasable Programmable Read Only Memory and the capacity of the memory is 64 Kbytes. The time clock cycle is 1 MHz.

The application program is a part of the executive program that consists of various subroutines that generate the logic to perform miscellaneous functions.

These subroutines consist of auto/manual logic, command request logic, synchronizing logic, trouble/disable logic and memory flag logic. The application program stored in ROM is executed to actuate some specific components, such as pumps and valves in nuclear power plants. The memory is tested periodically by a memory test routine to detect the memory error through the checksum technique. Memory flag logic raises a flag and initiates repair when a memory fault is detected.

An application program (auto/manual) is selected for application of developed methodology to the control system. The selected program is a simple logic algorithm, which leads to the result "yes" or "no." In addition, inputs of the program have only two kinds of values, 0 or 1. The program performs its function with values of thirteen input variables. The values of seven output variables are produced after program execution. Therefore, operational profiles of the software are determined by thirteen inputs and the application software is modeled with the logic network (Figure 3.10).

The failure rates of all hardware components are assumed to be 10^{-7}/h for the application to control system. The state probability of the system is shown in Figure 3.11, when any fault-handling techniques are not considered and the input domain of software is always $\boldsymbol{I}_0$; or, $(TR_{ij}, UF_{ij}, DF_{ij}) = (0, 1, 0)$ and $\boldsymbol{I} = \boldsymbol{I}_0$. This result is equal to the result calculated by the part count method proposed in MIL-

Table 3.5. Information on control system hardware

Component	Description
CPU	8085
Memory	64K EPROM(TS27C64A, 8192 × 8)
Time clock	1 MHz

HNBK-217. The simplification of the model proposed in this study is reduced to the combinatorial model in which the system and the components are treated in binary states, good/bad.

The state probability is shown in Figure 3.12 when the fault-handling techniques are considered at only the hardware component level and input domain of software is always $\boldsymbol{I}_0$; or, $(TR_{3j}, UF_{3j}, DF_{3j}) = (0.8, 0.1, 0.1), j = 0, 1, 2$, and $\boldsymbol{I} = \boldsymbol{I}_0$. The state probability of the system goes to a steady-state value when sufficient time passes. The steady-state value depends on the values, TR_{3j}, UF_{3j}, and DF_{3j}.

The state probability is shown in Figure 3.13 when any fault-handling techniques are not considered and the input domain of software is always $\boldsymbol{I}_3$; or, $(TR_{3j}, UF_{3j}, DF_{3j}) = (0, 1, 0), j = 0, 1, 2$, and $\boldsymbol{I} = \boldsymbol{I}_3$. The values of state probability show a different result from that of Figure 3.11 although the shape of state probability is similar to that of Figure 3.11. Thus, the operational profile of software affects the state probability of the system, as expected. The model used in this case study includes not only the coverage model of fault-handling techniques implemented in digital systems but also the model that considers the interaction between hardware and software. This model is simplified to the conventional model that treats the system in binary states since the proposed model is a general combinatorial model. This model covers the detailed interaction of system components, if the information on the coverage of fault-handling techniques implemented in system is obtained through fault injection experiments.

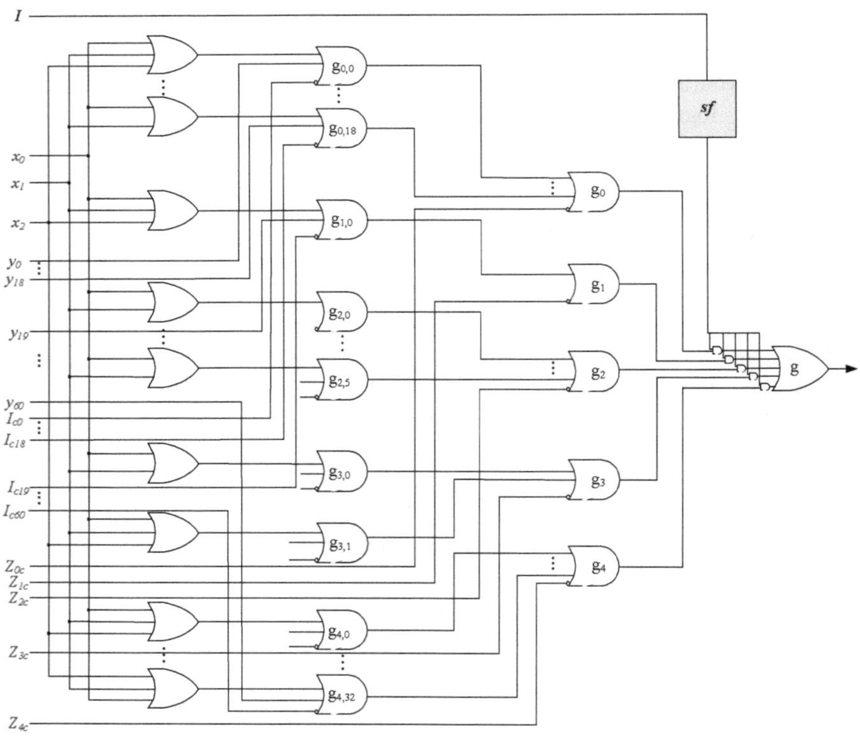

Figure 3.10. Logic network of the application software

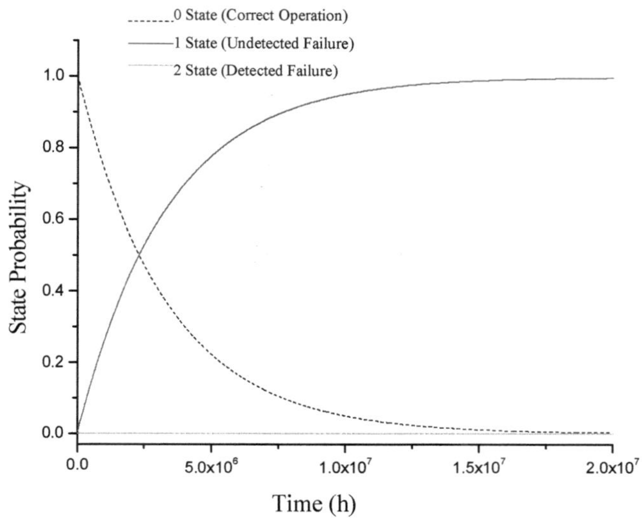

Figure 3.11. State probability of the system without fault-handling techniques

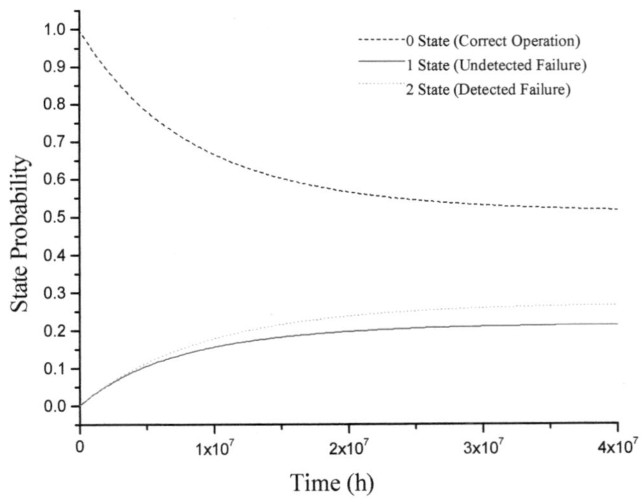

Figure 3.12. State probability of the system with fault-handling techniques of hardware components

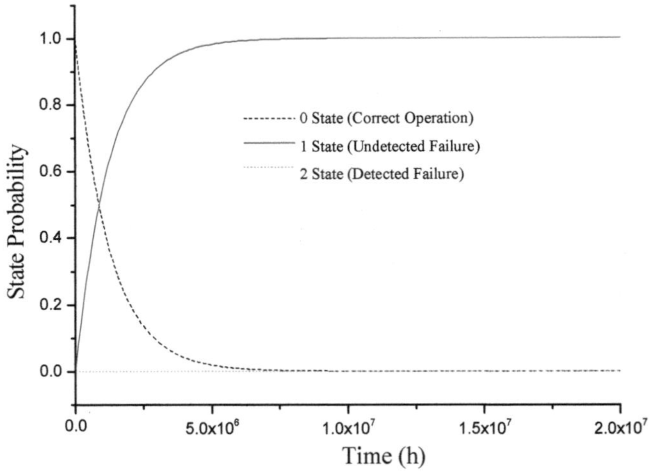

Figure 3.13. State probability of the system with consideration of software operational profile but without consideration of fault-handling techniques

This modeling method is particularly attractive for embedded systems in which small-sized software is implemented, since it is laborious to apply this method to systems with large software.

The work for obtaining the coverage of each fault-handling technique implemented in the components is required for the application of this model. One method to do this is to obtain coverage by fault injection experiment. A more general analytical approach is left for future work.

3.3 Case Study 3: Risk Assessment of Safety-critical Digital System

There are important, complex, correlated, and unresolved PRA issues in digital systems safety analyses. The characteristics of digital systems listed in Chapter 2 are summarized from the viewpoint of PRA modeling with the following factors:

- Modeling the multi-tasking of digital systems
- Estimating software failure probability
- Estimating the effect of software diversity and V&V efforts
- Estimating the coverage of fault-tolerant features
- Modeling the CCF in hardware
- Modeling the interactions between hardware and software
- Failure mode of digital system
- Environmental effects
- Digital-system-induced initiating events including human errors

The aim of this section is to examine the analysis framework of safety of digital systems in the context of the PRA and assess the effect of factors listed above.

3.3.1 Procedures for the PRA of Digital I&C System

System configuration and operating environment are investigated carefully in order to develop a PRA model. The PRA model aims at quantifying the risk due to failure of a safety function. The first step is identifying the hazard of a system failure. Only the fail-to-hazard failure modes are considered. Many safety-critical I&C systems are designed using the fail-to-safe philosophy. Clear identification of hazard status from safe status is required using fail-to-safe design for cases, such as a safety-critical network failure.

Understanding the failure mechanism of a safety function is important. Reasons for a specific safety function failure are categorized into two groups: mechanical actuator failure and signal generation failure (Figure 2.12). In a conventional analysis, mechanical failures are treated as the main contributors of risk because the conventional analysis treats the signal generation system as independent basic events or as simple fault trees. In the case of an analog signal-processing system, every signal maintains a fully independent processing channel. The situation with digital equipment is different. Microprocessors and software technologies make the digital system multi-functional and the system performs several functions sequentially or conditionally. This multi-tasking feature causes a risk concentration and deteriorates the system reliability.

The function of an I&C system is usually initiated by the input signals. Thus, the availability and validity of an input signal are important. Redundant input signals or an operator manual input are available in some cases. The development of a PRA model requires in-depth analysis for an input availability, including a document survey, a simulation, and an expert judgment [23]. For example, multiple trip parameters are processed in the digitalized signal-processing system for the reactor trip signal generation in nuclear power plants. Whether the diverse means for signal generation can be credited under a specific situation must be decided in this phase. Which trip parameters will exceed the setpoints must also be decided in this specific situation.

The processing mechanism of the digital system, which consists of many redundant components and utilizes a network communication and self-monitoring algorithms, is investigated. Self-monitoring or fault-tolerant mechanisms effectively enhances system availability. Quantification of the coverage of advanced algorithms is important. The treatment of software failure is also an important topic. The common-cause failure group is carefully identified with a consideration of the development/operation environment.

A human operator plays the role of a backup for an automated digital I&C system. The dependency of an automated system with an operator is considered for quantification of failure probability in a manual action. For example, an instrumentation sensor failure causes a concurrent failure of both signal generation mechanisms. Malfunction of an automated system is also an error-forcing context for the human operator [24].

3.3.2 System Layout and Modeling Assumptions

The reactor protection system (RPS) is one of the most important safety-critical digital systems in a nuclear power plant. Many RPSs, including those in Korean standard nuclear plants, adopt a four-channel layout to satisfy the single failure criterion and improve the plant availability. The schematic diagram of a typical four-channel RPS includes a selective two-out-of-four voting logic (Figure 3.14). The RPS has four channels which are located in electrically and physically isolated rooms. The RPS automatically generates reactor trip and engineered safety feature (ESF) actuation signals whenever monitored processes reach predefined setpoints.

The bistable processor (BP) module in each channel receives analog and digital inputs from sensors or from other processing systems through analog input (AI) and digital input (DI) modules. The BP module determines the trip state by comparing input signals with predefined trip setpoint. The logic-level trip signals generated in the BP module of any channel are transferred to coincidence processor (CP) modules of all the channels through hardwired cables or data links. The RPS includes multiple number of BPs in a channel for redundancy. BPs in a channel are connected with process variables in different order. The trip logic is also executed among redundant BPs in different order for providing diversity.

Each CP module performs two-out-of-four voting for each process input using signals from four BP modules in four channels. The CP module produces an output signal using a dedicated digital output (DO) module. A halt of the CP module causes the heart beat signal to a watchdog timer to stop, then the watchdog timer forces the RPS trip and initiates a trip signal.

A schematic diagram of a typical four-channel digital RPS, signal flow in RPS, and the structure of a selective two-out-of-four logic which initiates the interposing relay are illustrated in Figures 3.14 to 3.16. The faulttree model is built by using the faulttree analysis software tool, KwTree, which was developed by Korea Atomic Energy Research Institute (KAERI) as part of an integrated PRA software package, KIRAP. KIRAP consists of a faulttree analysis tool, a cutset generator, an uncertainty analysis tool, and basic event analysis tools [25].

Assumptions used in the model [26] are summarized as:

- All failure modes are assumed to be hazardous since there is not sufficient enough information about failure modes of digital systems.
- The effect from other components, such as trip circuit breakers, interposing relays, sensors, and transducers, is out of scope. This analysis concentrates on the digital system. Failure rates of non-RPS components are ignored for simplicity in analysis.
- Watchdog timers monitor the status of final output generation processors. The coverage of timer-to-processor monitoring is much lower than that of processor-to-processor monitoring because the processor-to-processor monitoring method uses more sophisticated algorithms. Watchdog timers are assumed to be able to detect software failures with the same coverage as in the case of hardware failures.

- Every processor is assumed to contain identical software program. Software failure induces the CCF of processors.
- The fail-to-hazard probability of the network or serial communications is ignored.
- The fail-to-hazard probability of the inter-system data bus and the back plane of PLC is ignored.
- The components are assumed to be tested at least once per month, or at a periodic test interval (T) of 730 hours. Component unavailability (Q) is half of the product of failure rate (λ) and periodic test interval: $Q = \lambda T/2$.

Investigating the quantitative relationship between important factors and PRA results is the aim of this case study. PRA results provide useful insights, even though the adequacy of these assumptions is not guaranteed and the model requires further refinement.

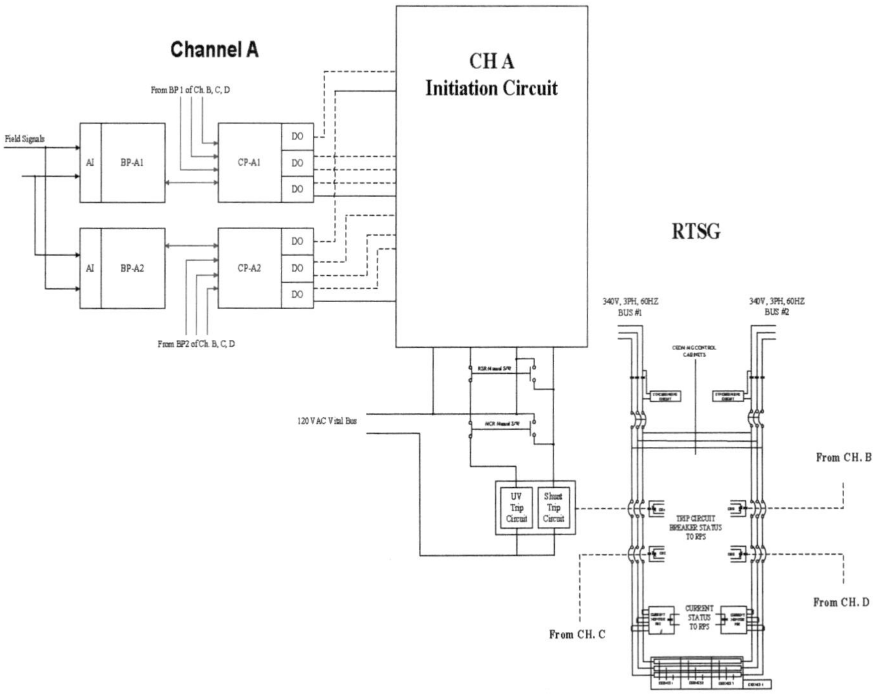

Figure 3.14. Schematic diagram of a typical RPS

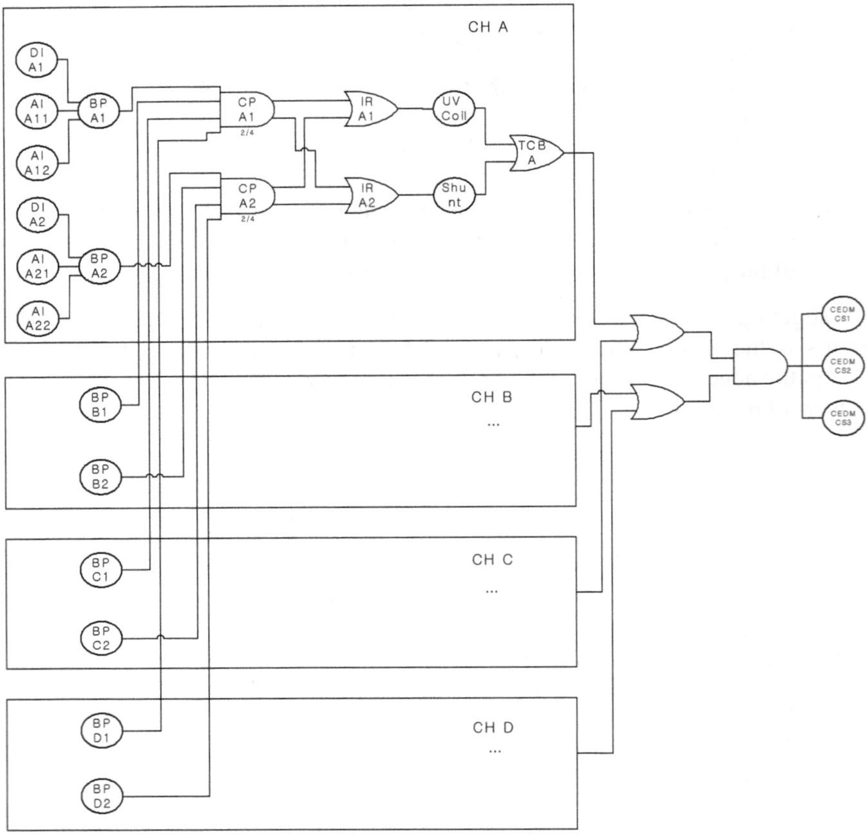

Figure 3.15. The signal flow in the typical RPS

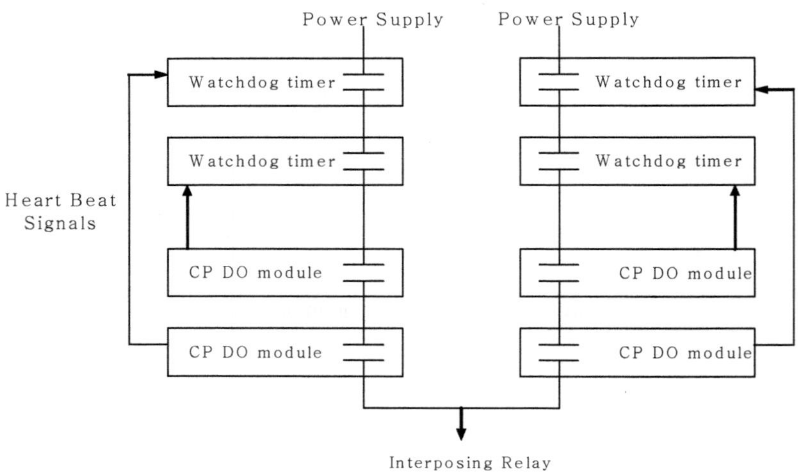

Figure 3.16. The detailed schematic diagram of watchdog timers and CP DO modules

3.3.3 Quantification

The analytic process is required to find critical factors and the explanation of the relationship between factors and PRA results. The result of the faulttree analysis is expressed in the form of a probability sum as:

$$\text{System Unavailability} = q_1 + q_2 + \ldots + q_i + \ldots + q_n \tag{3.15}$$

$$q_i = p_1 \times p_2 \times \ldots \times p_j \times \ldots \times p_m \tag{3.16}$$

where q_i is the probability of cutset i and p_j is the probability of basic event j. The probability of a basic event in the fault tree is the failure probability of a corresponding component. Cutset is defined as a set of system events that, if they all occur, would cause system failure [27].

The cutsets of multi-channel protection systems are categorized into two groups [26]: cutsets which include dependent events that result in multiple failures caused by the same reason, and cutsets which consist of the possible combinations of independent events which make all channels unavailable. The first group is divided into three groups: (1) cutsets which disturb collecting input signals, (2) cutsets which disturb generating proper output signals, and (3) cutsets which cause the distortion of processing results.

Four-channel redundancy makes the probabilities of almost whole "possible combinations of independent basic events" negligible, because the order of failure probabilities of safety-grade digital modules is very low (usually less than 10^{-3} per demand). Thus, the cutsets in group 2 are negligible. The cutsets which contain CCF events become main contributors to system unavailability. CCF is the failure of multiple components at the same time. CCF probability is much higher than the probability of combinations of different basic events in the case of the example system, because the value of CCF probability usually equals several percent of the independent failure probability, while the multiplication of the low probabilities (usually less than 10^{-3}) of independent failures is a thousand or more times lower than the independent failure probability.

Analyses of several possible design alternatives provide similar results. Details of the analysis result depend upon the design concept of the system. But, every dominant cutset of the four-channel digital protection system consists of CCF probabilities of digital modules and the error probability of a human operator. Conceptually, the dominant cutsets of a multi-channel digital protection system are expressed mathematically as:

$$q_1 = \Pr(\text{OP}) \times \Pr(\text{AI CCF})$$

$$q_2 = \Pr(\text{OP}) \times \Pr(\text{DO CCF})$$

$$q_3 = \Pr(\text{OP}) \times \Pr(\text{PM CCF}) \times \Pr(\text{WDT CCF})$$

$$q_4 = \Pr(\text{OP}) \times \Pr(\text{PM CCF}) \times \{\Pr(\text{WDT } a_1) \times \Pr(\text{WDT } a_3) \ldots\}$$

$$q_5 = \Pr(\mathrm{OP}) \times \Pr(\mathrm{PM\ CCF}) \times \{\Pr(\mathrm{WDT}\ a_1) \times \Pr(\mathrm{DO}\ a_3) \ldots\} \quad (3.17)$$

where:

Pr(OP) = the probability that a human operator will fail to manually initiate the reactor trip

Pr(AI CCF) = the probability of the CCF of analog input modules

Pr(DO CCF) = the probability of the CCF of digital output modules

Pr(PM CCF) = the probability of the CCF of processor modules

Pr(WDT CCF) = the probability of the CCF of watchdog timers

Pr(WDT α) = the probability that the watchdog timer α will fail to initiate the reactor trip

Pr(DO β) = the probability that the digital output module β will fail to initiate the reactor trip

The first and second cutsets (denoted by q_1 and q_2) of Equation 3.17 correspond to the probability of simultaneous failures of a human operator and all input/output modules, each of which belongs to groups 1 and 2 respectively. The third cutset, q_3, implies the probability of simultaneous failures of a human operator, all processor modules and all watchdog timers. The fourth and fifth cutsets, q_4 and q_5, of Equation 3.17 correspond to the probability of simultaneous failures of a human operator and all processor modules and the combined failures of watchdog timers and digital output modules. The cutsets of q_3, q_4 and q_5 are related to the processor module, and belong to group 3.

The processor module is the most complex part of a digital system. The reliability of this module is relatively lower than input/output modules. Software is installed in processor modules. Software failure is assumed to be included in the CCF event of processor modules. Installation of the same software in redundant systems might remove the redundancy effect. Therefore, the CCF of processor modules is a major obstruction to the proper working of digital protection systems.

However, most safety-critical applications, such as protection systems of nuclear plants, will reduce risk by adopting fault-tolerant mechanisms. Effective fault-tolerant mechanisms protect system safety so that it is not severely affected by the failure probability of processor modules. Relatively low fault detection probability is expected in the case of watchdog timer applications. The failure probability of a watchdog timer includes the detection probability of a processor module's fault.

Pr(OP) is not directly correlated to the digital system. The effect of Pr(WDT α) and Pr(DO β) on the system safety is relatively small. Critical variables are determined based on Equation 3.17: Pr(AI CCF), Pr(DO CCF), Pr(PM CCF), and Pr(WDT CCF).

The relationships between factors mentioned earlier in Equation 3.17 are summarized as:

- Modeling the multi-tasking of digital systems: N/A (should be explicitly modeled)
- Estimating software failure probability: q_3
- Estimating the effect of software diversity and V&V efforts: q_3

- Estimating the coverage of fault-tolerant features: q_4 and q_5
- Modeling the CCF in hardware: All (q_1, q_2, q_3, q_4 and q_5)
- Modeling the interactions between hardware and software: q_3
- Failure mode of digital system: All
- Environmental effects: All
- Digital-system-induced initiating events including human errors: N/A (should be inspected from the plant-wide viewpoint)

3.3.4 Sensitivity Study for the Fault Coverage and the Software Failure Probability

The sensitivity of the PRA result along with critical variables mentioned in Section 3.2.3 is quantitatively examined in this section. Equation 3.17 is derived using static methodology and the fault tree method. As a result, the complex and dynamic features of digital systems are not fully reflected. The lack of failure data is another weak point of analysis. Intuition from Equation 3.17 is helpful in designing a safer system. A systematic analysis and quantitative comparison between design alternatives are expected to support decision-making for design improvement [28].

Three factors are considered in this sensitivity study: The CCF group, the software failure probability, and the watchdog timer coverage. Pr(AI CCF), Pr(DO CCF), Pr(PM CCF), and Pr(WDT CCF) are the most critical variables. The reasons for parameter extraction from each critical variable are as follows.

CCF probabilities of input/output modules, Pr(AI CCF) and Pr(DO CCF), depend on the system design because the CCF component group, that is the set of components affected by the same failure cause, varies by hardware design. Three design alternatives are assumed: (1) a system which uses identical input modules and identical output modules; (2) a system which uses two kinds of input modules and identical output modules; and (3) a system which uses two kinds of input modules and two kinds of output modules. A separate faulttree model is established for each design alternative to perform sensitivity studies.

The CCF probability of processor modules, Pr(PM CCF), depends on the hardware failure probability of the processor module, the software failure probability, the diversity of processor modules, and the interaction effect between hardware and software. Identical processor modules containing the same software are assumed to be used. The interaction effect between hardware and software is ignored. Pr(PM CCF) depends on the hardware and software failure probability. Software failure probability is treated in a probabilistic manner because of the randomness of the input sequences ("error crystals in software" concept), which is the most common justification for the apparent random nature of software failure. This case study adopts the "error crystal" concept, and uses 0.0, 1.0×10^{-6}, 1.0×10^{-5}, and 1.0×10^{-4} for values of software failure probability.

Pr(WDT CCF) depends on the failure probability of the contained relay and the fault coverage of watchdog timers. Pr(WDT CCF) mainly depends on the fault coverage of a watchdog timer, since the reliability variation of safety-grade relays used in watchdog timers is negligible. The failure rate of the watchdog device and the failure rate of the microprocessor determines the system unavailability related to microprocessors, if the watchdog mechanism is perfect. The system unavailability is expressed as the probability sum of covered faults and uncovered faults. The reliability of a watchdog device is much higher than that of a microprocessor-based device, because of its simplicity. Then fault coverage is the dominant contributor to systems unavailability. Several discrete values of 0.3 (poor coverage), 0.4, 0.6, 0.7, and 1.0 (perfect coverage) are used for the value of the coverage factor.

A total of 60 ($3 \times 5 \times 4$) calculations were performed. Only two trip parameters (steam generator level and pressurizer pressure) were considered. The CCF probability between different kinds of hardware devices was ignored, even if they were used for the same purpose. The results of calculation for a typical four-channel RPS design are graphically illustrated in Figures 3.17 to 3.19. The best unavailability of 4.80×10^{-9} was obtained from the system which had diverse input/output modules, perfect software and 100% fault coverage while the system which had identical input/output modules, poor software and poor fault coverage showed the worst result of 1.60×10^{-5}. A more detailed explanation of this sensitivity study is found in reference 26.

These factors remarkably affect the system safety as shown by the quantitative assessment results. The value of each factor changes the system unavailability up to several thousand times. Inappropriate consideration of these three important factors causes unreasonable assumptions and severely distorts the analysis results. Values used in the analysis must be in a reasonable range, even though the effects of these factors are hard to accurately quantify. The more realistic PRA results result in more reasonable and accurate engineering decisions.

RPS unavailability is estimated based on conventional failure probability data of digital components. A software failure probability under 1×10^{-5} and a fault coverage of a watchdog timer over 0.7 are hard to practically prove. System unavailability of the design with identical input/output/processor modules was around 3.7×10^{-6}. The diverse input/output modules resulted in the reduction of system unavailability to 4.6×10^{-7}. However, the designer must consider the economical aspects, such as development and maintenance costs. Design is the art of trade-off.

CCF events are one of the main contributors to a multi-channel system's unavailability because the CCF implies the concurrent failure of redundant backups. Unsystematic treatment of the CCF is responsible for much of the uncertainty about risks from operating nuclear power plants [29]. The importance of precise CCF modeling of digital equipment is emphasized because designers provide various redundancies and diversities throughout separated systems. For example, in the RPS of the Korean standard nuclear plant, there are 16 processors and 16 digital output modules which do the identical function of local coincidence logic. These huge redundant systems will simultaneously lose their function in the event of CCFs of processors and digital output modules. Designers should be careful to avoid the CCF since even the products from different vendors do not guarantee the independence of faults. Analysts also should be careful to make an assumption of independence.

Software failure in digital safety-critical systems induces severe problems on assessing system safety. The redundancy of hardware modules is useless in the case of software failure, if the same software is installed in redundant systems. Software failure also cannot be detected by a pure hardware-based monitoring mechanism. Prediction of software reliability using a conventional model is much harder than for hardware reliability. A testing-based approach requires a huge number of well-designed test cases [30]. The use of diverse versions of software in a multi-processor system makes the analysis more complex. Software diversity increases reliability compared with single versions, but this increase is much less than what completely independent failure behavior would imply. The independence assumption is often unreasonable in practice [31]. There is no generally accepted method to estimate the degree of dependence. Therefore, the degree of dependence must be estimated for each particular case.

The effort for proving complete software is compensated by large coverage of a sophisticated monitoring mechanism. A system which has 10^{-4} as the software failure probability and 0.4 as the fault coverage has 7.2×10^{-6} as the system unavailability (Figure 3.18). The software failure probability must be proved below 10^{-6}, if the system unavailability must be reduced to 1.6×10^{-6} by using the low failure probability of software; 10^{-6} with 90% confidence level implies 2.30×10^{6} tests without failure while 10^{-4} implies 2.30×10^{4} tests [32]. The designer also achieves the target system unavailability by improving the fault coverage of watchdog timers from 0.4 to 0.6. Then the designer should check and compare the costs of alternatives.

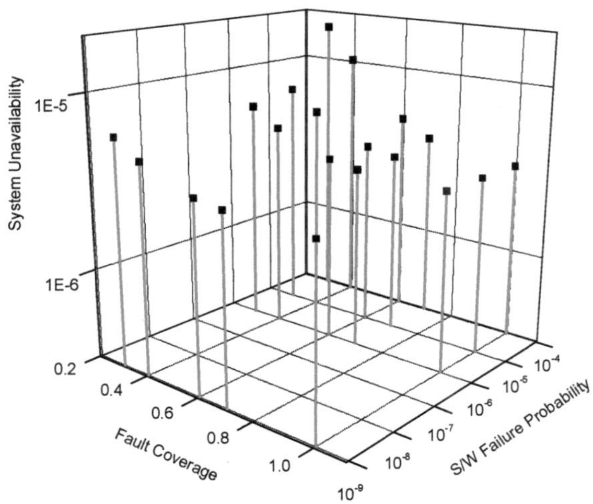

Figure 3.17. System unavailability along fault coverage and software failure probability when identical input and output modules are used

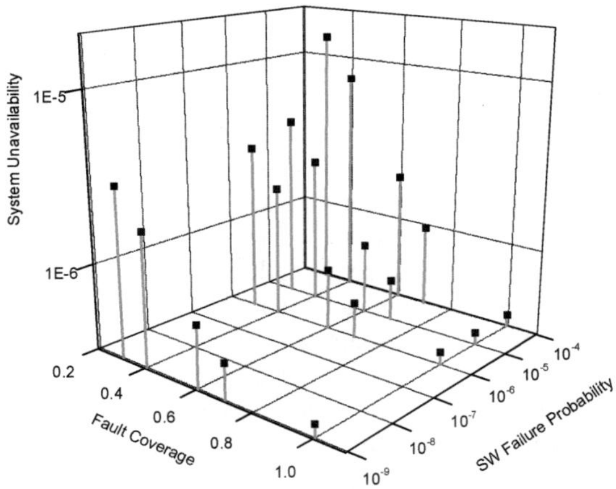

Figure 3.18. System unavailability along fault coverage and software failure probability when two kinds of input modules and the identical output modules are used

There are limitations in this analysis. First, all failure modes are assumed to be hazardous. The more precise estimation of failure modes is helpful in obtaining more realistic analysis results. Second, the result might be more realistic if coverage for software failures is considered separately from coverage for hardware failures. However, there is no available research regarding the coverage for the software failure. Third, the diversity of software versions mentioned above is not considered.

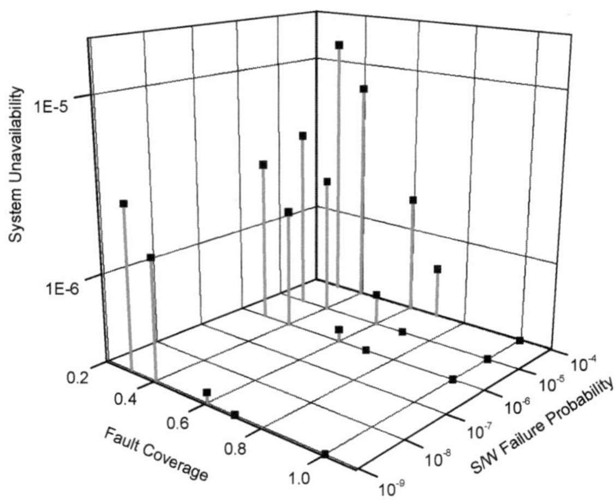

Figure 3.19. System unavailability along fault coverage and software failure probability when two kinds of input modules and two kinds of output modules are used

3.3.5 Sensitivity Study for Condition-based HRA Method

Quantification as part of a human reliability assessment involves the derivation of a probability distribution for basic events modeled in a PRA. Each HEP consists of one unsafe action (UA) of which the probability is affected by error forcing contexts (EFC). The HEP (H) in a specific accident scenario is calculated as [33, 34]:

$$H = \sum_{i} \Pr(\text{UA} \mid \text{EFC}_i) \Pr(\text{EFC}_i) \tag{3.18}$$

Two kinds of EFC are considered in this sensitivity study: alarms and sensor indications. The failure of display/actuation devices is not considered as an EFC for simplicity. The effect of independent failure of redundant equipment on system unavailability is relatively small. Some alarms are generated by the automatic system, of which a failure is also a reason for a signal generation failure. Both signal generation failure reasons must be considered: automated system failure and manual actuation failure. Sensor failures are independent of the accident scenario. For sensors (S) and automatic systems (A), The failure of an automatic system implies the failure of a safety signal generation and the loss of alarms. The signal generation failure probability (F) is calculated based on the HEP of Equation 3.18:

$$F = H = \sum_{i} \sum_{j} \Pr(\text{UA} \mid A_i, S_j) \Pr(A_i \mid S_j) \Pr(S_j) \tag{3.19}$$

Human operator performance is affected by the automated signal generation system. The failure of a system is one of signal failure reasons. Automatic systems consist of many components and input sensors. There are several redundant channels, and each channel processes input signals from corresponding sensors. Furthermore, there are complicated voting processes and monitoring mechanisms in order to avoid the loss of safety function in the case of a single component failure. The status of instrumentation sensors affects the performance of both automatic systems and human operators in a complicated manner. The construction of PRA model for all the combinations of EFCs, case by case, would require an infeasibly large amount of effort.

The following steps (condition-based HRA, CBHRA) are required to take into account the HEP issue with conditional events in a more effective manner based on the fault tree method:

1. Conducting an investigation into possible EFCs
2. Selecting important EFCs
3. Developing a set of conditions in consideration of selected EFCs
4. Estimating the HEP for each condition
5. Constructing a fault tree which includes one human error (HE) event for each manual action
6. Obtaining minimal cutsets (MCS) by solving the fault tree

7. Post-processing of MCSs

The purpose of steps (1) to (3) is the development of the EFC groups. Possible EFC combinations are categorized into several groups (*n* groups) in order to treat them in a practical manner, since the consideration of all the EFC combinations in a separate manner is very complicated. Steps (5) and (6) are the same steps as in a conventional PRA approach. The MCSs must be categorized into several sets with the viewpoint of the HE event. The number of MCS sets equals that of the HE events used in step (5).

Step (7) implies a substitution of the HE event in a set of MCSs with the EFC-group-specified HE event with consideration of the other events in each MCS. For example, the event of "the manual reactor trip failure (MRTF)" is substituted by one of the possible EFC-group-specified HE events: "MRTF given EFC group 1," "MRTF given EFC group 2,"..., or "MRTF given EFC group *n*."

The manual implementation of step (7) is expected to require much effort. Therefore, an automatic conditioning with a PRA software package is recommended. Automatic conditioning is enabled based on logical rules, such as "if there are more than three sensor-failure events in the MCS, then substitute the basic HE event with the HE event given no alarm and no indication," or "if there is no sensor failure, then substitute the basic HE event with the HE event given no alarm and all indications."

On the other hand, an investigation into the event of AUTO_SUCCESS is necessary for the EOC to distinguish the groups. Generally, when a negation gate is used in the fault tree model, obtaining the corresponding MCSs is difficult because the usual software packages require many resources and a long processing time to solve the negation logics. The model of a single EOC event is preferable to that of multiple EOC events for practical use. The probability of an AUTO_SUCCESS event is assumed to be unity when the automated signal processing channels are highly reliable.

This case study considers a single parameter safety function, such as an auxiliary feedwater actuation signal in nuclear power plants. The automatic feedwater makeup signal is generated when a signal-processing system detects that the water level in the steam generator is less than the setpoint.

The availabilities of automated safety signal, indication of the parameter, and alarm are tabulated based on the status of the automated system and the instrumentation sensors. The results of a single-parameter safety function in consideration of two-out-of-four voting logic are shown in Table 3.6. The bold entries are the EOC area in which the safety signals are automatically generated. The operator is expected not to interrupt them. The other entries indicate the EOO area in which the operator is expected to actively play the role of a backup for the automated system. There are two EOO conditions and one EOC condition for the single-parameter functions. The delicate quantification of the HEP in each condition, especially the EOC probability, is beyond the scope of this analysis.

The operator is assumed to spend a certain portion of the available diagnosis time overcoming the lack of information. That is, the operator is assumed to consume the given time for gathering the information from the other information sources. Thirty percent of the diagnosis time is assumed to remain in the case of

Table 3.6. The conditions of a human error in the case of the 4-channel single-parameter functions (O: available, X: unavailable)

Status of instrumentation	Status of the automated system	
	Normal	Abnormal
3 or more channels available	**Auto. signal: O** **Indication: O** **Alarm: O** **<Condition 1>**	Auto. signal: X Indication: O Alarm: X <Condition 2>
2 channels available	**Auto. signal: O** **Indication: X** **Alarm: O** **< Condition 1*>**	Auto. signal: X Indication: X Alarm: X <Condition 3>
1 or no channel available	Auto. signal: X Indication: X Alarm: X <Condition 3>	Auto. signal: X Indication: X Alarm: X <Condition 3>

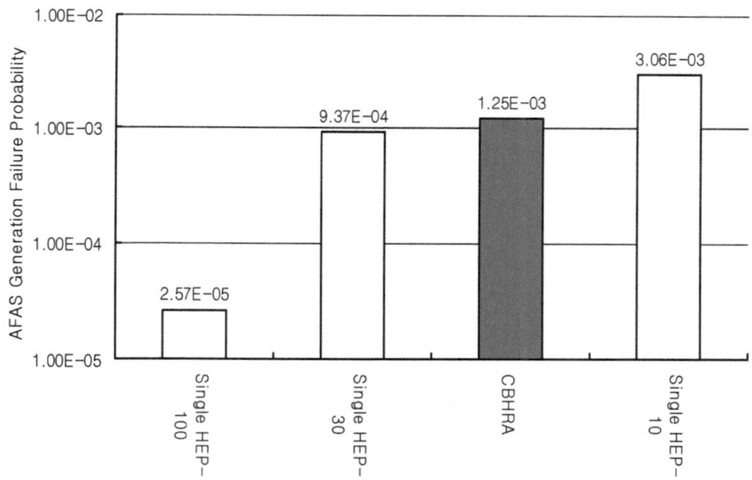

Figure 3.20. Comparison among single HEP methods and the CBHRA method for AFAS generation failure probabilities. Single HEP-100, 30, and 10 means that the single HEP method is used and the HEP is calculated based on the assumption that 100%, 30%, and 10% of the diagnosis time is available, respectively. For the CBHRA, 30% and 10% is assumed to be available for conditions 2 and 3, respectively.

condition 2 when the operator recognizes the situation under the trip/actuation alarms unavailable condition. Just 10% is assumed to remain for condition 3. The result of CBHRA is calculated, based on the HEPs for the conditions 1 to 3 in Table 3.6. The results of calculation for a typical four-channel RPS design are graphically illustrated in Figure 3.20. The other results in Figure 3.20 are calculated using the conventional single-human-error-event method. The CBHRA result, 1.25×10^{-3}, is significantly higher than the conventional result, 2.57×10^{-5}. This difference is caused by a consideration of information availability. The CBHRA considers the lack of information as an EFC, while a conventional analysis assumes that all the information is delivered to the operator. The result also demonstrates the merit of the CBHRA method due to a more sophisticated treatment of the EFCs.

3.4 Concluding Remarks

Digital systems have been installed in safety-critical applications, such as nuclear power plants, and their safety effect evaluation has become an important issue. The multi-tasking feature of digital I&C equipment increases the risk factor because it affects the actuation of safety functions by several mechanisms. The system modeling framework is overviewed in this chapter by using case studies.

The risk model should be properly developed to reasonably assess the risk from individual digital modules, by considering signal-processing paths and fault detection algorithms, which vary along the circuit and imbedded program designs. The various kinds of the CCF and the fault tolerance features should be carefully treated. Software failure is treated as a kind of CCF of processor modules in addition to the traditional CCFs. Human operator failure and automated signal generation failure also have interdependency.

Some valuable insights are derived from this quantitative study. The framework explained in these case studies is useful in characterizing the bounds of system unavailability.

References

[1] National Research Council (1997) Digital Instrumentation and Control Systems in Nuclear Power Plants, National Academy Press, Washington, D.C

[2] Kang HG, Jang SC, and Lim HG (2004) ATWS Frequency Quantification Focusing on Digital I&C Failures, Journal of Korea Nuclear Society, Vol. 36

[3] Laprie JC, Arlat J, Beounes C, and Kanoun K (1990) Definition and Analysis of Hardware-and-Software-Fault-Tolerant Architectures, IEEE Computer, Vol. 23, pp. 39–50

[4] Yau M, Apostolakis G, and Guarro S (1998) The Use of Prime Implicants in Dependability Analysis of Software Controlled Systems, Reliability Engineering and System Safety, No. 62, pp. 23–32

[5] Thaller K and Steininger A (2003) A Transient Online Memory Test for Simultaneous Detection of Functional Faults and Soft Errors in Memories, IEEE Trans. Reliability, Vol. 52, No. 4

[6] Bolchini C (2003) A Software Methodolgy for Detecting Hardware Faults in VLIW Data Paths, IEEE Trans. Reliability, Vol. 52, No. 4

[7] Nelson VP (1990) Fault-Tolerant Computing: Fundamental Concepts, IEEE Computer, Vol. 23, pp. 19–25

[8] Fenton NE and Neil M (1999) A Critique of Software Defect Prediction Models, IEEE Trans. Software Engineering, Vol. 25, pp. 675–689
[9] Butler RW and Finelli GB (1993) The Infeasibility of Quantifying the Reliability of Life-Critical Real-Time Software, IEEE Trans. Software Engineering, Vol. 19, pp. 3–12
[10] Choi JG and Seong PH (1998) Software Dependability Models Under Memory Faults with Application to a Digital system in Nuclear Power Plants, Reliability Engineering and System Safety, No. 59, pp. 321–329
[11] Goswami KK and Iyer RK (1993) Simulation of Software Behavior Under Hardware Faults, Proc. on Fault-Tolerant Computing Systems, pp. 218–227
[12] Laprie JC and Kanoun K (1992) X-ware Reliability and Availability Modeling, IEEE Trans. Software Eng., Vol. 18, No. 2, pp. 130–147
[13] Vemuri KK and Dugan JB (1999) Reliability Analysis of Complex Hardware-Software Systems, Proceedings of the Annual of Reliability and Maintainability, pp. 178–182
[14] Doyle SA, Dugan JB and Patterson-Hine FA (1995) A Combinatorial Approach to Modeling Imperfect Coverage, IEEE Trans. Reliability, Vol. 44, No. 1, pp. 87–94
[15] Davio M, Deshamps JP, and Thayse A (1978) Discrete and Switching Functions, McGraw-Hill
[16] Janan X (1985) On multistate system analysis, IEEE Trans. Reliability, Vol. R-34, pp. 329–337
[17] Levetin G (2003) Reliability of Multi-State Systems with Two Failure-modes, IEEE Trans. Reliability, Vol. 52, No. 3
[18] Levetin G (2004) A Universal Generating Function Approach for the Analysis of Multi-state Systems with Dependent Elements, Reliability Engineering and System Safety, Vol. 84, pp. 285–292
[19] Kaufman LM, Johnson BW (1999) Embedded Digital System Reliability and Safety Analysis, NUREG/GR-0020
[20] Siewiorek DP (1990) Fault Tolerance in Commercial Computers, IEEE Computer, Vol. 23, pp. 26–37
[21] Veeraraghavan M and Trivedi KS (1994) A Combinatorial Algorithm for Performance and Reliability Analysis Using Multistate Models, IEEE Trans. Computers, Vol. 43, No. 2, pp. 229–234
[22] Beizer B (1990) Software Testing Techniques, Van Notrand Reinhold
[23] Kang HG and Jang SC (2006) Application of Condition-Based HRA Method for a Manual Actuation of the Safety Features in a Nuclear Power Plant, Reliability Engineering and System Science, Vol. 91, No. 6
[24] American Nuclear Society (ANS) and the Institute of Electrical and Electronic Engineers (IEEE), 1983, PRA Procedures Guide: A Guide to the Performance of Probabilistic Risk Assessments for Nuclear Power Plants, NUREG/CR-2300, Vols. 1 and 2, U.S. Nuclear Regulatory Commission, Washington, D.C
[25] Han SH et al. (1990) PC Workstation-Based Level 1 PRA Code Package KIRAP, Reliability Engineering and Systems Safety, Vol. 30
[26] Kang HG and Sung T (2002) An Analysis of Safety-Critical Digital Systems for Risk-Informed Design, Reliability Engineering and Systems Safety, Volume 78, No. 3
[27] McCormick NJ (1981) Reliability and Risk Analysis, Academic Press, Inc. New York
[28] Rouvroye JL, Goble WM, Brombacher AC, and Spiker RE (1996) A Comparison Study of Qualitative and Quantitative Analysis Techniques for the Assessment of Safety in Industry, PSAM3/ESREL'96
[29] NUREG/CR-4780 (1988) Procedures for Treating Common Cause Failures in Safety and Reliability Studies
[30] HSE (1998) The use of computers in safety-critical applications, London, HSE books

[31] Littlewood B and Strigini L (1993) Validation of Ultrahigh Dependability for Software Based Systems, Communications of ACM, Vol. 36, No. 11

[32] Kang HG and Sung T (2001) A Quantitative Study on Important Factors of the PSA of Safety-Critical Digital Systems, Journal of Korea Nuclear Society, Vol. 33, No. 6

[33] US Nuclear Regulatory Commission (USNRC) (2000) Technical Basis and Implementation Guidelines for a Technique for Human Event Analysis (ATHEANA), Washington, D.C., NUREG-1624 Rev. 1

[34] Forester J, Bley D, Cooper S, Lois E, Siu N, Kolaczkowski A, and Wreathall J (2004) Expert Elicitation Approach for Performing ATHEANA Quantification, Reliability Engineering and System Safety, Vol. 83

Part II

Software-related Issues and Countermeasures

4

Software Faults and Reliability

Han Seong Son[1] and Man Cheol Kim[2]

[1]Department of Game Engineering
Joongbu University
#101 Daehak-ro, Chubu-myeon, Kumsan-gun, Chungnam, 312-702, Korea
hsson@joongbu.ac.kr

[2]Integrated Safety Assessment Division
Korea Atomic Energy Research Institute
1045 Daedeok-daero, Yuseong-gu, Daejeon, 305-353, Korea
charleskim@kaeri.re.kr

Software, unlike hardware, does not fail, break, wear out over time, or fall out of tolerance [1]. Hardware reliability models are based on variability and the physics of failure (Chapter 1), but are not applied to software since software is not physical. For example, it is not possible to perform the equivalent of accelerated hardware stress testing on software. Consequently, different paradigms must be used to evaluate software reliability, which raises a few issues for software reliability engineers.

Software reliability issues in safety modeling of digital control systems are introduced in Section 2.3. Issues considered are quantification of software reliability, assessment of software lifecycle management, diversity, and hardware–software interactions. These issues are directly related to software faults. This chapter discusses software reliability issues in view of software faults. Software faults themselves are discussed in Section 4.1. Software reliability is a part of overall system reliability, particularly from the viewpoint of large-scale digital control systems. Integrating software faults into system reliability evaluation, such as probabilistic risk assessment, is important (Chapter 2). This involves quantitative and qualitative software reliability estimation (Section 4.2 and 4.3). Software reliability includes issues related to software reliability improvement techniques (Chapter 5).

4.1 Software Faults

Software takes inputs from other systems (software, hardware, or humans) and produces outputs that are used by either humans, or other software and hardware. Software runs on a computer platform that interacts with the environment.

Software failures may originate within the software or from the software interface with the operational environment. Software faults are classified into software functional faults (faults within software) and software interaction faults (input/output faults and support faults) [2]. Support faults are related to failures in computing resource competition and computing platform physical features. An abnormal operation of the hardware platform may cause failure of the software.

There have been two different views of software faults. Software faults may be random or systematic. Random failures may occur at any time. It is not possible to predict when a particular component will fail. A statistical analysis is performed to estimate the probability of failure within a certain time period by observing a large number of similar components. Failures caused by systematic faults are not random and cannot be analyzed statistically. Such failures may be predictable. Once a systematic fault has been identified, its likely effect on the reliability of the system is studied. However, unidentified systematic faults represent a serious problem, as their effects are unpredictable and are not normally susceptible to a statistical analysis. Software functional faults and some software interaction faults correspond to the "systematic" view, while other software interaction faults correspond to the "random" view.

4.1.1 Systematic Software Fault

Software faults are not random but systematic [3]. Software failure is caused by either an error of omission, an error of commission, or an operational error. Systematic software faults are tightly coupled with humans. An error of omission is an error that results from something that was not done [3]:

- Incomplete or non-existent requirements
- Undocumented assumptions
- Not adequately taking constraints into account
- Overlooking or not understanding design flaws or system states
- Not accounting for all possible logic states
- Not implementing sufficient error detection and recovery algorithms

Software designers often fail to understand system requirements from functional domain specialists. This results in an error of omission. Domain experts tend to take for granted things which are familiar to them, but which are usually not familiar to the person eliciting the requirements [4]. This is also one of the main reasons for errors of omission. A ground-based missile system is an example [3]. The launch command could be issued and executed without verifying if the silo hatch had been first opened during simulated testing and evaluation of the system. The fact that everyone "knows" that you are supposed to do something may cause an error of omission in a requirement elicitation process.

Errors of commission are caused by making a mistake or doing something wrong in a software development process. Errors of commission include [3]:

- Logic errors
- Faulty designs
- Incorrect translation of requirements in software

- Incorrect handling of data
- Faulty system integration

A typical example for logic errors is the inadequate use of logic constructs – CASE constructs or IF/THEN/ELSE constructs – resulting in an unintended output [3]. Different logic constructs have different evaluation and execution mechanisms. A software designer should be careful to use logic constructs so that unplanned events do not occur.

Operational errors result from the incorrect use of a product, which can be accidental or intentional incorrect usage. Examples of operational errors include [3]:

- Induced or invited errors
- Illegal command sequences
- Using a system for a purpose or in an environment for which it was not intended
- Inhibiting reliability features
- Not following operational reliability procedures

Designers can minimize the opportunity for induced or invited errors by incorporating comprehensive human factors and engineering practices [1]. Extensive error detection and recovery logic prevents the execution of illegal command sequences. Adequate documentation of intended operational environment and procedures reduces the likelihood of accidental operational errors.

4.1.2 Random Software Fault

Some reliability engineers believe that the cause of a software failure is not random but systematic. However, software faults take an almost limitless number of forms because of the complexity of the software within a typical application. The complex process involved in generating software causes faults to be randomly distributed throughout the program code. The effect of faults cannot be predicted and may be considered to be random in nature. Unknown software faults are sufficiently random to require statistical analysis.

Software faults become failures through a random process [5]. Both the human error process and the run selection process are dependent on many time-varying variables. The human error process introduces faults into software code and the run selection process determines which code is being executed in a certain time interval. A few methods for the quantification of the human error process are introduced in Chapter 8. The methods may be adopted to analyze the random process of software faults.

Some software interaction faults (*e.g.*, support faults with environmental factors) are involved in the viewpoint that software faults are random. There exists a software masking effect on hardware faults (Section 2.3.3). A substantial number of faults do not affect the outputs of a software-based system. A random hardware masking effect of software faults exists. The randomness is understood more easily by considering the aging effect on hardware. This effect induces slight changes on hardware. The system may produce faulty outputs by some kinds of software, but it may not by other kinds of software.

4.1.3 Software Faults and System Reliability Estimation

A system fails to produce reliable responses when the system has faults. Systems may fail for a variety of reasons. Software faults are just one of the reasons. Random or systematic software faults are issues of decision making, particularly in large-scale digital control systems. A reliability engineer who considers random software faults will easily incorporate them into system reliability estimation or directly estimate software reliability based on quantitative software reliability models (Section 4.2).

Software faults are integrated into PRA to statistically analyze system reliability and/or safety [2]. Software failure taxonomy, developed to integrate software into the PRA process, identifies software-related failures (Section 4.1). Software failure events appear as initiating and intermediate events in the event sequence diagram or event tree analysis or even as elements of fault trees. The software PRA three-level sub-model includes a special gate depicting propagation between failure modes and the downstream element. The downstream element is the element that comes after the software in an accident scenario or after the software in a fault tree.

Quantification approaches for the three-level PRA sub-model are being developed. The first approach is based on past operational failures and relies on public information. Only quantification of the first level is performed by modeling the software and the computer to which a failure probability is assigned. PRA analysts use this information to quantify the probability of software failure when no specific information is available in the software system. The second approach pursues target quantification of the second level using expert opinion elicitation. The expert opinion elicitation approach is designed to identify causal factors that influence second-level probabilities and to quantify the relationship between such factors and probabilities. Analysts, who have knowledge of the environment in which the software is developed, are able to assess the values taken by these causal factors and hence quantify the unknown probabilities once such a causal network is built.

A reliability engineer considers systematic software faults and evaluates software reliability based on qualitative models (Section 4.3). A holistic model and BBN are used to evaluate the effect of systematic software faults on reliability of a whole system as well as qualitative models. A holistic model is introduced in Section 4.4. BBN is discussed in Section 2.3.2.

4.2 Quantitative Software Reliability Models

There has been a debate within industry, academia, and the international standards community concerning the quantification of software reliability. Some promote qualitative assessment of software reliability while others promote quantitative assessment. Quantitative models focus on product issues. Qualitative software reliability models (Section 4.3) focus on process issues [1].

4.2.1 A Classification of Quantitative Software Reliability Models

Software reliability models are classified according to software development lifecycle phases [6]:

- Early Prediction Models – These models use characteristics of the software development process from requirements to tests and extrapolate this information to predict the behavior of software during operation [7]. These include Phase-based Model, Rome Laboratory Model, Raleigh Model, Musa Prediction Model, Industry Data Collection, and Historical Data Collection.
- Software Reliability Growth Models – These models capture failure behavior of software during testing and extrapolate to determine its behavior during operation. These models use failure data information and trends observed in failure data to derive reliability predictions. Software reliability growth models are classified as Concave models and S-shaped models [6]. Musa Basic Model, Goel Okumoto NHPP Model, Musa Okumoto NHPP Model, Musa Poisson Execution Time Model, Jelinski Moranda Model, Littlewood Verall Model, Weibull Model, and Raleigh Model are representative Concave models. S-shaped models include the Yamada S-shaped and Gompertz models.
- Input-Domain-Based Models – These models use properties of software input domain to derive a correctness probability estimate from test cases that executed properly [8]. Nelson Model, Tsoukalas Model, and Weiss & Weyuker Model fall into this class. Input-domain-based models are used in the validation phase, but cannot be used early in the software development lifecycle.
- Architecture-Based Models – These models emphasize *SA and derive reliability estimates by combining estimates obtained for different modules of software [9]. The architecture-based software reliability models are further classified into State-based, Path-based, and Additive models. Heterogeneous Software Reliability Model, Laprie Model, Gokhale *et al.* Model, and the Gokhale Reliability Simulation Approach are State-based models. Path-based models include Shooman Model, Krishnamurthy and Mathur Model, and Yacoub Cukic Ammar Model. Everett Model and Xie Wohlin Model fall into the Additive model class.
- Hybrid Black Box Models – These models combine the features of input-domain-based models and software reliability growth models. Input-Domain-based Software Reliability Growth Model is a representative hybrid black box model.
- Hybrid White Box Models – These models use selected features from both white box models and black box models. These models are considered in hybrid white box models since these models consider the architecture of the system for reliability prediction. A time/structure-based model for

estimating software reliability has been proposed. This model is a hybrid white box model.

All these models are appropriately selected, considering many criteria, to estimate software reliability. The reliability model selection is a new decision-making problem. Criteria used for software reliability model selection have been proposed [6], including lifecycle phase, output desired by the user, input required by model, trend exhibited by the data, validity of assumptions according to data, nature of the project, structure of the project, test process, and development process. All criteria are tightly related to issues of software reliability quantification (Section 4.2.3).

4.2.2 Time-related Software Reliability Models *Versus* Non-time-related Software Reliability Models

There is a debate on whether or not software reliability models should be time related, in addition to the quantitative versus qualitative debate. Quantitative software reliability models are divided into time-related models and non-time-related models. Quantitative time-related models have been extensively studied [10 – 11]. A definition of hardware reliability is "the probability of operating failure free for a specified time under a specified set of operating conditions." Hardware reliability models are quantitative and time related. Traditional software reliability models from the 1980s are based on the number of errors found during testing and the amount of time it took to discover them. These models use various statistical techniques, many borrowed from hardware reliability assessments, to estimate the number of errors remaining in the software and to predict how much time will be required to discover them [11, 12]. Thus, these models are also quantitative and time-related.

A main argument against a time-related software reliability model is [13]: "Software is susceptible to pattern failures that are not discovered until particular program or data sequences are processed." This argument is rephrased as "no one knows when the failure will occur through a processing of the defective sequence," making the time factor irrelevant. Non-time-related quantitative models have been reviewed [14]. The amount of execution time or test time has no bearing on the functionality of software. Software testing or mission simulation consists of verifying software output under varying input conditions. Simulation time is dependent on the verification rate, which is not consistent among tests.

A quantitative non-time-related software reliability model focusing on the effectiveness of the test suite and meeting established reliability goals has been developed [14]. The model is derived in part by using the Taguchi design of experiments technique. The first step is to determine what constitutes an effective test matrix by examining factors such as nominal, upper, and lower level operating specifications. Next, test effectiveness is calculated as the ratio of the number of input combinations in the matrix to the number of total possible input combinations. The third step is the measurement of success probability as number of successful tests divided by number of input combinations in the test matrix. The results are plotted and compared against customer expectations and reliability goals. The process is repeated and corrective action is taken until reliability goals are met.

A reliability engineer for a large-scale digital control system decides which one is more appropriate for the system between a time-related model and non-time-related model. Characteristics of the system and reliability-related information are investigated thoroughly in order to make a decision.

4.2.3 Issues in Software Reliability Quantification

Important issues directly involved in software reliability quantification are reliability modeling, field reliability prediction, reliability objective, operational profile, and reliability estimation based on rare events. Reliability modeling and field reliability prediction are related to the accuracy of models and the reliability data (Sections 4.2.3.1 and 4.2.3.2). Reliability objective and operational profile issues are directly coupled with expected software reliability data (Section 4.2.3.2). Rare-event-based reliability estimation is discussed in Section 4.2.3.3.

4.2.3.1 Accuracy of Models

Reliability modeling and field reliability prediction fall into a model accuracy issue. Models consist of two components: parameter estimation and prediction. Parameter estimation inputs either failure counts in time intervals or time between failures, and produces estimates of parameters related to failure rate or failure intensity, where failure rate is defined as the ratio of the number of failures of a given category to a given unit of measure [15]. Failure intensity has been defined [16]. Predictions are made of future software reliability once the model has been fitted with parameters. This does not guarantee accurate predictions, even with achieving a good model fit. Model accuracy is only validated by comparing predictions with future software reliability. A good fit with historical data is required in order to obtain accurate model parameter estimates (*e.g.*, accurate estimates of failure rate parameters).

4.2.3.2 Software Reliability Data

Another important aspect of software reliability quantification is the variety of data that is necessary to support reliability modeling and prediction. This data is also used to make empirical assessments of reliability. There are three types of software reliability data [16]:

- Error: A human action that produces an incorrect result (*e.g.*, an incorrect action on the part of a programmer or an operator).
- Fault: An incorrect step, process, or data definition in a computer program.
- Failure: The inability of a system or component to perform its required functions within specified performance requirements.

The three types are related in chronological sequence:

Error → Fault → Failure

Examples of the use of this data are [17]:

- Estimate parameters for reliability models from technique details.

- Empirically analyze trends for reliability growth or decrease.
- Empirically assess software reliability during test and operation against predetermined reliability objectives.
- Use reliability data to decide where and how much to inspect and test.

The first two uses of software reliability data are directly related to reliability models. Optimally selected failure data improve the accuracy of parameter estimation and prediction [17]. Not all failure data should be used to estimate model parameters and to predict failures using software reliability models. Old data may not be as representative of current and future failure processes as is recent data. More accurate predictions of future failures are obtained by excluding or giving lower weight to earlier failure counts. There must be a criterion for determining the optimal value of the starting failure count interval in order to use the concept of data aging. The mean square error criterion is used to determine the amount of historical failure data to use for reliability prediction [17].

Reliability data for the assessment of software reliability during test and operation are related to reliability objectives. Reliability objectives and the concept of necessary reliability are based on operational data [10]. Most large-scale digital control systems have large programs. The number of possible input sequences is huge in a large program with many inputs. Combined with the large number of possible program paths (in some cases, infinite), a significant number of execution sequences in the program are generated. Failure data are dynamically obtained from the complicated execution of programs. Statistical and analytical reliability models are employed to set a software reliability objective in terms of faults and/or failure densities that remain at the beginning of operational use using the failure data accumulated by many operations. Fault density is much easier to determine than remaining faults. Accurate models for remaining faults as a function of both fault density and observed failure intensity have been validated.

Final use involves operational profile, such as regression testing using failure data. Regression testing is performed after making a functional improvement or repair to the program. Its purpose is to determine if the change has adversely affected other aspects of the program. It is performed by rerunning the subset of the program's test cases – determined by the operational profile – that pertains to the change [5, 10]. Regression testing is important because changes and error corrections tend to be much more error prone than the original program code (in much the same way that most typographical errors in newspapers are the result of last-minute editorial changes, rather than errors in the original copy) [18].

4.2.3.3 Required Number of Tests if Failure Occurs or Not

The probability of failure is predicted to be zero when random testing reveals no failures. This approach will not differentiate between no failures after two tests and no failures after two billion tests. Testing once and finding no failures may cause one to think that it is reliable. Testing twice and finding no failures may cause one to think more confidently that it is reliable. This approach, named "Bayesian," is adoptable for a highly reliable system. The Bayesian framework incorporates prior assumptions into the analysis of testing results, and provides a mathematical framework for incorporating information other than random testing results into probability of failure estimates [19]. The Bayesian estimate of failure probability and the variance of the estimate are important in this framework in that the variance can be utilized as a factor to assure confidence.

All known faults are removed in software projects. Faults found must be identified and removed if there is a failure during operational testing. Tests for a safety-critical system should not fail during the specified number of test cases or a specified period of working. Numbers of fault-free tests to satisfy this requirement are calculated to ensure reliability (Section 2.3.1).

The number of additional failure-free tests for the software reliability to meet the criteria needs to be predetermined if failure occurs during the test. An approach based on a Bayesian framework was suggested to deal with this problem (Section 2.3.1) [20]. The test is assumed as an independent Bernoulli trial with probability of failure per demand, *p*, to derive the probability distribution function. From this assumption, the distribution of *p* can be derived and then the Bayesian framework is introduced to use prior knowledge. Prior knowledge is used as trials before failure occurs. The equation for the reliability requirement is obtained by the Bayesian approach:

$$\Pr(\text{no failures in the next } n_0 \text{ demands}) \geq 1 - \alpha \tag{4.1}$$

where $1 - \alpha$ is the confidence level. The mean and variance for number of failures R_f in the next n_f demands based on the Bayesian predictive distribution, if *r* failures have been met in the past *n* demands, are calculated as:

$$\mathrm{E}(R_f) = n_f \left[\frac{a + r}{a + b + n} \right] \tag{4.2}$$

$$\mathrm{Var}(R_f) = n_f \left[\frac{a + r}{a + b + n} \right] \left[1 - \frac{a + r}{a + b + n} \right] \left[\frac{a + b + n + n_f}{a + b + n + 1} \right] \tag{4.3}$$

where *a* (>0) and *b* (>0) represent prior knowledge. An observer represents a belief about the parameter of interest with values in the Bayesian framework. The uniform prior with $a = b = 1$ is generally used when no information about the system and its development process is available.

Error is corrected once failure occurs. The correction is always assumed to be perfect in this calculation. One of these three would result in: (1) error is corrected completely, (2) error is corrected incompletely, (3) error is treated incorrectly, so introduces more errors. These cases are treated differently.

4.2.4 Reliability Growth Models and Their Applicability

The reliability of software is directly estimated by using software reliability growth models, such as Jelinski–Moranda model [21] and the Goel–Okumoto's non-homogeneous Poisson process (NHPP) model [22], provided test results or failure histories are available. Software faults are detected and removed with testing effort expenditures in the software development process. The number of faults remaining in the software system is decreased as the test goes on. A mathematical tool that describes software reliability is a software reliability growth model. Software reliability growth models cannot be applied to large-scale safety-critical software systems due to a small number of expected failure data from the testing. The possibilities and limitations of practical models are discussed.

Unavailability due to software failure is assumed not to exceed 10^{-4}, which is the same requirement as that used for proving the unavailability requirement of programmable logic comparators for the Wolsung NPP unit 1 in Korea. The testing period is assumed to be one month, which is the assumption that is used in the unavailability analysis for the digital plant protection system of the Ulchin NPP unit 5 and 6 in Korea. Based on these data, the required reliability of the safety-critical software is calculated as:

$$\frac{\lambda T}{2} \leq U \tag{4.4}$$

$$\lambda \leq \frac{U}{2T} = \frac{10^{-4}}{2 \times 1\,\text{month}} = 2.78 \times 10^{-7}\,\text{hr}^{-1} \tag{4.5}$$

where:
U : required unavailability
λ : failure rate (of the software)
T : test period

Software reliability growth models are summarized and categorized into two groups [5]: (1) binomial-type models and (2) Poisson-type models. Well-known models are the Jelinski–Moranda [21] and Goel–Okumoto NHPP models [22]. The two representative models are applied to the example failure data, which are selected from the work of Goel and Okumoto [22]. The criteria for the selection of the example data are reasonability (the failure data can reasonably represent the expected failures of safety-critical software) and accessibility (other researchers can easily get the example failure data). Software reliability growth models are found to produce software reliability results after 22 failures through the analysis of the example failure data. The change in the estimated total number of inherent software faults (which is a part of software reliability result) was calculated by two software reliability growth models (Figure 4.1). Time-to-failure data (gray bar) represents the time-to-failure of observed software failures. For example, the 24th failure was observed 91 days after the occurrence, when correct repair of the 23rd software failure was implemented. The estimated total number of software inherent faults in the Jelinski–Moranda model and the Goel–Okumoto NHPP model are represented with a triangle-line and an x-line, respectively. The number of already observed failures is represented by the straight line. The triangle-line and the x-line should not be below the straight line because the total number of inherent software faults should not be less than the number of already observed failures.

There are several limitations for software reliability growth models that are applied to a safety-critical software system. One of the most serious limitations is the expected total numbers of inherent software faults calculated by the software

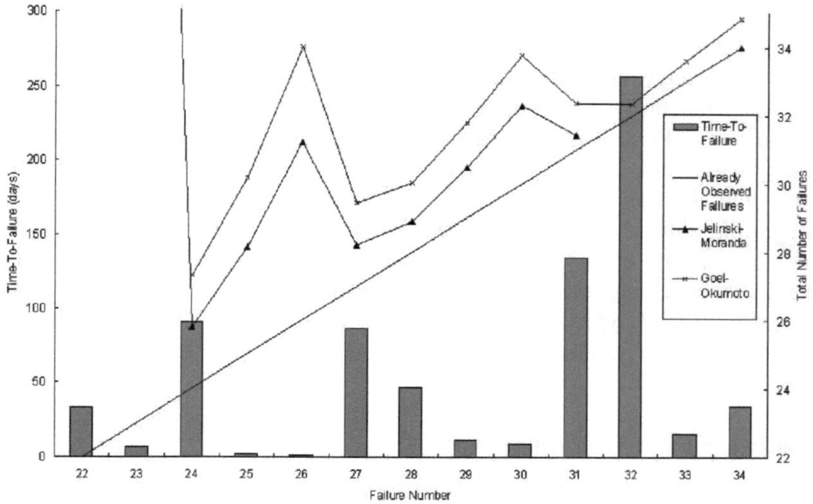

Figure 4.1. Estimated total numbers of inherent software faults calculated by Jelinski–Moranda model and Goel–Okumoto NHPP model

reliability growth models that are highly sensitive to time-to-failure data. After long time-to-failures, such as shown in the 24th failure, 27th failure, and 31st failure, drastic decreases in the estimated total number of inherent software faults are observed for both software reliability growth models (Figure 4.1). This sensitivity to time-to-failure data indicates that the resultant high software reliability (Equation (4.6)) could be a coincidence in the calculation process. One of other limitations is that, although at least 20 failure data are needed, we cannot be sure that the amount of failure data is revealed during the development and testing of a safety-critical software system.

4.3 Qualitative Software Reliability Evaluation

Qualitative software reliability models focus on process issues. Process issues concern how a software product is developed through the software lifecycle. Software reliability/risk assessment provides a warning to software managers of impending reliability problems early in the software lifecycle (*i.e.*, during requirements analysis). More efficient software management is possible by using risk factors to predict cumulative failures and values of the risk factor thresholds where reliability significantly degrades. Management is able to better schedule and prioritize development process activities (*e.g.*, inspections, tests) with advance warning of reliability problems. Some examples of software risk factors are attributes of requirement changes that induce reliability risk, such as memory space and requirements issues. Reliability risk due to memory space involves the amount of memory space required to implement a requirements change (*i.e.*, a requirements change uses memory to the extent that other functions do not have sufficient memory to operate effectively, and failures occur). Requirements issues mean conflicting requirements (*i.e.*, a requirements change conflicts with another requirements change, such as requirements to increase the search criteria of a web site and simultaneously decrease its search time, with added software complexity, causing failures). Process issues like requirements change are involved in software reliability evaluation. Thus, qualitative software reliability evaluation is useful in software reliability engineering.

Integrating software faults for probabilistic risk assessment has demonstrated that software failure events appear as initiating events and intermediate events in the event sequence diagram or event tree analysis, or even as elements of the fault trees, which are all typical analysis techniques of PRA [2]. This means that qualitative software evaluation methods are useful for quantitative system reliability assessment.

4.3.1 Software Fault Tree Analysis

Software Fault Tree Analysis (SFTA) is used in software safety engineering fields. SFTA is the method derived from Fault Tree Analysis (FTA) that has been used for system hazard analysis and successfully applied in several software projects. SFTA forces the programmer or analyst to consider what the software is not supposed to do. SFTA works backward from critical control faults determined by the system fault tree through the program code or the design to the software inputs. SFTA is applied at the design or code level to identify safety-critical items or components and detects software logic errors after hazardous software behavior has been identified in the system fault tree.

A template-based SFTA is widely used [4]. Templates are given for each major construct in a program, and the fault tree for the program (module) is produced by composition of these templates. The template for IF-THEN-ELSE is depicted in Figure 4.2. The templates are applied recursively, to give a fault tree for the whole module. The fault tree templates are instantiated as they are applied (*e.g.*, in the above template the expressions for the conditions would be substituted, and the event for the THEN part would be replaced by the tree for the sequence of statements in the branch). SFTA goes back from a software hazard, applied top down, through the program, and stops with leaf events which are either "normal events" representing valid program states, or external failure events which the program is intended to detect and recover from. The top event probability is determined if FTA is applied to a hardware system and the hardware failure event probabilities are known. This is not the case for software reliability – the logical contribution of the software to the hazard is analyzed.

Performing a complete SFTA for large-scale control systems is often prohibitive. The analysis results become huge, cumbersome, and difficult to relate to the system and its operation. Software is more difficult to analyze for all functions, data flows or behavior as the complexity of the software system increases. SFTA is applied at all stages of the lifecycle process. SFTA requires a different fault tree construction method (i.e., a set of templates) for each language used for software requirement specification and software design description.

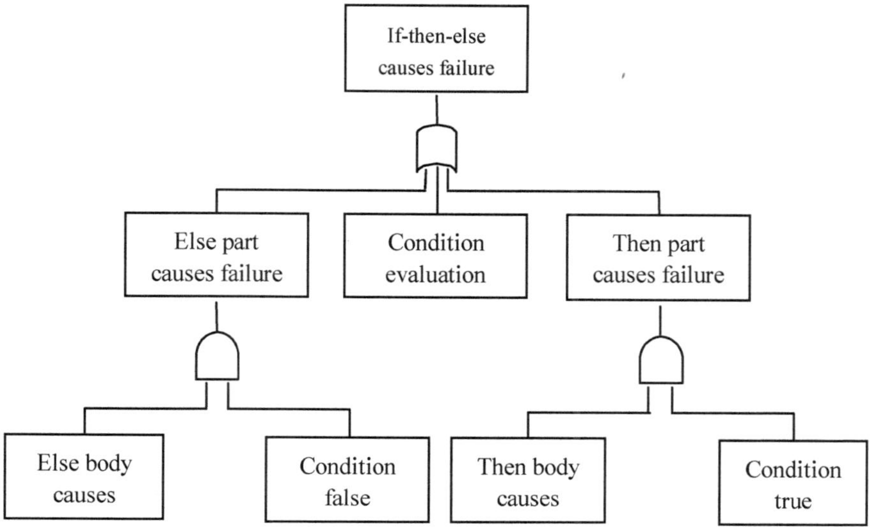

Figure 4.2. An example of software fault tree template

This makes SFTA labor-intensive. SFTA being applied top-down has advantages that it can be used on detailed design representations (*e.g.*, statecharts and Petri nets, rather than programs, especially where code is generated automatically from such representations). More appropriate results are given and less effort is required in constructing the trees. A set of guidelines needs to be devised for applying SFTA in a top-down manner through the detailed design process. Guidelines are also needed for deciding at what level it is appropriate to stop the analysis and rely on other forms of evidence (*e.g.*, of the soundness of the code generator). Techniques need to be developed for applying SFTA to design representations as well as programs.

4.3.1.1 Quality of Software Fault Trees

The quality of fault trees is crucial to the evaluation of software reliability because poor fault trees may lead to a wrong estimation of software reliability. Fault tree analysis has a fundamental limitation since it is informal in nature, although widely used in industry [23]. Graphical notations help the analyst to organize the thought process systematically. However, the technique offers no help in investigating causal events and the relationship among them. The result is not guaranteed to be repeatable and analysis may contain flaws when different experts apply fault tree analysis. An inspection technique [24] is used to detect errors of fault trees, but it is informal in nature.

An interesting method for the improvement of the fault tree quality using formal methods (Section 5.1) was suggested [25]. The method was proposed to provide formal (wherever possible) automated and qualitative assistance to informal and/or quantitative safety analysis. The correctness of fault trees was validated by a model-checking technique among formal methods. A real-time model checker UPPAAL [26] was used to validate the correctness of fault trees.

The property patterns are found to be particularly useful when validating the correctness of fault trees, although property specification accepted by UPPAAL is an arbitrarily complex temporal logic formula [27]:

- $\forall\Box\,(\neg pN)$: Let pN be the temporal logic formula semantically equivalent to the failure mode described in the fault tree node N. The property $\forall\Box(\neg pN)$ determines if the system can ever reach such a state. If the model checker returns TRUE, the state denoted by pN will never occur, and the system is free from such a hazard. It means that a safety engineer has thought a logically impossible event to be feasible, and the model checker found an error in the fault tree. If, on the other hand, the property is not satisfied, such a failure mode is indeed possible, and the model checker generates detailed (but partial) information on how such a hazard may occur. Detailed analysis of the counterexample may provide assurance that safety analysis has been properly applied. The counterexample may also reveal a failure mode which the human expert had failed to consider.
- $\forall\Box\,((B_1 \wedge \ldots \wedge B_n) \rightarrow A)$ / $\forall\Box\,((B_1 \vee \ldots \vee B_n) \rightarrow A)$: This pattern is used to validate if AND/OR connectors, used to model relationship among causal events, are correct. The refinement of the fault tree was done properly if the model checker returns true. Otherwise, there are two possibilities: (1) the gate connector is incorrect; or (2) failure modes in the lower level fault tree nodes are incorrect. A counterexample can provide insight as to why the verification failed and how the fault tree might be corrected.

A reactor shutdown system at the Wolsong nuclear power plant is required to continually monitor the state of the plant by reading various sensor inputs (*e.g.*, reactor temperature and pressure) and generating a trip signal should the reactor be found in an unsafe state [28]. The primary heat transport low core differential pressure (PDL) trip condition has been used as an example, among the six trip parameters, because it is the most complex trip condition and has time-related requirements. The trip signal can be either an immediate trip or a delayed trip; both trips can be simultaneously enabled. Delayed trip occurs if the system remains in certain states for over a period of time. High-level requirements for PDL trip were written in English in a document called the Program Functional Specification (PFS) as:

> *If the D/I is open, select the 0.3% FP conditioning level. If ϕLOG < 0.3% FP – 50mV, condition out the immediate trip. If $\phi LOG \geq$ 0.3% FP, enable the trip. Annunciate the immediate trip conditioning status via the PHT ΔP trip inhibited (ϕLOG < 0.3% FP) window D/O.*
>
> *If any ΔP signal is below the delayed trip setpoint and fAVEC exceeds 70% FP, open the appropriate loop trip error message D/O. If no PHT DP delayed trip is pending or active, then execute a delayed trip as follows:*
>
> *Continue normal operation without opening the parameter trip D/O for normally three seconds. The exact delay must be in the range [2.7, 3.0] seconds.*

> *Once the delayed parameter trip has occurred, keep the parameter trip D/O open for one second (± 0.1 seconds), and then close the parameter trip D/O once all DP signals are above the delayed trip setpoint or fAVEC is below 70% FP.*

Additional documents, including Software Requirements Specification (SRS) and Software Design Documentation (SDD), are used when performing fault tree analysis. Detailed and technical insight about the system are provided by these documents, which were thoroughly reviewed by a group of technical experts and government regulators before an operating license was granted. Fault tree (Figure 4.3) was initially developed by a group of graduate students who were majoring in software engineering, had previously reviewed the shutdown system documents, and performed independent safety analysis. They are also familiar with technical knowledge of software safety, in general, and fault tree analysis in particular. They possessed in-depth knowledge on how the trip conditions work. The fault tree was subsequently reviewed and revised by a group of domain experts in nuclear engineering who concluded that the fault tree appeared to be correct.

The top-level event, derived from the results of preliminary hazard analysis, is given as "PDL trip fails to clear digital output (D/O) in required time." The fault tree node had been refined into three causal events connected by OR gate. Failure modes described in some nodes (*e.g.*, 2 and 4) were further refined.

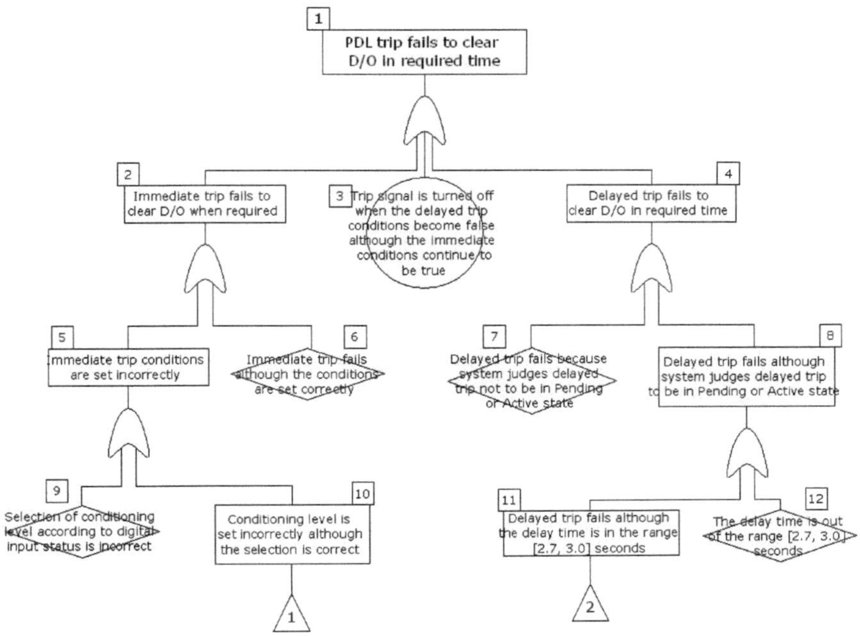

Figure 4.3. A part of fault tree of Wolsong PDLTrip

The validation of fault trees consists of the following steps:

- Translate functional requirements into a set of concurrent timed automata. Variables used in the timed automata follow the convention used in the four-variable approach, and prefixes m_, c_, and k_ represent monitored variables, controlled variables, and constant values, respectively. For example, functional requirement "If ϕ_{LOG} < 0.3% FP – 50mV, condition out the immediate trip" is captured by the rightmost transition of Figure 4.4 labeled "If m_PDLCond == k_CondSwLo and f_Flog < 2689, then f_PDLCond:= k_CondOut." For the PDL trip alone, the complete specification consisted of 12 concurrent timed automata. There were about 215 feasible states, clearly too many to fully inspect manually.
- Derive properties to be verified using one of the two patterns described earlier.
- Run UPPAAL to perform model checking.

Domain knowledge is needed to correctly rewrite the failure mode in the temporal logic formula. In this example, the formula (f_PDLSnrI == k_SnrTrip AND f_PDLCond == k_CondIn) denotes the activation of immediate trip condition. Delayed trip is canceled when the PDLDly process moves from the waiting state to the normal state and the value of f_PDLTrip becomes k_NonTrip in some states other than the initial state (e.g., denoted by having clock variable z > 0). The temporal logic formula corresponding to the absence of system state corresponding to the fault tree node 3 is given as follows:

$\forall\Box(\neg p3)$ whereas p3 corresponds to
(f_PDLSnrI == k_SnrTrip and f_PDLCond == k_CondIn) and
(f_PDLDly == k_InDlyNorm and f_PDLTrip == k_NotTrip and z > 0) (4.7)

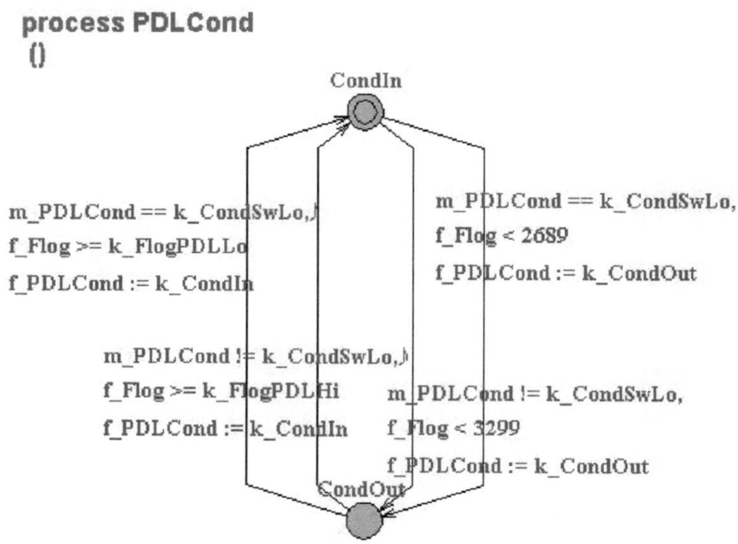

Figure 4.4. Timed automata for PDLCond trip condition

UPPAAL concluded that the property was not satisfied, and a counterexample, shown in terms of a simulation trace, was generated (Figure 4.5). Each step can be replayed. The tool graphically illustrates which event took place in a certain configuration. The simulation trace revealed that the property does not hold if the trip signal is (incorrectly) turned off (*e.g.*, becomes NotTrip) when the immediate trip condition becomes false while delayed trip condition continues to be true. This is possible because two types of trips have the same priority. While the failure mode captured in node 3 is technically correct when analyzed in isolation, model checking revealed that it was incomplete and that it must be changed to "Trip signal is turned off when the condition of one trip becomes false although the condition of the other continues to be true." (Or, two separate nodes can be drawn.) Analysis of the simulation trace provided safety analysts an interactive opportunity to investigate details of subtle failure modes humans forgot to consider.
Node 12 describes a failure mode where the system incorrectly clears a delayed trip signal outside the specified time range of [2.7, 3.0] seconds. UPPAAL accepts only integers as the value of a clock variable, *z* in this example. Using 27 and 30 to indicate the required time zone, a literal translation of the failure mode shown in the fault tree would correspond to:

$$\forall\Box(\neg p12) \text{ where p12 is } ((z < 27 \text{ or } z > 30) \text{ and f\_PDLTrip == k\_NotTrip}) \quad (4.8)$$

Model-checking of this formula indicated that the property does not hold, and an analysis of the counterexample revealed that the predicate p12 does not hold when *z* is equal to zero (*i.e.*, no time passed at all). This is obviously incorrect, based on domain-specific knowledge of how delayed trip is to work, and it quickly reminds

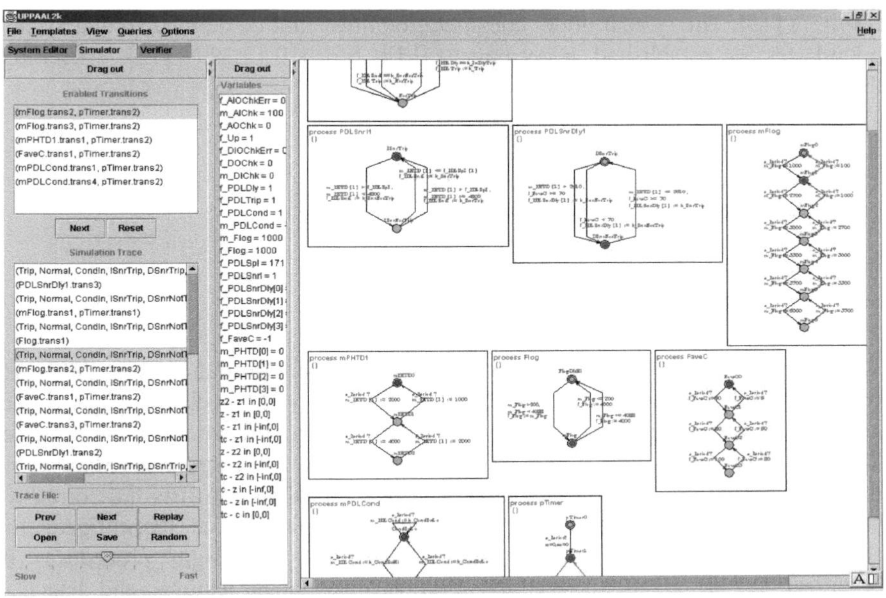

Figure 4.5. Screen dump of the UPPAAL outputs

a safety analyst that the failure mode, as it is written, is ambiguous in that the current description of the failure mode fails to explicitly mention that the system must be in the waiting state, not the initial system state, before the delayed trip timer is set to expire. That is, the property needs to be modified as:

$$\forall\Box(\neg p12)$$
where p12 is (f_PDLSnrDly == k_SnrTrip and f_FaveC >= 70)
and ($z < 27$ or $z > 30$) and (f_PDLTrip == k_NotTrip) (4.9)

The following clause in the PFS provides clues as to how the formula is to be revised: "If any ΔP signal is below the delayed trip setpoint and fAVEC exceeds 70% FP, open the appropriate loop trip error message D/O. If no PHT DP delayed trip is pending or active, then execute a delayed trip as follows: Continue normal operation without opening the parameter trip D/O for the normal three seconds. The exact delay must be in the range [2.7, 3.0] seconds." Model-checking of the revised property demonstrated that the property is satisfied meaning that fault tree node 12 is essentially correct, although it initially contained implicit assumptions. Thus, the application of a model-checking technique helped a reliability/safety engineer better understand the context in which specified failure mode occurs and therefore conduct a more precise reliability/safety analysis.

4.3.2 Software Failure Mode and Effect Analysis

Failure mode and effect analysis (FMEA) is an analytical technique, which explores the effects of failures or malfunctions of individual components in a system, such as software (*e.g.*, "If this software, in this manner, what will be the result?") [15]. The system under consideration must first be defined so that system boundaries are established, according to IEEE Std 610.12.1990 [15]. Thereafter, the essential questions are:

1. How can each component fail?
2. What might cause these modes of failure?
3. What would be the effects if the failures did occur?
4. How serious are these failure modes?
5. How is each failure mode detected?

Both hardware FMEA and software FMEA identify design deficiencies. Software FMEA is applied iteratively throughout the development lifecycle. Analysts collect and analyze principle data elements such as the failure, the cause(s), the effect of the failure, the criticality of the failure, and the software component responsible identifying each potential failure mode. Software FMEA also lists the corrective measures required to reduce the frequency of failure or to mitigate the consequences. Corrective actions include changes in design, procedures or organizational arrangements (*e.g.*, the addition of redundant features, and detection methods or a change in maintenance policy).

The criticality of the failure is usually determined based on the level of risk. The level of risk is determined by the multiplication of probability of failure and severity. Probability of failure and severity is categorized in Table 4.1 and Table 4.2, respectively. A risk assessment matrix is prepared usually depending on system characteristics and expert opinions. FMEA is used for single point failure modes (e.g., single location in software) and is extended to cover concurrent failure modes. It may be a costly and time-consuming process but once completed and documented it is valuable for future reviews and as a basis for other risk assessment techniques, such as fault tree analysis. The output from software FMEA is used as input to software FTA.

4.3.3 Software Hazard and Operability Studies

Hazard and operability studies (HAZOP) is a well-established system analysis technique. HAZOP has been used for many years in assessing the safety of proposed process plants, and suggesting design modifications to improve safety [12]. It asks if deviations from design intent possibly occur and what may be the causes and consequences of the deviations. HAZOP examines all the components and the interfaces of the system under analysis. HAZOP is particularly powerful for analyzing the interactions between the system components. One of the prominent features of HAZOP is the use of guidewords used to prompt the analyses. HAZOP is considered as an alternative to FMEA, or vice versa. But HAZOP complements FMEA. FMEA starts with a failure mode of a component and analyzes the effects of the failure mode. FMEA is inductive, in that it works from the specific to the general. HAZOP works both backward and forward. It starts with a particular fault and moves backwards to find out possible causes of the fault. It moves forward to analyze the consequences of the fault at the same time.

Over recent years there have been many research projects for adapting HAZOP to software. Software HAZOP has been aimed at hazard identification and exploratory analysis and derived requirements activities. The major work on software HAZOP has led to the development of a draft Interim Defense Standard 00-58 in the UK [29]. The draft follows the traditional HAZOP style, suggesting team structures for HAZOP meetings as used in the process industry.

Table 4.1. Category of probability of failure mode

Level	Probability	Description	Individual failure mode
A	10^{-1}	Frequent	Likely to occur frequently
B	10^{-2}	Probable	Likely to occur several times in the life of a component
C	10^{-3}	Occasional	Likely to occur sometime in the life of a component
D	10^{-4}	Remote	Unlikely to occur but possible
E	10^{-5}	Improbable	So unlikely that occurrence may not be experienced

Table 4.2. Severity category for software FMEA

Category	Degree	Description
I	Minor	Failure of component – no potential for injury
II	Critical	Failure will probably occur without major damage to system or serious injury
III	Major	Major damage to system and potential serious injury to personnel
IV	Catastrophic	Failure causes complete system loss and potential for fatal injury

HAZOP is based on deviations from design intent. What the deviations (applications of guidewords to flows) mean is usually fairly clear in process plant. However, what deviations are possible is not always clear in software. Here an important technical issue rises that the design representations are the basis of software HAZOP. Ideally, design representation should reflect the underlying computational model of the system so that any deviation which is meaningfully described in the design representation has physical meaning. The HAZOP technique has low predictive accuracy and is of limited effectiveness in the exploratory analysis, if this ideal is not met. So, a set of guidelines needs to be devised for assessing the suitability of a design (or requirements) representation for software HAZOP, and for adapting a representation to reflect the underlying computational model, if necessary. Integrating formal methods with software HAZOP is an interesting challenge.

4.4 Concluding Remarks

Issues in software reliability, related to handling software faults, have been examined. Software is a complex intellectual product. Inevitably, some errors are made during requirements formulation as well as during designing, coding, and testing the product. The development process for high-quality software includes measures that are intended to discover and correct faults resulting from these errors, including reviews, audits, screening by language-dependent tools and several levels of test. Managing these errors involves describing, classifying, and modeling the effects of the remaining faults in the delivered product, helping to reduce their number and criticality.

Software reliability is considered in view of the software lifecycle. There are two categories of software reliability: pre-release software reliability and post-release software reliability [2]. Pre-release software reliability is an assessment of design integrity. It measures the robustness of the design features and procedures, evaluating how well preventive measures have been incorporated. A product may be marketed if and when the pre-release software reliability assessments are deemed accurate and adequate. Actual data concerning performance is also collected and analyzed once a software product is released; this information represents post-release software reliability. Post-release software reliability is an analysis of the type and source of errors found, once a software product has been fielded. It determines "what went wrong" and what corrective measures are needed, such as "lessons learned." This information is then fed back into the continuous process improvement cycle.

The combination of quantitative and qualitative software reliability evaluations is very useful. A software project may adopt quantitative reliability measures for pre-release software reliability and qualitative reliability measures for post-release software reliability, or *vice versa.* Both qualitative reliability and quantitative reliability are assessed based on the information collected throughout the development lifecycle. Some pioneering work has been done in developing a holistic model for software reliability [30–31]. Current software reliability estimation and prediction techniques do not take into account a variety of factors which affect reliability. The success (of these models) relates only to those cases where the reliability being observed is quite modest and it is easy to demonstrate that reliability growth techniques are not plausible ways of acquiring confidence that a program is ultra-reliable. A holistic model integrates many different sources and types of evidence to assess reliability [32]:

- Product – metrics about the product, its design integrity, behavior, failure modes, failure rates, and so forth
- Process – metrics about how the product was developed and its reliability assessed
- Resources – metrics about the resources used to develop the systems, such as the people and their qualifications, the tools used and their capabilities and limitations
- Human computer interaction – metrics about the way people interact with the system, which could be derived from formal scenario analysis and a HAZOP analysis

Qualitative and quantitative information is collected and analyzed throughout the development lifecycle. This model integrates metrics to yield a comprehensive software reliability assessment.

The reliability of digital and/or software-based systems is much different from that of analog and/or hardware-based systems. There is as yet no certified method to assess software reliability. Therefore, the software development process based on various quality improvement methods, such as formal V&V, test, and analysis methods, is essential. The methods should all be integrated as a basis of the reliability evaluation. Various quality improvement methods, which are described in Chapter 5, are used as a process metric. The results from an optimized number of reliability tests are used as a product metric.

References

[1] Leveson NG (1995) Safeware: system safety and computers. Addison–Wesley
[2] Li B, Li M, Ghose S, Smidts C (2003) Integrating software into PRA. Proceedings of

the 14th ISSRE, IEEE Computer Society Press
[3] Herrmann DS (2002) Software safety and reliability. IEEE Computer Society Press, ISBN 0-7695-0299-7
[4] Rugg G (1995) Why don't customers tell you everything you need to know? or: why don't software engineers build what you want? Safety Systems, Vol. 5, No. 1, pp. 3–4, Sep.
[5] Musa JD, et al. (1987) Software reliability: measurement, prediction, application. McGraw–Hill, New York
[6] Asad CA, Ullah MI, Rehman MJ (2004) An approach for software reliability model selection. Proceedings of the 28th Annual International Computer Software and Applications Conference (COMSAC'04), IEEE
[7] Smidts C, Stoddard RW, Stutzke M (1998) Software reliability models: an approach to early reliability prediction. IEEE Transactions on Reliability, Vol. 47, No. 3, pp. 268–278
[8] Gokhale SS, Marinos PN, Trivedi KS (1996) Important milestones in software reliability modeling. In Proceedings of Software Engineering and Knowledge Engineering (SEKE '96), Lake Tahoe, NV, pp. 345–352
[9] Gokhale SS, Wong WE, Trivedi KS, Horgan JR (1998) An analytical approach to architecture-based software reliability prediction. In IEEE Int. Computer Performance and Dependability Symposium, pp. 13–22, Sept.
[10] Musa J (1999) Software reliability engineering. McGraw–Hill
[11] Lyu M (ed.) (1996) Handbook of software reliability engineering. McGraw–Hill/ IEEE Computer Society Press
[12] Storey N (1996) Safety-critical computer systems. Addison–Wesley
[13] Rees RA (1994) Detectability of software failure. Reliability Review, Vol. 14, No. 4, pp. 10–30, Dec.
[14] Bieda F (1996) Software reliability: a practitioner's model. Reliability Review, Vol. 16, No. 2, pp. 18–28, June
[15] IEEE Std 610.12.1990 (1990) IEEE Standard glossary of software engineering terminology. IEEE, New York, March
[16] American National Standards Institute/American Institute of Aeronautics and Astronautics (1993) Recommended practice for software reliability. R-013-1992
[17] Schneidewind NF, Keller TW (1992) Application of reliability models to the space shuttle. IEEE Software, Vol. 9, No. 4, pp. 28–33, July
[18] Myers GJ (1979) The art of software testing. John Wiley and Sons, New York
[19] Miller KW, et. al. (1992) Estimating the probability of failure when testing reveals no failures. IEEE Transactions on Software Engineering, Vol. 18, No. 1, pp. 33–43
[20] Littlewood B, Wright D (1997) Some conservative stopping rules for the operational testing of safety-critical software. IEEE Transactions on Software Engineering, Vol. 23, No. 11, pp. 673–683
[21] Jelinski Z, Moranda PB (1972) Software reliability research (W. Freiberger, Editor). Statistical Computer Performance Evaluation, Academic, New York, p. 465
[22] Goel AL, Okumoto K (1979) Time-dependent error-detection rate model for software reliability and other performance measures. IEEE Transactions on Reliability, Vol. R-28, No. 3, p. 206
[23] Kocza G, Bossche A (1997) Automatic fault-tree synthesis and real-time trimming, based on computer models. Proc. Ann. Reliability and Maintainability Symp., pp. 71–75
[24] WWW Formal Technical Review(FTR) Archive
[25] http://www.ics.hawaii.edu/~johnson/FTR/
[26] Cha SD, Son HS, Yoo JB, Jee EK, Seong PH (2003) Systematic evaluation of fault

trees using real-time model checker UPPAAL. Reliability Engineering and System Safety, Vol. 82, pp. 11–20
[27] Bengtsson J, Larsen KG, Larsson F, Pettersson P, Yi W (1995) UPPAAL – a tool suite for automatic verification of real-time systems. In Proceedings of the 4th DIMACS Workshop on Verification and Control of Hybrid Systems, New Brunswick, New Jersey, October
[28] Pnueli A (1977) The temporal logic of programs. In Proceedings of the 18th IEEE Symposium on Foundations of Computer Science, pp. 46–77
[29] AECL CANDU (1993) Program functional specification, SDS2 programmable digital comparators, Wolsong NPP 2,3,4. Technical Report 86-68300-PFS-000 Rev.2, May
[30] DEF STAN 00-58 (1996) HAZOP studies on systems containing programmable electronics. UK Ministry of Defence, (interim) July
[31] Littlewood B (1993) The need for evidence from disparate sources to evaluate software safety. Directions in Safety-Critical Systems, Springer–Verlag, pp. 217–231
[32] Herrmann DS (1998) Sample implementation of the Littlewood holistic model for assessing software quality, safety and reliability. Proceedings Annual Reliability and Maintainability Symposium, pp. 138–148

5

Software Reliability Improvement Techniques

Han Seong Son[1] and Seo Ryong Koo[2]

[1]Department of Game Engineering
Joongbu University
#101 Daehak-ro, Chubu-myeon, Kumsan-gun, Chungnam, 312-702, Korea
hsson@joongbu.ac.kr

[2]Nuclear Power Plant Business Group
Doosan Heavy Industries and Construction Co., Ltd.
39-3, Seongbok-Dong, Yongin-Si, Gyeonggi-Do, 449-795, Korea
seoryong.koo@doosan.com

Digital systems offer various advantages over analog systems. Their use in large-scale control systems has greatly expanded in recent years. This raises challenging issues to be resolved. Extremely high-confidence in software reliability is one issue for safety-critical systems, such as NPPs. Some issues related to software reliability are tightly coupled with software faults to evaluate software reliability (Chapter 4). There is not "one right answer" as to how to estimate software reliability. Merely measuring software reliability does not directly make software more reliable, even if there is a "proper answer" for estimation of software reliability. Software faults should be carefully handled to make software more reliable with as many reliability improvement techniques as possible. However, software reliability evaluation may not be useful. Software reliability improvement techniques dealing with the existence and manifestation of faults in software are divided into three categories:

- Fault avoidance/prevention that includes design methodologies to make software provably fault-free
- Fault removal that aims to remove faults after the development stage is completed. This is done by exhaustive and rigorous testing of the final product
- Fault tolerance that assumes a system has unavoidable and undetectable faults and aims to make provisions for the system to operate correctly, even in the presence of faults

Some errors are inevitably made during requirements formulation, designing, coding, and testing, even though the most "thorough" fault avoidance techniques are applied. No amount of testing can certify software as fault-free, although most bugs, which are deterministic and repeatable, can be removed through rigorous and extensive testing and debugging. The remaining are usually bugs which elude detection during testing. Fault avoidance and fault removal cannot ensure the absence of faults. Any practical piece of software can be presumed to contain faults in the operational phase. Designers must deal with these faults if the software failure has serious consequences. Hence, fault tolerance should be applied to achieve more dependable software. Fault tolerance makes it possible for the software system to provide service even in the presence of faults. This means that prevention and recovery from imminent failure needs to be examined.

Formal methods, such as fault avoidance techniques, verification and validation, such as fault removal techniques, and fault tolerance techniques, such as block recovery and diversity, are discussed.

5.1 Formal Methods

Formal methods use mathematical techniques for the specification, design, and analysis of computer systems, that are based on the use of formal languages that have very precise rules. There are various definitions of formal methods in the literature. For example, Nancy Leveson states:

> *A broad view of formal methods includes all applications of (primarily) discrete mathematics to software engineering problems. This application usually involves modeling and analysis where the models and analysis procedures are derived from or defined by an underlying mathematically-precise foundation* [1].

The main purpose of formal methods is to design an error-free software system and increase the reliability of the system. Formal methods treat components of a system as mathematical object modules and model them to describe the nature and behavior of the system. Mathematical models are used for the specifications of the system so that formal methods can reduce the ambiguity and uncertainty which are introduced to the specifications by using natural language. Formal models are systematically verified, proving whether the user's requirements are properly reflected in them or not, by virtue of their mathematical nature.

The definition of formal methods provides a more concrete understanding that has two essential components, formal specification and formal verification [2]. Formal specification is based on a formal language, which is a set of strings for a well-defined alphabet [3]. Rules are given for distinguishing strings, defined over the alphabet, which belongs to the language, from other strings that do not. Users lessen ambiguities and convert system requirements into a unique interpretation with rules. Formal verification includes a process for proving whether the system design meets the requirements. Formal verification is performed using mathematical proof techniques, since formal languages treat system components as mathematical objects. Formal methods support formal reasoning about formulae in formal languages. The completeness of system requirements and design are verified with formal proof techniques. In addition, system characteristics, such as safety, liveness, and deadlock, are proved manually or automatically with the techniques.

Formal methods include but are not limited to specification and verification techniques based on process algebra, model-checking techniques based on state machines, and theorem-proving techniques based on mathematical logic.

5.1.1 Formal Specification

There exist many kinds of formal specification methods. Formal specifications are composed using languages based on graphical notations, such as state diagrams, or languages that are based on mathematical systems, such as logics and process algebra. Which language is appropriate to the specified system requirements determines choice of formal methods. The level of rigor is another factor to be considered for choice of formal methods. Formal methods are classified based on Rushby's identification of levels of rigor in the application of formal methods [3]

- Formal methods using concepts and notation from discrete mathematics (Class 1)
- Formal methods using formalized specification languages with some mechanized support tools (Class 2)
- Formal methods using fully formal specification languages with comprehensive support environments, including mechanized theorem proving or proof checking (Class 3)

Notations and concepts derived from logic and discrete mathematics are used to replace some of the natural language components of requirements and specification documents in Class 1. This means that a formal approach is partially adopted, and proofs, if any, are informally performed. The formal method in this class incorporates elements of formalism into an otherwise informal approach. The advantages gained by this incorporation include the provision of a compact notation that can reduce ambiguities. A systematic framework, which can aid the mental processes, is also provided.

A standardized notation for discrete mathematics is provided to specification languages in Class 2. Automated methods of checking for certain classes of faults are usually provided. Z, VDM, LOTOS, and CCS are in this class. Proofs are informally conducted and are referred to as rigorous proofs (rather than formal proofs). Several methods provide explicit formal rules of deduction that permit formal proof, even if manual.

Class 3 formal methods use a fully formal approach. Specification languages are used with comprehensive support environments, including mechanized theorem proving or proof checking. The use of a fully formal approach greatly increases the probability of detecting faults within the various descriptions of the system. The use of mechanized proving techniques effectively removes the possibility of faulty reasoning. Disadvantages associated with these methods are the considerable effort and expense involved in their use, and the fact that the languages involved are generally very restrictive and often difficult to use. This class includes HOL, PVS, and the Boyer–Moore theorem prover.

The formal methods in Class 1 are appropriate when the objective is simply to analyze the correctness of particular algorithms or mechanisms [4]. Class 2 methods are suitable if the nature of the project suggests the use of a formalized specification together with manual review procedures. The mechanized theorem proving of Class 3 is suggested where an element of a highly critical system is crucial and contains many complicated mechanisms or architectures.

The main purpose of formal specification is to describe system requirements and to design the requirements, so that they can be implemented. A formal specification can be either a requirement specification or a design specification. The design specification primarily describes how to construct system components. The requirement specification is to define what requirements the system shall meet. Design specification is generated for the purpose of implementing the various aspects of the system, including the details of system components. Design specification is verified as correct by comparing with requirement specification.

5.1.2 Formal Verification

Formal verification proves or disproves the correctness of intended functions, algorithms, or programs underlying a system with respect to formal specification or property. For example, a formal process to check whether a design specification is well satisfied with a requirement specification is a formal verification activity. There are two approaches to formal verification. The first approach is model checking [3]. Model-checking is a technique for verifying finite-state systems. Verification can be performed automatically in model-checking, and thus is preferable to deductive verification. The model-checking procedure normally uses an exhaustive search of the state space of the system to determine if a specification is true or not. A verification tool generates a counterexample which is traced to a failure path if a deviation exists between the system and the specification.

The second approach is logical inference, such as proof checking and theorem proving [3]. Proof checking checks the steps of a proof produced by an engineer, whereas theorem proving discovers proofs without human assistance. A proof begins with a set of axioms, which are postulated as true, in all cases. Inference rules state that if certain formulae, known as premises, are derivable from the axioms, then another formula, known as the consequent, is also derivable. A set of inference rules must be given in each formal method. A proof consists of a sequence of well-defined formulae in the language in which each formula is either an axiom or derivable by an inference rule from previous formulae in the sequence. The last formula in the sequence is said to be proven. When all the properties are proven this means that an implementation is functionally correct; that is, it fulfills its specification.

5.1.3 Formal Methods and Fault Avoidance

Software testing involves verification and validation activities and is an effective approach to a fault-free software system. Software testing alone cannot prove that a system does not contain any defects, since testing demonstrates the presence of faults but not their absence. A software design process that ensures high levels of quality is also very important. Formal methods support this design process to ensure high levels of software quality by avoiding faults that can be formally specified and verified.

An important advantage of formal methods is the performance of automated tests on the specification. This not only allows software tools to check for certain classes of error, but also allows different specifications of the system to be compared to see if they are equivalent. The development of a system involves an iterative process of transformation in which the requirements are abstracted through various stages of specification and design, that ultimately appear as a finished implementation. Requirements, specification, and levels of design are all descriptions of the same system, and thus are functionally equivalent. It is possible to prove this equivalence, thereby greatly increasing the fault avoidance possibility in the development process, if each of these descriptions is prepared in a suitable form.

Fault avoidance is accomplished with formal methods through automation lessoning for the possibility of human error intervention. Formal methods have inspired the development of many tools. Some tools animate specifications, thereby converting a formal specification into an executable prototype of a system. Other tools derive programs from specifications through automated transformations. Transformational implementation suggests a future in which many software systems are developed without programmers, or at least with more automation, higher productivity, and less labor [5, 6]. Formal methods have resulted in one widely agreed criterion for evaluating language features: how simply can one formally evaluate a program with a new feature? The formal specification of language semantics is a lively area of research. Formal methods have always been an interest of the Ada community, even before standardization [7, 8]. A program is automatically verified and reconstructed in view of a formal language.

The challenge is to apply formal methods for projects of large-scale digital control systems. Formal specifications scale up much easier than formal verifications. Ideas related to formal verifications are applicable to projects of any size, particularly if the level of formality is allowed to vary. A formal method provides heuristics and guidelines for developing elegant specifications and for developing practically useful implementations and proofs in parallel. A design methodology incorporating certain heuristics that support more reliable and provable designs has been recommended [9]. The Cleanroom approach was developed, where a lifecycle of formal methods, inspections, and reliability modeling and certification are integrated in a social process for producing software [10, 11]. Formal methods are a good approach to fault avoidance for large–scale projects.

Fault avoidance capability of formal methods is demonstrated in the application of the formal method NuSCR (Nuclear Software Cost Reduction), which is an extension of the SCR-style formal method [12]. The formal method and its application are introduced in Chapter 6. NuSCR specification language was originally designed to simplify the complex specification techniques of certain requirements in the previous approach. The improved method describes the behavior of history-related and timing-related requirements of a large-scale digital control system by specifying them in automata and timed-automata, respectively [13, 14]. NuSCR is very effective in determining omitted input and/or output variables in a software specification and pinpointing ambiguities in a specification by virtue of improvements as well as formality (Chapter 6). Omitted variables in the specification are easily found in the project because of the NuSCR feature that all the inputs and outputs shall be specified. Application reports that NuSCR helps to determine ambiguous parts and then change the specification to the precise one.

5.2 Verification and Validation

Verification and validation (V&V) is a software-engineering discipline that helps to build quality into software. V&V is a collection of analysis and testing activities across the full lifecycle and complements the efforts of other quality-engineering functions.

V&V comprehensively analyzes and tests software to determine that it correctly performs its intended functions, to ensure that it performs no unintended functions, and to measure its quality and reliability. V&V is a systems engineering discipline to evaluate software in a systems context. A structured approach is used to analyze and test the software against all system functions and against hardware, user, and other software interfaces.

Software validation is "establishing by objective evidence that all software requirements have been implemented correctly and completely and are traceable to system requirements." Software validation is essentially a design verification function as defined in FDA's Quality System Regulation (21 CFR 820.3 and 820.30), and includes all verification and testing activities conducted throughout the software life cycle. Design validation encompasses software validation, but goes further to check for proper operation of the software in its intended use environment.

Verification is defined in FDA's Quality System Regulation (21 CFR 820.3) as "confirmation by examination and provision of objective evidence that specified requirements have been fulfilled." Software verification confirms that the output of a particular phase of development meets all input requirements for that phase.

Verification involves evaluating software during each life cycle phase to ensure that it meets the requirements set forth in the previous phase. Validation involves testing software or its specification at the end of the development effort to ensure that it meets its requirements (that it does what it is supposed to). Maximum benefit is derived by synergism and treating "V&V" as an integrated definition, while "verification" and "validation" have separate definitions. Ideally, V&V parallels software development and yields several benefits:

- High-risk errors are uncovered early, giving the design team time to evolve a comprehensive solution rather that forcing a makeshift fix to accommodate development deadlines.
- Management is evaluated with continuous and comprehensive information about the quality and progress of the development effort.

- An incremental preview of system performance is given to the user, with the chance to make early adjustments.

There are four types of V&V:

- *Inspection*
  Typical techniques include desk checking, walkthroughs, software reviews, technical reviews, and formal inspections (*e.g.*, Fagan approach).
- *Analysis*
  Mathematical verification of the test item, which includes estimation of execution times and estimation of system resources.
- *Testing*
  This is also known as "white box" or "logic driven" testing. Given input values are traced through the test item to assure that they generate the expected output values, with the expected intermediate values along the way. Typical techniques include statement, condition, and decision coverage.
- *Demonstration*
  This is also known as "black box" or "input/output driven" testing. Given input values are entered and the resulting output values are compared against the expected output values. Typical techniques include error guessing, boundary-value analysis, and equivalence partitioning.

Four types of V&V are used at any level in software products. The most effective way to find anomalies at the component level is inspection. Inspection is not applicable at the system level (details of code are not examined when performing system level testing). Testing logically utilizes techniques and methods that are most effective at a given level.

V&V for software components is very expensive. Most projects need to avoid making statements like "all paths and branches will be executed during component testing." These statements result in a very expensive test program, since all code requires labor-intensive testing performed on it. V&V develops rules for determining V&V method(s) needed for each of the software functions to minimize costs. Very low complexity software function, which is not on the safety-critical list, may need informal inspections (walkthrough) performed. Other complicated functions require white box testing, since it is difficult to determine how the functions work. Inspections should be performed before doing white box testing for a given module, as it is less expensive to find the errors early in the development.

V&V is embraced as the primary way of proving that the system does what is intended. The resulting V&V effort is effective in fault removal and thus has become a significant part of software development. Demonstrating that the system is implemented completely without faults uses a requirements traceability matrix (RTM), which documents each of the requirements traced to design items, code, unit, integration and system test cases. RTM is an effective way of documenting implementation – what are the requirements, where are they implemented, and how have you tested them.

5.2.1 Lifecycle V&V

Lifecycle refers to the "start-to-finish" phases of system development. The software development lifecycle encompasses: requirements, design, implementation, integration, field installation, and maintenance. A software lifecycle provides a systematic approach to the development and maintenance of a software system. A well-defined and well-implemented lifecycle is imperative for the successful application of V&V techniques.

There are two types of lifecycle models: the sequential model and the iterative model. The sequential model is a once-through sequence of steps and does not provide formal feedback from later phases to prior phases. The iterative model, on the other hand, involves repeated feedback cycling through lifecycle phases. Generally, the sequential model is used where requirements are well known and not subjected to change. The iterative lifecycle is appropriate when the requirements are not well known, or are undergoing changes, and/or there are significant technical issues/questions about how the software can be implemented to meet those requirements [15].

5.2.1.1 Requirements Phase

The purpose of requirements verification is to identify whether the requirements specified would correctly and completely describe a system that satisfies its intended purpose. The functional and performance requirements for the system are established from the viewpoint of the plant engineering, licensing, operations, and maintenance staff, at this phase. The requirements stage is critical to the overall success of the development procedure. Each software requirement is identified and evaluated with respect to software quality attributes, including correctness, consistency, completeness, understandability, accuracy, feasibility, traceability, and testability during requirements verification.

Requirements tracing is an important V&V technique that begins during the requirements specification stage of the development lifecycle and continues through the development process. A software requirement is traceable if its origin is clear, is testable, and facilitates referencing to future development steps. Backward traceability is established by correlating software requirements to applicable regulatory requirements, guidelines, and operational concept, or any other preliminary system concept documentation. Forward traceability to design elements and code modules is established by identifying each requirement with a unique name or number.

5.2.1.2 Design Phase

A software design description is produced at this stage. A description of the overall system architecture contains a definition of the control structures, algorithms, equations, and data inputs and outputs for each software model. The description is evaluated in view of software quality attributes such as correctness, completeness, consistency, accuracy, and testability. Verification of compliance with any applicable standards is also performed.

Requirements tracing continues during design verification by mapping documented design items to system requirements. This ensures that the design meets all specified requirements. Non-traceable design elements are identified and evaluated for interference with required design functions. Design analysis is performed to trace requirement correctness, completeness, consistency, and accuracy.

5.2.1.3 Implementation Phase

Detailed design of software is translated into source code during the coding and implementation phase. This activity also creates supporting data files and databases. The purpose of implementation verification is to provide assurance that the source code correctly represents the design. Source code is analyzed to obtain equations, algorithms, and logic for comparison with the design. This process detects errors made during the translation of detailed design to code. Information gained during analysis of the code, such as frequently occurring errors and risky coding structures and techniques, is used in finalizing test cases and test data.

The source code is traced to design items and evaluated for completeness, consistency, correctness, and accuracy during evaluation of this stage. Detailed test cases and test procedures are generated and documented using the knowledge gained about the program through its structure and detected deficiencies.

5.2.1.4 Validation (Testing) Phase

The whole system is evaluated against original requirements specification during the system validation phase. Validation consists of planned testing and evaluation to ensure that the final system complies with the system requirements.

Validation is more than just testing since analysis is involved. A test is a tool used by the validation team to uncover previously undiscovered specification, design, or coding errors throughout the development process. Validation uses testing plus analysis to reach objectives. The analysis is the design of test strategies, procedures, and evaluation criteria, based on knowledge of the system requirements and design. This proves system acceptability in an efficient fashion.

Tests must be defined to demonstrate that all testable requirements have been met. Test cases and test procedures are evaluated for completeness, correctness, clarity, and repeatability. Requirements tracing continues during validation by tracing test cases to requirements. This ensures that all testable requirements are covered. Expected results specified in test cases are verified for correctness against the requirements and design documentation.

5.2.2 Integrated Approach to V&V

V&V is an integrated definition that synergistically derives maximum benefit. An integrated environment (IE) approach to support software development and V&V processes has been proposed [16]. The IE approach has been utilized for NPP safety-critical systems based on a programmable logic controller (PLC). The IE approach focuses on V&V processes for PLC software. The IE approach achieves better integration between PLC software development and the V&V process for NPP safety-critical systems. System specification based on formalism is supported by V&V activities, such as completeness, consistency, and correctness, in the IE approach. Software engineers avoid using them despite practical benefits, if software development methods are complicated and hard to use. System specification of the IE approach focuses on ease of use and understanding.

The IE approach supports optimized V&V tasks for safety-critical systems based on PLC throughout the software lifecycle. Software requirements inspection, requirements traceability, and formal specification and analysis are integrated in this approach for more efficient V&V. The IE approach allows project-centered configuration management. All documents and products in this approach are systematically managed for configuration management. Major features of the IE approach are: document evaluation, requirements traceability, formal requirements specification and V&V, and effective design specification and V&V (Figure 5.1). Document analysis based on Fagan inspection [17] is supported throughout the software lifecycle for the document evaluation. Sentence comparison based on inspection results is supported by using RTM for requirements traceability. Document evaluation and traceability analysis are major tasks of the concept and requirements phases. The user requirements are described and evaluated through documentation in the concept phase, which is the initial phase of a software development project. The requirements phase is the period in the software lifecycle when requirements such as functional and performance capabilities of a software product are defined and documented. The IE approach adopts NuSCR, which is a formal method that is suitable for NPP systems, for formal requirements specification and V&V [18]. Formal requirements specification and analysis is performed by using the NuSCR method in the IE approach.

Effective design specification and V&V are supported in the IE approach. The IE approach also adopts the NuFDS (nuclear FBD-style design specification) approach, which is a suitable design method for NPP systems [19, 20]. NuFDS supports a design consistency check using ESDTs (extended structured decision tables), architecture analysis, and model-checking for design analysis. The software design phase is a process of translating software requirements into software structures that are built in the implementation phase. A well-formed

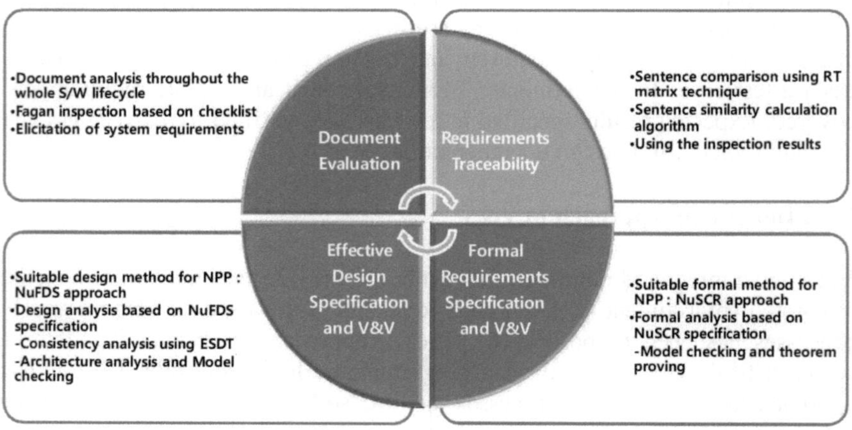

Figure 5.1. Major features of IE approach

design specification is very useful for coding during the implementation phase in that an implementation product, such as code, can be easily translated from design specifications. The function block diagram (FBD), among PLC software languages, is considered an efficient and intuitive language for the implementation phase. The boundary between design phase and implementation phase is not clear in software development based on PLC languages. The level of design is almost the same as that of implementation in PLC software. It is necessary to combine the design phase with the implementation phase in developing a PLC-based system. Coding time and cost are reduced by combining design and implementation phases for PLC application. The major contribution of the NuFDS approach is achieving better integration between design and implementation phases for PLC applications. The IE approach provides an adequate technique for software development and V&V for the development of safety-critical systems based on PLC. The function of the interface to integrate the whole process of the software lifecycle and flow-through of the process are the most important considerations in this approach. The scheme of the IE approach is shown in Figure 5.2. The IE approach can be divided into two categories: IE for requirements [16], which is oriented in the requirements phase, and IE for design and implementation [19, 20], which is oriented in the combined design and implementation phase. The NuSEE toolset was developed for the efficient support of the IE approach. NuSEE consists of four CASE tools: NuSISRT, NuSRS, NuSDS, and NuSCM (Chapter 6). The integrated V&V process helps minimize some difficulties caused by difference in domain knowledge between the designer and analyzer. Thus, the V&V process is more comprehensive by virtue of integration. V&V is more effective for fault removal if the software development process and the V&V process are appropriately integrated.

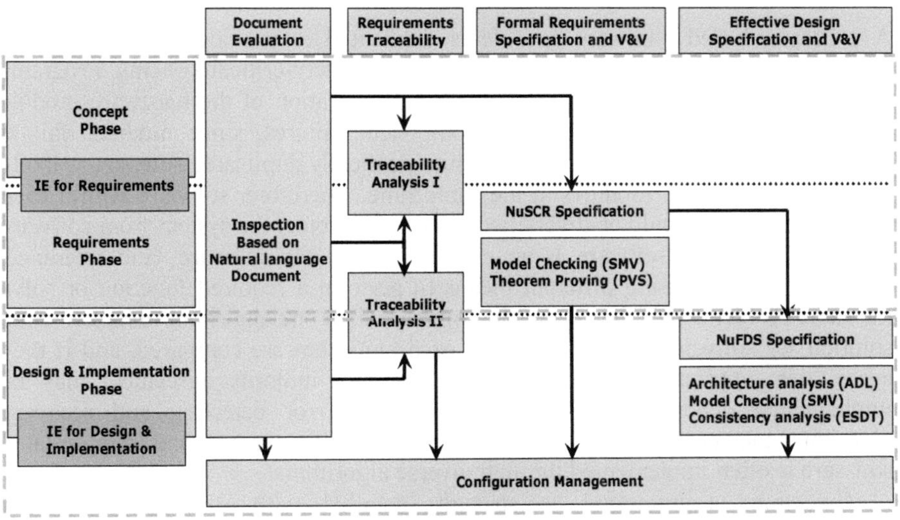

Figure 5.2. Overall scheme of IE approach

5.3 Fault Tolerance Techniques

Software fault tolerance, like hardware fault tolerance using diverse redundancy, provides protection against systematic faults associated with software and hardware [4]. It means and also refers to the tolerance of faults (of whatever form) by the use of software. Some software fault tolerance techniques falling within this definition are used to tolerate faults within software, while others are used to deal with both hardware and software faults.

Traditional hardware fault tolerance has tried to solve a few common problems which plagued earlier computer hardware, such as manufacturing faults. Another common hardware problem is transient faults from diverse sources. These two types of faults are effectively guarded against using redundant hardware of the same type. However, redundant hardware of the same type will not mask a design fault. Software fault tolerance is mostly based on traditional hardware fault tolerance. For example, *N*-version programming closely parallels *N*-way redundancy in the hardware fault tolerance paradigm. Recovery blocks are modeled after the current *ad hoc* method being employed in safety critical software [21]. The *ad hoc* method uses the concept of retrying the same operation, hoping that the problem would be resolved when the second try occurred. Software fault tolerance cannot sufficiently mask design faults, like hardware fault tolerance. Software fault tolerance and hardware fault tolerance needs to evolve to solve a design fault problem, as more large-scale digital control systems are designed and built, especially safety critical systems.

5.3.1 Diversity

A processor-based hardware module is duplicated to provide redundancy in a large-scale digital control system, particularly in safety-critical systems. Programs within the module must also be duplicated. Duplication of the hardware module provides protection against random component failures, since modules fail at different times. However, a problem within identically duplicated software is likely to affect all identical modules at the same time. Therefore, software within each hardware module should be diversified in order to protect the system from software faults. Diversity in software is an essential factor for software fault tolerance. Diversity refers to using different means to perform a required function or solve the same problem. This means developing more than one algorithm to implement a solution for software. The results from each algorithm are compared, and if they agree, then appropriate action is taken. Total or majority agreement may be implemented depending on system criticality. Error detection and recovery algorithms take control if the results do not agree. Safety-critical and safety-related software is often implemented through diverse algorithms.

N-version programming is strongly coupled with diversity. *N*-version programming provides a degree of protection against systematic faults associated with software through the use of diversity. *N*-version programming involves using several different implementations of a program [22]. These versions all attempt to implement the same specification and produce the same results. Different versions may be run sequentially on the same processor or in parallel on different processors. Various routines use the same input data and their results are compared. The unanimous answer is passed to its destination in the absence of disagreement between software modules. The action taken depends on the number of versions used if the modules produce different results. Disagreement between the modules represents a fault condition for a duplicated system. However, the system can not tell which module is incorrect. This problem is tackled by repeating the calculations in the hope that the problem is transient. This approach is successful if the error had been caused by a transient hardware fault which disrupted the processor during the execution of the software module. Alternatively, the system might attempt to perform some further diagnostics to decide which routine is in error. A more attractive arrangement uses three or more versions of the software. Some form of voting to mask the effects of faults is possible in this case. Such an arrangement is a software equivalent of the triple or N-modular redundant hardware system. The high costs involved usually make them impractical, although large values of N have attractions from a functional viewpoint.

The main disadvantages of N-version programming [4] are processing requirements and cost of implementation. The calculation time, for a single processor system, is increased by a factor of more than N, compared to that of a single version implementation. The increase beyond a factor of N is caused by the additional complexity associated with the voting process. This time overhead for a N-processor system may be removed with the cost of additional hardware. Software development costs tend to be increased by a factor of more than N, in either case, owing to the cost of implementing the modules and the voting software. This high development cost restricts the use of this technique to very critical applications where the cost can be tolerated.

5.3.2 Block Recovery

Another essential factor is recovery from software faults. Diverse implementation does not guarantee the absence of common faults. Design features which provide correct functional operation in the presence of one or more errors are required in addition to diversity. Block recovery is one of the design features [22]. The block recovery method uses some form of error detection to validate the operation of a software module. An alternative software routine is used if an error is detected. This scheme is based on the use of acceptance tests. These tests may have several components and may, for example, include checks for runtime errors, reasonability, excessive execution time, and mathematical errors. It is necessary to demonstrate that each module achieves the functionality set out in its specification during software development. Such an approach is also used to devise runtime tests which will demonstrate that a module has functioned correctly.

Systems using the block recovery approach require duplication of software modules. A primary module is executed in each case, followed by its acceptance test. Failure of the test will result in the execution of an alternative redundant module, after which the acceptance test is repeated. Any number of redundant modules may be provided to give increased fault tolerance. Execution proceeds to the next software operation as soon as execution of one of the versions of the module results in a successful test. The system must take appropriate action if the system fails the acceptance test for all of the redundant modules, when an overall software failure is detected.

There are three main types of block recovery: backward block recovery, forward block recovery, and *n*-block recovery. The system is reset to a known prior safe state if an error is detected with backward block recovery. This method implies that internal states are saved frequently at well-defined checkpoints. Global internal states or only those for critical functions may be saved.

The current state of the system is manipulated or forced into a known future safe state if an error is detected with forward block recovery. This method is useful for real-time systems with small amounts of data and fast-changing internal states.

Several different program segments are written which perform the same function in *n*-block recovery. The first or primary segment is executed first. An acceptance test validates the results from this segment. The result and control is passed to subsequent parts of the program if the test passes. The second segment, or first alternative, is executed if the test fails. Another acceptance test evaluates the second result. The result and control is passed to subsequent parts of the program if the test passes. This process is repeated for two, three, or *n* alternatives, as specified.

5.3.3 Perspectives on Software Fault Tolerance

Software fault tolerance becomes more necessary for modern computer technologies. Current software fault tolerance methods cannot adequately compensate for all faults, especially design faults. Recovery blocks may be a good solution to transient faults. However, recovering faults faces the same inherent problem that *N*-version programming does in that it does not offer (sufficient) protection against design faults. Design faults need to be effectively dealt with in the next generation of software fault tolerance methods, since most software faults are design faults which are the result of human error in interpreting a specification or correctly implementing an algorithm. For example, it is possible for a limited class of design faults to be recovered by using distributed *N*-version programming. Memory leaks, which are a design fault, can cause a local heap to grow beyond the limits of the computer system. Distributed heaps could run out of memory at different times and still be consistent with respect to a valid data state by using distributed *N*-version programming or one of its variants [23].

Some next-generation software fault tolerance methods need to address how to resolve the increasing cost problem of building correct software. These methods should be cost effective to be applied to safety-critical large-scale digital control systems.

The performance of software fault tolerance depends on the capabilities of fault diagnostics and fault-masking mechanisms. Software fault tolerance is achieved by software and hardware. Improved performance of software fault tolerance is acquired when software and hardware are adequately integrated into fault diagnostics and masking mechanisms. The integration of software and hardware increases protection against design faults and decreases the cost of building correctly functioning systems.

5.4 Concluding Remarks

Three categories of software reliability improvement techniques (fault avoidance, fault removal, and fault tolerance) have been discussed in this chapter. Fault-free and fault-tolerant software are produced by these techniques, increasing the reliability of software. Fault-free software is software that conforms to its specification. The following is crucial for developing fault-free software:

- Production of a precise (preferably formal) system specification (fault avoidance)
- Use of information hiding and encapsulation (fault avoidance)
- Extensive use of reviews in the development process (fault avoidance and/or fault removal)
- Careful planning of system testing (fault removal)

A fault-tolerant system needs software fault tolerance in order to create a system that is reliable. Particularly, the system inevitably requires software fault tolerance, to ensure reliability of the system that will operate throughout its life.

Application of improved techniques results in more reliable software. What is the qualitative and/or quantitative limit of software reliability improvement? This question from Chapter 4 reminds us that reliability improvement techniques described in this chapter should be considered to evaluate software reliability based on process metrics.

References

[1] Leveson NG (1990) Guest Editor's Introduction: Formal Methods in Software Engineering. IEEE Transactions in Software Engineering, Vol. 16, No. 9

[2] Wing JM (1990) A Specifier's Introduction to Formal Methods. Computer, Vol. 23, No. 9

[3] Rushby J (1993) Formal Methods and the Certification of Critical Systems. Technical Report CSL-93-7, SRI International, Menlo Park, CA

[4] Storey N (1996) Safety-Critical Computer Systems. Addison–Wesley.

[5] Proceedings of the Seventh Knowledge-Based Software Engineering Conference, McLean, VA, September 20–23, 1992

[6] Agresti WW (1986) New Paradigms for Software Development. IEEE Computer Society

[7] London RL (1977) Remarks on the Impact of Program Verification on Language Design. In Design and Implementation of Programming Languages. Springer–Verlag

[8] McGettrick AD (1982) Program Verification using Ada. Cambridge University Press

[9] Gries D (1991) On Teaching and Calculation. Communications of the ACM, Vol. 34, No. 3

[10] Mills HD (1986) Structured Programming: Retrospect and Prospect. IEEE Software, Vol. 3, No. 6

[11] Dyer M (1992) The Cleanroom Approach to Quality Software Development. John Wiley & Sons

[12] AECL (1991) Wolsong NPP 2/3/4, Software Work Practice Procedure for the Specification of SRS for Safety Critical Systems. Design Document no. 00-68000-SWP-002, Rev. 0
[13] Hopcroft J, Ullman J (1979) Introduction to Automata Theory, Language and Computation. Addison–Wesley.
[14] Alur R, Dill DL (1994) A Theory of Timed Automata. Theoretical Computer Science Vol. 126, No. 2, pp. 183–236
[15] EPRI (1995) Guidelines for the Verification and Validation of Expert System Software and Conventional Software. EPRI TR-103331-V1 Research project 3093-01, Vol. 1
[16] Koo S, Seong P, Yoo J, Cha S, Yoo Y (2005) An Effective Technique for the Software Requirements Analysis of NPP Safety-critical Systems, Based on Software Inspection, Requirements Traceability, and Formal Specification. Reliability Engineering and System Safety, Vol. 89, No. 3, pp. 248–260
[17] Fagan ME (1976) Design and Code Inspections to Reduce Errors in Program Development. IBM System Journal, Vol. 15, No. 3, pp. 182–211
[18] Yoo J, Kim T, Cha S, Lee J, Son H (2005) A Formal Software Requirements Specification Method for Digital Nuclear Plants Protection Systems. Journal of Systems and Software, No. 74, pp. 73–83
[19] Koo S, Seong P, Cha S (2004) Software Design Specification and Analysis Technique for the Safety Critical Software Based on Programmable Logic Controller (PLC). Eighth IEEE International Symposium on High Assurance Systems Engineering, pp. 283–284
[20] Koo S, Seong P, Jung J, Choi S (2004) Software design specification and analysis (NuFDS) approach for the safety critical software based on programmable logic controller (PLC). Proceedings of the Korean Nuclear Spring Meeting
[21] Lyu MR, ed. (1995) Software Fault Tolerance: John Wiley and Sons, Inc.
[22] IEC, IEC 61508-7: Functional Safety of Electrical/Electronic/Programmable Electronic Safety-related Systems – Part 7: Overview of Techniques and Measures
[23] Murray P, Fleming R, Harry P, Vickers P (1998) Somersault Software Fault-Tolerance. HP Labs whitepaper, Palo Alto, California

6

NuSEE: Nuclear Software Engineering Environment

Seo Ryong Koo[1], Han Seong Son[2] and Poong Hyun Seong[3]

[1]Nuclear Power Plant Business Group
Doosan Heavy Industries and Construction Co., Ltd.
39-3, Seongbok-Dong, Yongin-Si, Gyeonggi-Do, 449-795, Korea
seoryong.koo@doosan.com

[2]Department of Game Engineering
Joongbu University
#101 Daehak-ro, Chubu-myeon, Kumsan-gun, Chungnam, 312-702, Korea
hsson@joongbu.ac.kr

[3]Department of Nuclear and Quantum Engineering
Korea Advanced Institute of Science and Technology
373-1 Guseong-dong, Yuseong-gu, Daejeon, 305-701, Korea
phseong@kaist.ac.kr

The concept of software V&V throughout the software development lifecycle has been accepted as a means to assure the quality of safety-critical systems for more than a decade [1]. The Integrated Environment (IE) approach is introduced as one of the countermeasures for V&V (Chapter 5). Adequate tools are accompanied by V&V techniques for the convenience and efficiency of V&V processes. This chapter introduces NuSEE (Nuclear Software Engineering Environment), which is a toolset to support the IE approach developed at Korea Advanced Institute of Science and Technology (KAIST) [2]. The software lifecycle consists of concept, requirements, design, implementation, and test phases. Each phase is clearly defined to separate the activities to be conducted within it. Minimum V&V tasks for safety-critical systems are defined for each phase in IEEE Standard 1012 for "Software Verification and Validation" (Figure 6.1) [3]. V&V tasks are traceable back to the software requirements. A critical software product should be understandable for independent evaluation and testing. The products of all lifecycle phases are also evaluated for software quality attributes, such as correctness, completeness, consistency, and traceability. Therefore, it is critical to define an effective specification method for each software development phase and V&V task based on the effective specifications during the whole software lifecycle.

One single complete V&V technique does not exist because there is no adequate software specification technique that works throughout the lifecycle for safety-critical systems, especially for NPP I&C systems. There have been many attempts to use various specification and V&V techniques, such as formal

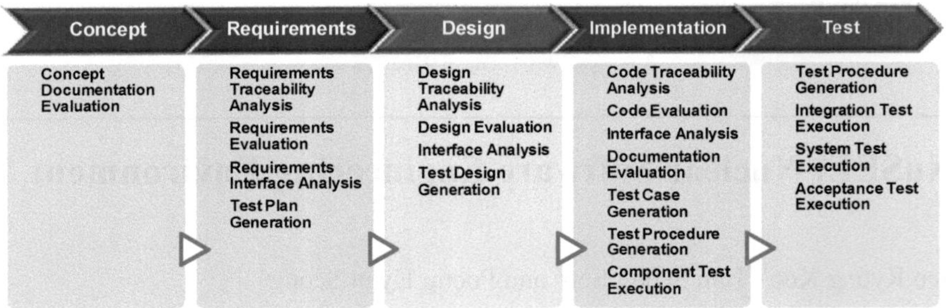

Figure 6.1. Software V&V tasks during the lifecycle

specification and analysis, software inspection, traceability analysis, and formal configuration management in NPP software fields. However, most are extremely labor-intensive, and their users require tool support.

The IE approach for software specification and V&V is in accordance with the above software V&V tasks during the entire software lifecycle for safety-critical systems. The NuSEE toolset was developed to support and integrate the entire software lifecycle for NPP safety-critical systems systematically implementing this IE approach. The NuSEE toolset also achieves optimized integration of work products. NuSEE provides effective documentation evaluation and management, formal specification and analysis, and systematic configuration management. NuSEE consists of four major tools: NuSISRT (Nuclear Software Inspection Support and Requirements Traceability) for the concept phase, NuSRS (Nuclear

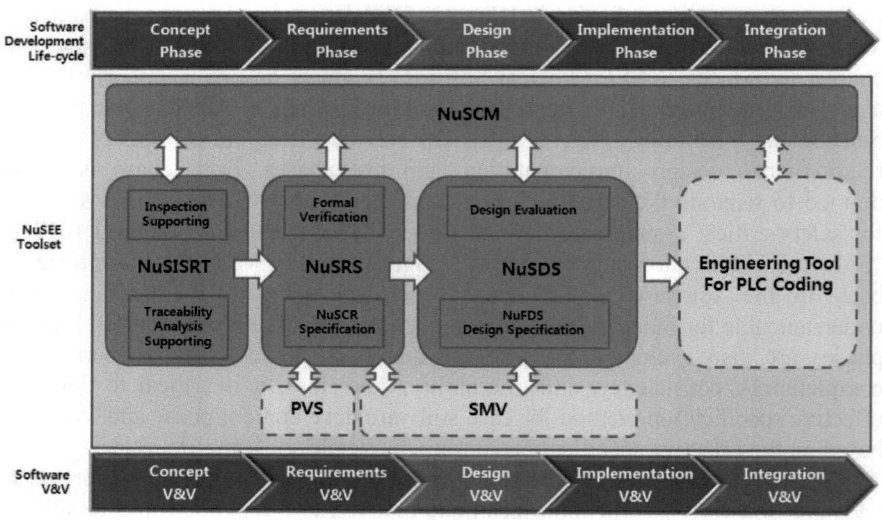

Figure 6.2. Overall features of NuSEE

Software Requirements Specification and analysis) for the requirement phase, NuSDS (Nuclear Software Design Specification and analysis) for the design phase, and NuSCM (Nuclear Software Configuration Management) for configuration management.

Features of the NuSEE toolset are shown in Figure 6.2. Each tool supports each phase of the software development lifecycle and software V&V process. The tools are integrated in a straightforward manner through the special features of the interface. Potential errors are found at an early point throughout the software lifecycle using the NuSEE toolset. Software engineers fix them with the lowest cost and the smallest impact on system design.

6.1 NuSEE Toolset

6.1.1 NuSISRT

NuSISRT (Nuclear Software Inspection Support and Requirements Traceability) is a PC-based tool designed to manage requirements. NuSISRT in the IE approach supports all software lifecycle phases as well as the concept phase. Inspection, based on documents written in natural language, is believed to be an effective software V&V technique that is extensively used for NPP I&C systems. Inspection provides a great increase in both productivity and product quality by reducing development time and by removing more defects than is possible without its use. The outstanding feature of inspection is that it is applied to the whole lifecycle.

Requirements traceability analysis capability is integrated into the software inspection support tool in NuSISRT. This is also considered as a major method for software V&V. The capabilities of structural analysis and inspection meeting support are integrated in NuSISRT. NuSISRT comprises tools for document evaluation, traceability analysis, structural analysis, and inspection meeting support. NuSISRT has three kinds of view to systematically support the IE approach: inspection view, traceability view, and structure view. NuSISRT also has a web page for inspection meetings.

6.1.1.1 Inspection View

The support of document evaluation with inspection view is the main function of NuSISRT. This view has an extracting function that reads a text file and copies paragraph numbers and requirement text to a NuSISRT file. Any text data that is convertible to ".txt" format can be read. This view also supports the manual addition of individual requirements and imports requirements from text data with various formats. Inspection view permits users to associate database items with each other by defining attributes; the attributes attached to individual database items provide a powerful means of identifying subcategories or database items. Inspection view of NuSISRT supports the parent-child links for managing requirements. Peer links between items in a database and general documents are also supported. Peer links provide an audit trail that shows compliance for quality standards or contractual conditions.

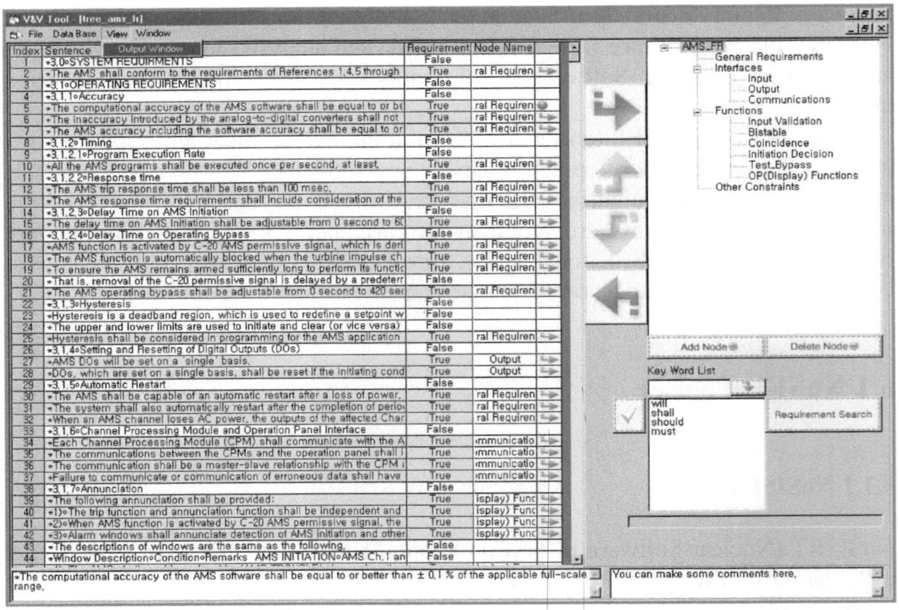

Figure 6.3. Inspection view of NuSISRT

There are many sentences in a requirements document, but all of them are not requirements. Adequate requirement sentences have to be elicited for more effective inspection. A software requirements inspection based on checklists is performed by each inspector using inspection view of NuSISRT (Figure 6.3). The view reads source documents, identifies requirements, and extracts the requirements. Inspection view automatically extracts requirements based on a set of keywords defined by the inspector. The requirements found are then highlighted (Figure 6.3). The inspector also manually identifies requirements. Inspection view enables the production of a user-defined report that shows various types of inspection results. The user builds up the architecture of the desired reports in the right-hand window of this view. NuSISRT directly supports software inspection with this functional window if the user writes down checklists in the window. The requirements to be found by the tool are located in a suitable checklist site using the arrow buttons in the window. Each inspector examines the requirements and generates the inspection result documents with the aid of NuSISRT.

6.1.1.2 Traceability View

Traceability view of NuSISRT supports a requirements traceability analysis between two kinds of system documents. This view provides mechanisms to easily establish and analyze traceability between requirements sentences through the visual notification of change in the Requirements Traceability Matrix (RTM). This capability allows users to pinpoint its impact across the project and assess the coverage for V&V. An identification number of requirements is assigned to each requirement sentence elicited from inspection view for the traceability analysis. The relation between source requirements and destination requirements is described in the RTM for the requirements traceability analysis. Traceability between documents is analyzed using the results pertaining to the relation. An example of requirements traceability analysis using traceability view of NuSISRT is illustrated in Figure 6.4. Traceability view of NuSISRT supports the parent/child links to manage requirements and the peer links between items in the database and general documents to provide an audit trail (Figure 6.5). The column number of the matrix represents a requirement of the source document and the row number of the matrix represents that of the destination document in Figure 6.5. The relationships between source and destination are expressed through the matrix window with linked and/or unlinked chains. The linked chains indicate source requirements that are reflected onto destination requirements. The unlinked chains represent source and destination requirements that are changed. Therefore, it is necessary to verify the change between the source and destination documents. The question marks denote difficulties in defining traceability between requirements. Another analysis is required to verify requirements in this case.

Traceability view has an additional function to calculate the similarity between two requirements using similarity calculation algorithms in order to more efficiently support traceability analysis [4]. Traceability view automatically represents the similarity by percentage through this function (Figure 6.6). This similarity result is helpful to the analyzer. A traceability analysis between two documents is performed through traceability view in this way.

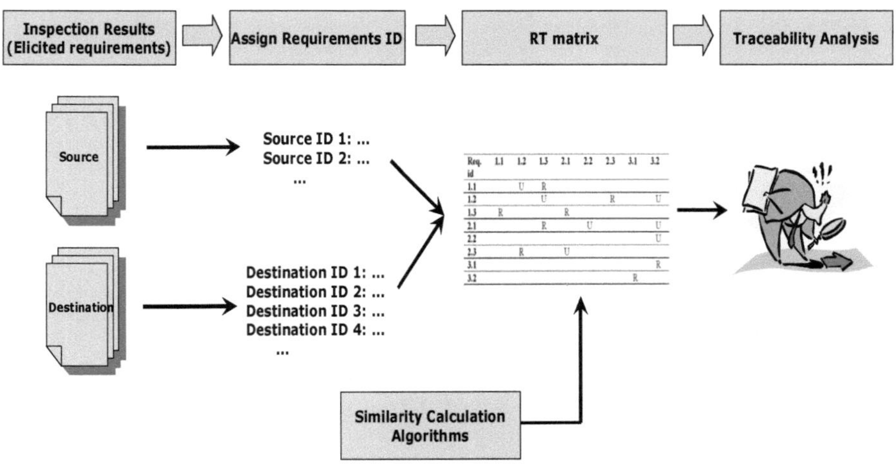

Figure 6.4. Schematic diagram of requirements traceability

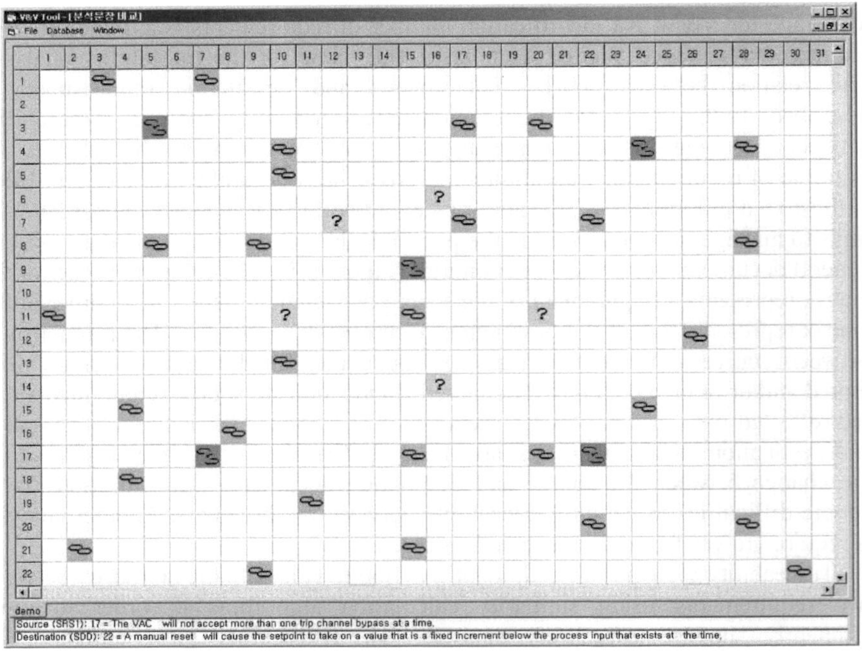

Figure 6.5. Traceability view of NuSISRT

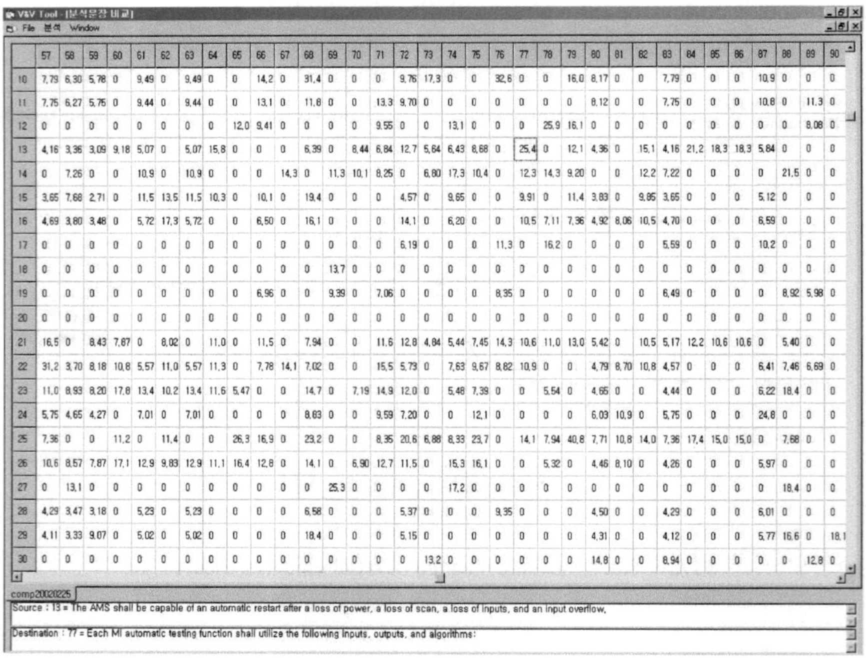

Figure 6.6. An example of similarity calculation

6.1.1.3 Structure View

Structure view of NuSISRT enables effective translation requirements into NuSRS as one of the interfacing functions in the IE approach. Users analyze system development documents in view of the system's structure through structure view (Figure 6.7). These analysis results then help generate a formal specification from a natural language document in the requirements phase. Inputs/outputs and functions are essentially defined in the structural analysis of systems through structure view. The IE approach proposes an input-process-output structure. Several tabular forms help users easily build up the input-process-output structure in structure view. This structure is represented in the right-hand side window as a tree. Structure view generates a result file written in XML language, which is then translated to NuSRS, after the structure analysis. FOD can be drawn automatically in NuSRS with this file.

6.1.2 NuSRS

Several formal methods are effective V&V harnesses [5–8], but are difficult to properly use in safety-critical systems because of their mathematical complexity. Formal specification lessens requirement errors by reducing ambiguity and imprecision and by clarifying instances of inconsistency and incompleteness. The Atomic Energy of Canada Limited (AECL) approach specifies a methodology and format for the specification of software requirements for safety-critical software

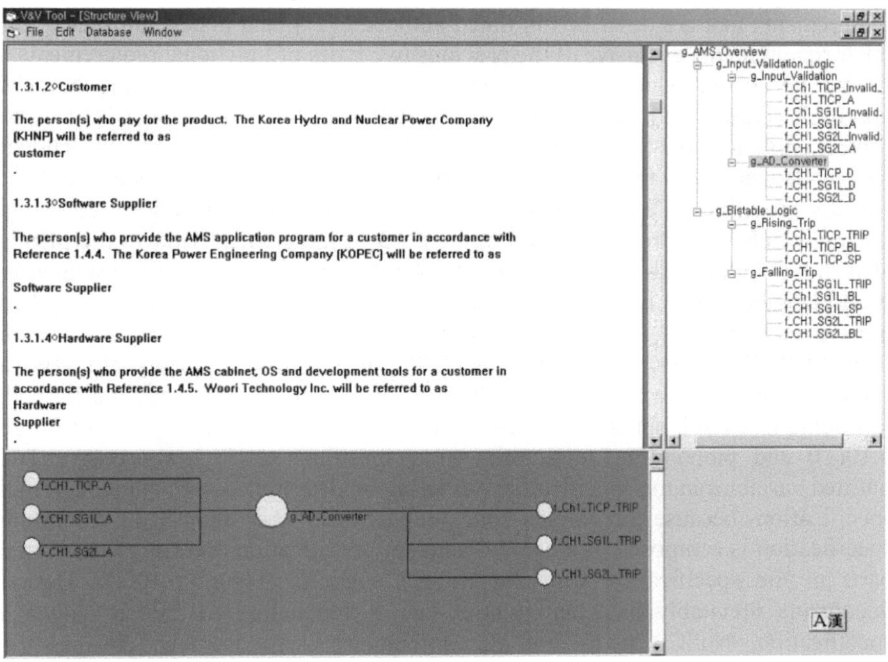

Figure 6.7. Structure view of NuSISRT

used in real-time control systems in nuclear systems. The approach is an SCR-style software requirements specification (SRS) method based on Parnas' four-variable method. A system reads environment states through monitored variables that are transformed into input variables. The output values of the output variables are calculated and changed into control variables. The AECL approach provides two different views of the requirements. A larger view, the Function Overview Diagram (FOD), where each of the functions in the FOD, is described by the smaller view of the Structured Decision Table (SDT). The AECL approach specifies all requirements of the nuclear control system in FOD and SDT notations. This is complex in cases where timing requirements and history-related requirements are considered. Difficulty with specification is modified in the NuSCR approach.

The NuSCR approach is a formal method that is an extension of the existing SCR-style AECL approach [9]. The NuSCR specification language was originally designed to simplify complex specification techniques of certain requirements in the AECL approach. The improved method describes the behavior of the history-related requirements and timing requirements of nuclear control systems by specifying them in automata and timed-automata, respectively. All specifications including history-related requirements and timing requirements are specified with only one type of function node in the FOD and with SDT tables in the existing AECL method. However, NuSCR uses three different types of nodes in the FOD to specify the properties derived from the requirements. The types consist of nodes that specify history-related requirements as are described in automata [10], timing requirements that are described in timed-automata [11], and nodes that specify all other requirements exclusive of the previous two types of functional requirements.

NuSRS is an editor for requirement specifications based on the NuSCR approach (Figure 6.8). An example of NuSCR specification of the NPP reactor protection system (RPS) is shown in Figure 6.9. NuSRS is a platform-independent tool made using java for formally specifying the SRS of a nuclear control system. NuSRS provides an environment to draw FOD and SDT and allows automata diagrams to be built from the nodes of the FOD. A hierarchical view of the SRS is also shown on the left side in Figure 6.8. NuSRS generates a result file written in XML language that includes all of the information in NuSRS, which is then transferred to NuSDS. Advantages of formal methods are shown by the application of NuSCR. Examples showing how NuSCR can improve the quality of software specification are shown in Figure 6.10. NuSCR is very effective, and not limited, to finding omitted input and/or output variables in a software specification (Figure 6.10(a)) and pinpointing ambiguities in a specification (Figure 6.10(b)). The omitted variables in the specification can easily be determined with NuSCR formal specification because all inputs and outputs shall be described. A formal specification is composed of a natural language specification revealing the omitted parts of the specification with the NuSCR supports (Figure 6.10(a)). Design documents inevitably have ambiguities due to the nature of natural language specification. NuSCR helps find the ambiguous parts and then changes the specification to the precise one in order to decrease discrepancies generated from these ambiguities (Figure 6.10(b)). The faults introduced into the system are avoided or prevented by using formal methods like NuSCR.

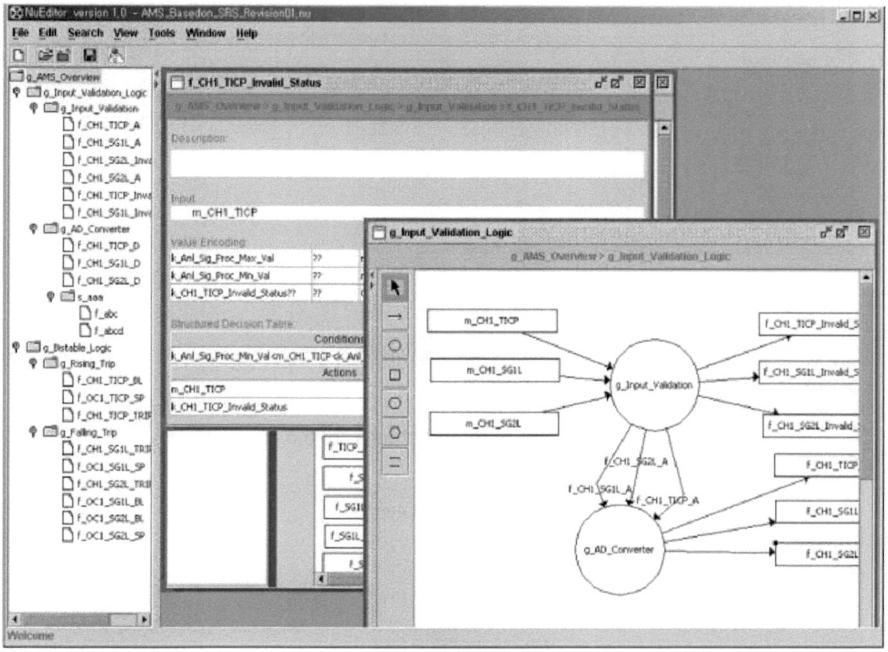

Figure 6.8. Editing windows of NuSRS

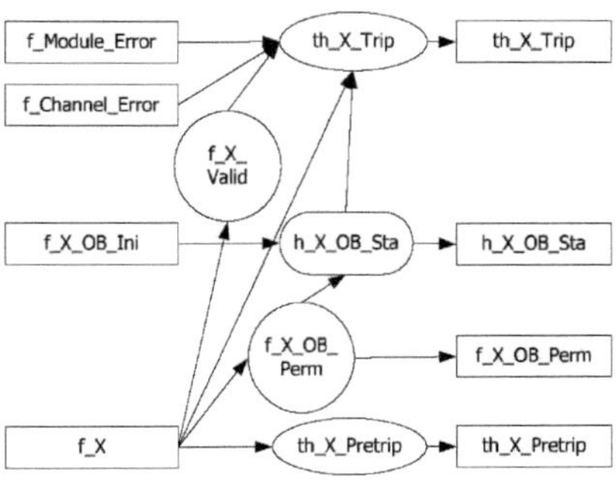

(a) FOD for *g_Fixed_Setpoint_Rising_Trip_with_OB*

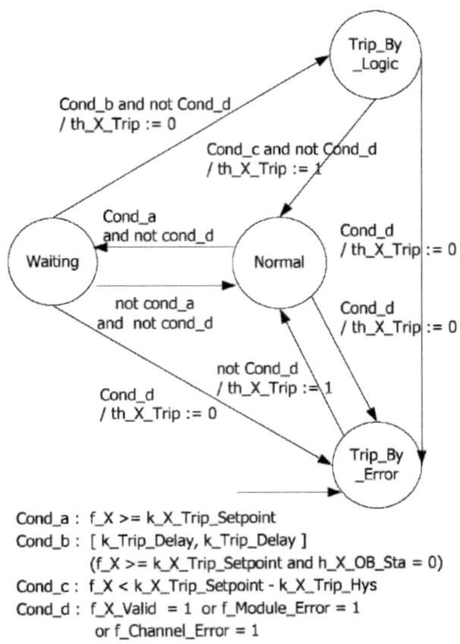

Conditions		
k_X_MIN <= f_X <= k_X_MAX	T	F
Actions		
f_X_Valid := 0	X	
f_X_Valid := 1		X

(b) SDT for function variable node *f_X_Valid*

(c) TTS for timed history variable node *th_X_Trip*

Figure 6.9. Part of NuSCR specification for the RPS

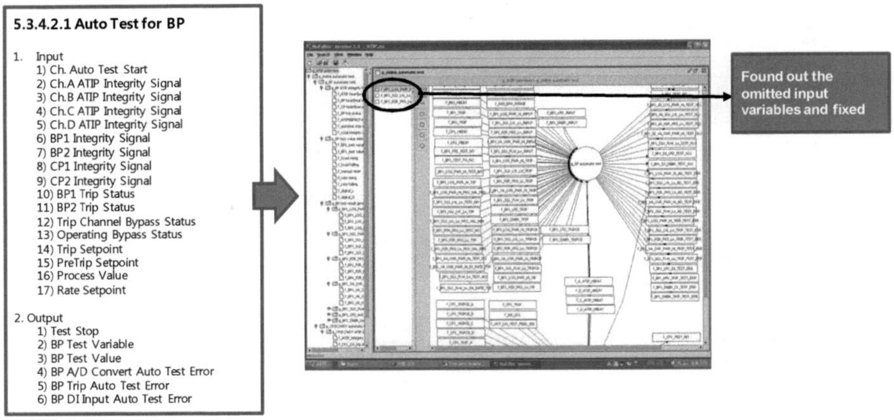

(a) Improvement of input/output variable completeness

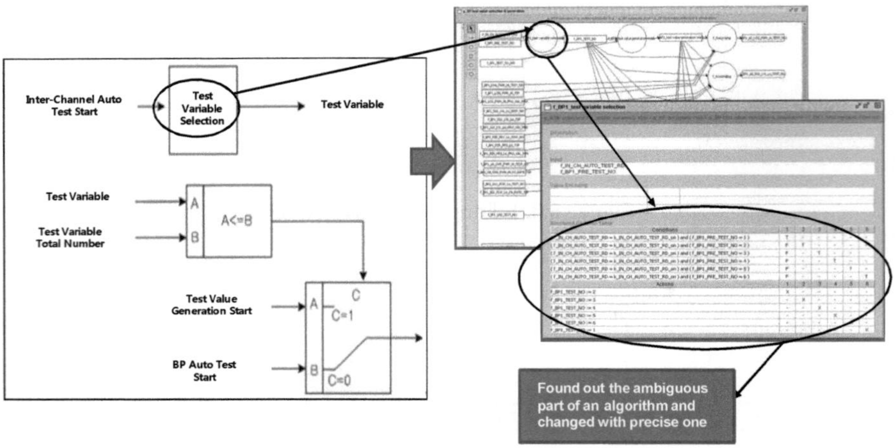

(b) Improvement of algorithm correctness

Figure 6.10. Partial application results of NuSCR for RPS

6.1.3 NuSDS

Software design is a process of translating problem requirements into software structures that are to be built in the implementation phase [12]. Generally, a software design specification (SDS) describes the overall system architecture and contains a definition of the control structure model. SDS should be evaluated in view of software quality attributes, such as correctness, completeness, consistency, and traceability. A well-constructed design specification is useful for the implementation phase, because an implementation product can be easily translated from good design specification. The most important task in the software design phase is to define an effective specification method.

NuFDS is a software design specification and analysis technique for safety-critical software based on a Programmable Logic Controller (PLC) [13]. NuFDS stands for nuclear FBD-style design specification. The function block diagram (FBD) is one of the PLC software languages [14]. NuSDS is a tool support based on the NuFDS approach. This tool is designed particularly for the software design specifications in nuclear fields. The specifications consist of four major parts: database, software architecture, system behavior, and PLC hardware configuration. A SDS is generated using these four specification features in NuSDS. The features of NuSDS are described in Figure 6.11. NuSDS fully supports design specifications according to the NuFDS approach. NuSDS partially supports design analysis based on the design specifications. NuSDS has been integrated with architecture description language and a model checker for design analysis. NuSDS translates the specifications into an input to model checking. NuSDS is also used in connection with other V&V tools.

A part of the bistable processor (BP) design specifications constructed by NuSDS is shown in Figure 6.12. The BP is a subsystem of a reactor protection system in an NPP. The I/O database specification of the BP is represented in Figure 6.12(a). The software architecture (*SA) specification of the BP using the architecture design block feature is shown in Figure 6.12(b). In the BP, the *SA is composed of H/W_Check_Module, Bistable_Module, Hearbeat_Module, and Comm_Module as its major architecture design blocks. Each major architecture contains sub-architectural modules. The FBD-style specification of the Signal_Check_Module of the BP is represented in Figure 6.12(c). This FBD-style specification addresses the interactions between the function blocks and the I/O variables. The hardware layout diagram for the PLC hardware configuration is shown in Figure 6.12(d). A basic verification is possible through the software

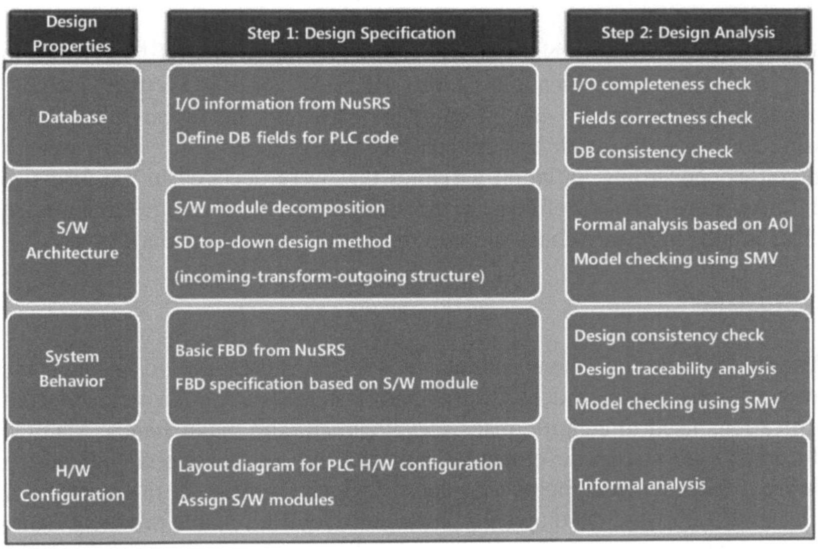

Figure 6.11. Features of NuSDS

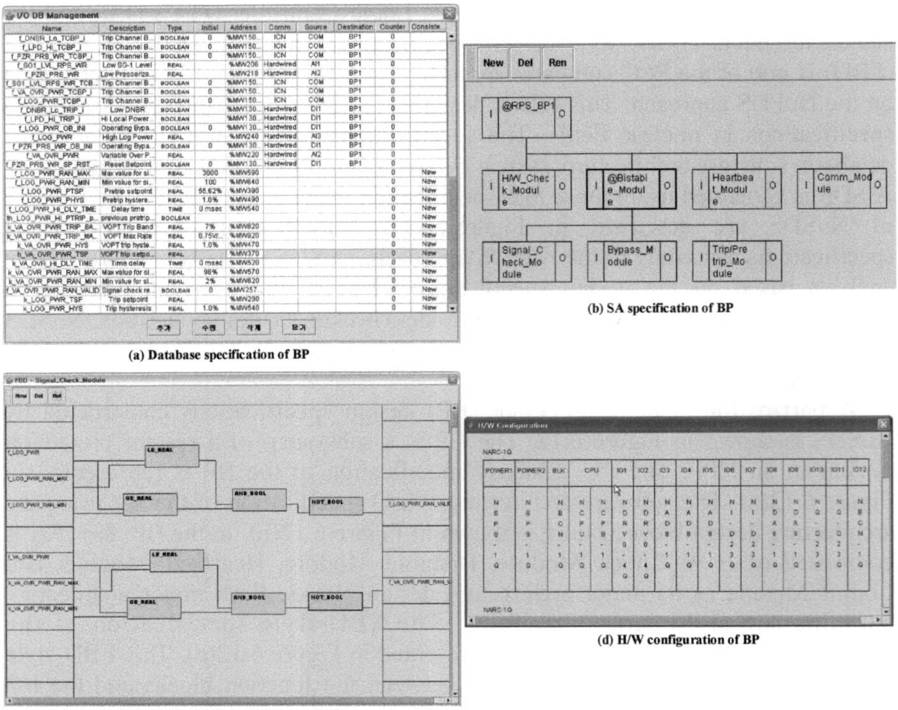

(a) Database specification of BP

(b) SA specification of BP

(c) FBD -style behavior specification of BP

(d) H/W configuration of BP

Figure 6.12. Software design specification of the BP

design specification using NuSDS. I/O errors and some missed *SAs were found during the design specification of the BP [13]. The I/O errors include the inconsistency between the natural language SRS and the formal SRS and some missing I/O variables, such as heartbeat-operation-related data. There were some ambiguities concerning initiation variables that were declared in the formal SRS. The *SAs were newly defined in the design phase since the communication module and the hardware check module were not included in the SRS.

6.1.4 NuSCM

Software configuration management (SCM) configures the form of a system (documents, programs, and hardware) and systematically manages and controls modifications used to compile plans, development, and maintenance. Many kinds of documents for system development and V&V processes are produced during the software lifecycle. Documents are controlled and governed to guarantee high quality in the software development phase and produce reliable products. Software quality management is highly valued in the development, modification, and maintenance phases. Requests in modification continue to be received even while operating the software. Specific corresponding plans are established in order to confront these requests. Deterioration in quality and declination in the life of the software will result if modification requests are not properly processed in the software maintenance phase. The risk of accidents due to software may increase, particularly in systems where safety is seriously valued. Many research institutes and companies are currently making attempts to automate systematic document management in an effort to satisfy high quality and reliability. NuSCM is a project-centered software configuration management system especially designed for nuclear safety systems. This integrated environment systematically supports the management of all system development documents, V&V documents and codes throughout the lifecycle. NuSCM also manages all result files produced from NuSISRT, NuSRS, and NuSDS for the interface between NuSCM and other tools. Web-based systems are being developed since most software systems are compatible and users can easily access regardless of location. NuSCM was also designed and embodied using the web. Document management and change request views in NuSCM are shown in Figure 6.13.

6.2 Concluding Remarks

The IE approach systematically supports a formal specification technique according to lifecycle and effective V&V techniques for nuclear fields, based on proposed specifications. NuSEE, integrated with NuSISRT, NuSRS, NuSDS, and NuSCM tools, supports the IE approach. NuSISRT is a special tool for software inspection and traceability analysis, that is used in all document-based phases as well as the concept phase. NuSRS and NuSDS support the generation of specifications, such as SRS in the requirement phase and SDS in the design phase for NPP software fields based on PLC applications. Formal analyses, such as theorem proving and model checking, are also supported. NuSCM is a project-centered software configuration management tool. NuSCM manages and controls

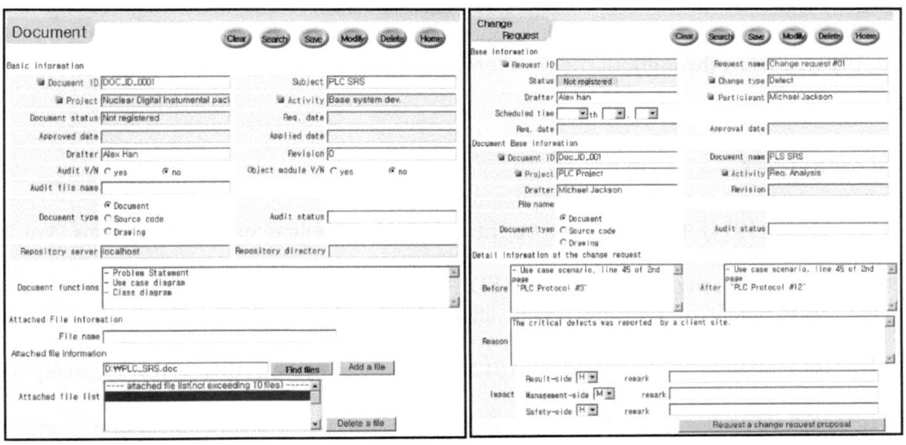

Figure 6.13. Document management view and change request view of NuSCM

Table 6.1. Summary of each tool

	S/W development life cycle	Main functions	Advantages
NuSISRT	● System concept phase ● Whole phases based on documents	● Documents inspection support ● Documents traceability analysis support ● System structure analysis support	● Systematic checklist management ● Reducing time of inspection work ● Minimize human error ● Effective traceability analysis ● Interface with NuSRS
NuSRS	● Software requirement phase	● Formal method (NuSCR) editing support ● Theorem proving (PVS) support ● Model checking (NuSMV) support	● Formal method for nuclear fields ● Effective system formal specification ● Formal requirement analysis ● Interface with NuSDS
NuSDS	● Software design phase	● System database, S/W architecture, system behavior, H/W configuration specification support ● Model checking support ● Traceability analysis support	● Optimal design technique for nuclear fields ● Effective system design specification ● Ease of PLC programming ● Formal design analysis
NuSCM	● Whole phases	● Project-centered configuration management support ● Change request form in nuclear fields support ● Source code management support	● CM technique for nuclear fields ● Various document styles support ● Interface with V&V tools

the modification of all system development and V&V documents. Resultant files from NuSISRT, NuSRS, and NuSDS are managed through NuSCM. The features of tools in the NuSEE toolset are summarized from the viewpoints of software development life-cycle support, main functions, and advantages (Table 6.1). The NuSEE toolset provides interfaces among the tools in order to gracefully integrate the various tools. The NuSEE toolset achieves optimized integration of work products throughout the software lifecycle of safety-critical systems based on PLC applications. Software engineers reduce the time and cost required for development of software. In addition, user convenience is enhanced with the NuSEE toolset, which is a tool for building bridges between specialists in system engineering and software engineering, because it supports specific system specification techniques that are utilized throughout the development lifecycle and V&V process.

References

[1] EPRI (1994) Handbook for verification and validation of digital systems Vol.1: Summary, EPRI TR-103291

[2] Koo SR, Seong PH, Yoo J, Cha SD, Youn C, Han H (2006) NuSEE: an integrated environment of software specification and V&V for NPP safety-critical systems. Nuclear Engineering and Technology

[3] IEEE (1998) IEEE Standard 1012 for software verification and validation, an American National Standard

[4] Yoo YJ (2003) Development of a traceability analysis method based on case grammar for NPP requirement documents written in Korean language. M.S. Thesis, Department of Nuclear and Quantum Engineering, KAIST

[5] Harel D (1987) Statecharts: a visual formalism for complex systems. Science of Computer Programming, Vol. 8, pp. 231–274
[6] Jensen K (1997) Coloured Petri nets: basic concepts, analysis methods and practical uses, Vol. 1. Springer–Verlag, Berlin Heidelberg
[7] Leveson NG, Heimdahl MPE, Hildreth H, Reese JD (1994) Requirements specification for process-control systems. IEEE Transaction on Software Engineering, Vol. 20, No. 9, Sept.
[8] Heitmeyer C, Labaw B (1995) Consistency checking of SCR-style requirements specification. International Symposium on Requirements Engineering, March
[9] Wolsong NPP 2/3/4 (1991) Software work practice procedure for the specification of SRS for safety critical systems. Design Document no. 00-68000-SWP-002, Rev. 0, Sept.
[10] Hopcroft J, Ullman J (1979) Introduction to automata theory, language and computation, Addison–Wesley
[11] Alur R, Dill DL (1994) A theory of timed automata. Theoretical Computer Science Vol. 126, No. 2, pp. 183–236, April
[12] Pressman RS (2001) Software engineering: a practitioner's approach. McGraw–Hill Book Co.
[13] Koo SR, Seong PH (2005) Software Design Specification and Analysis Technique (SDSAT) for the Development of Safety-critical Systems Based on a Programmable Logic Controller (PLC), Reliability Engineering and System Safety
[14] IEC (1993) IEC Standard 61131-3: Programmable controllers–Part 3, IEC 61131

Part III

Human-factors-related Issues and Countermeasures

7

Human Reliability Analysis in Large-scale Digital Control Systems

Jae Whan Kim

Integrated Safety Assessment Division
Korea Atomic Energy Research Institute
1045 Daedeok-daero, Yuseong-gu, Daejeon, 305-353, Korea
jhkim4@kaeri.re.kr

The reliability of human operators, which are basic parts of large-scale systems along with hardware and software, is introduced in Part III. A review of historic methods for human reliability analyses is presented in Chapter 7. The human factors engineering process to design a human–machine interface (HMI) is introduced in Chapter 8. Human and software reliability are difficult to completely analyze. The analysis may not guarantee the system against human errors. Strict human factors engineering is applied when designing human–machine systems, especially safety critical systems, to enhance human reliability. A new system for human performance evaluation, that was developed at KAIST, is introduced in Chapter 9. Measuring human performance is an indispensable activity for both human reliability analysis and human factors engineering.

The contribution of human error is in a range of 30–90% of all system failures, according to reports of incidents or accidents in a variety of industries [1]. Humans influence system safety and reliability over all the system lifespan, including design, construction, installation, operation, maintenance, and test to decommissioning [2]. Retrospective human error analysis investigates causes and contextual factors of past events. Prospective human error analysis (*i.e.*, human reliability analysis (HRA)) takes the role of predictive analysis of the qualitative and quantitative potential for human error, as well as a design evaluation of human–machine systems for system design and operation. The use of HRA for design evaluation is very limited. Most applications are an integral part of PRA by assessing the contribution of humans to system safety.

The major functions of HRA for PRA are to identify erroneous human actions that contribute to system unavailability or system breakdown, and to estimate the likelihood of occurrence as a probabilistic value for incorporation into a PRA model [3, 4]. Human errors for risk assessment are classified into three categories for risk assessment of nuclear power plants: pre-initiator human errors, human errors contributing to an initiating event, and post-initiator human errors [5]. Pre-initiator human errors refer to erroneous human events that occur prior to a reactor trip with undetected states and contribute to the unavailability or malfunction of important safety systems. Human errors contributing to an initiating event are human actions that induce unplanned reactor trips or initiating events. These are not dealt with separately from hardware-induced initiating events, but are statistically considered in an integrative manner in estimating the frequency of an initiating event for current PRAs. Post-initiator human errors occur during operator responses to emergency situations after reactor trip. The error domains that are treated in current HRAs are pre-accident human errors and post-accident human errors. A need for a method of analyzing human errors initiating a reactor trip event separately from the hardware-induced initiating events has been raised by PRA/HRA practitioners and developers. However, only partial applications have been performed for a few of power plants [6].

The development of HRA methods for use in risk assessment started in the early 1970s. HRA methods that appeared before the critiques of Dougherty [7] are called first-generation HRAs. Those that appeared afterward are called second-generation HRAs. The critiques raised by Dougherty have common perceptions among developers and practitioners of HRA on these HRA methods. The main focus of second-generation HRAs has been on post-initiator human errors. First-generation HRAs have biased emphasis on quantitative calculation of human error probabilities, which is deeply rooted in the quantitative demand of PRA. Major features of second-generation HRAs over first-generation HRAs are summarized:

1. Being capable of describing the underlying causes of specific erroneous human actions, or the context in which human errors occur
2. Being capable of identifying various kinds of error modes, including EOC, that might deteriorate the safety condition of a plant
3. Quantification of human error probability on the basis of error-producing conditions or context

This chapter surveys existing HRA methods involving first- and second-generation HRAs, representative first-generation HRA methods, including THERP [8], HCR [9], SLIM [10], and HEART [11] (Section 7.1), and representative second-generation HRA methods, including CREAM [12], ATHEANA [13], and the MDTA-based method [14, 15] (Section 7.2).

7.1 First-generation HRA Methods

7.1.1 THERP

THERP (technique for human error rate prediction) was suggested by Swain and Guttmann of Sandia National Laboratory [8]. THERP is the most widely used HRA method in PSA. A logical and systematic procedure for conducting HRA with a wide range of quantification data is provided in THERP. One of the important features of this method is use of the HRA event tree (HRAET), by which a task or an activity under analysis is decomposed into sub-task steps for which quantification data are provided, HEP is calculated and allotted for each sub-task step, and an overall HEP for a task or an activity is obtained by integrating all sub-task steps. Basic human error probabilities, including diagnosis error probability and execution error probabilities, uncertainty bounds values, adjusting factors with consideration of performance-shaping factors (PSFs), and guidelines for consideration of dependency between task steps are covered in Chapter 20 of the THERP handbook. The general procedure for conducting HRA using THERP is:

1. Assign a nominal HEP value for a task step on a branch of HRAET of a task or an activity.
2. Adjust the nominal HEP by considering PSFs.
3. Assess the dependencies between task steps.
4. Calculate the overall HEP by integrating all branches of the HRAET.
5. Assess the effect of recovery factors on overall HEP to obtain the final HEP.

7.1.2 HCR

The HCR (human cognitive reliability) model was suggested by Hannaman [9]. The non-response probability that operators do not complete a given task within the available time is produced by using the HCR model. Three major variables are used in calculating the non-response probability:

- The variable representing the level of human cognitive behavior (*i.e.*, skill, rule, and knowledge) defined by Jens Rasmussen [16]
- The median response time by the operator for completing a cognitive task
- Three PSF values: operator experience, level of stress, and level of HMI design

An event tree is provided to aid the decision for level of human cognitive behavior. The median response time is obtained through simulator experiments, expert judgments, or interviews of operators. The constant, K, is determined by integrating the levels of three PSFs using the equation:

$$K = (1 + K_1)(1 + K_2)(1 + K_3) \tag{7.1}$$

where K_1: the level of experience, K_2: the level of stress, and K_3: the level of HMI design.

The adjusted median response time, $T_{1/2}$, is represented as:

$$T_{1/2} = T_{1/2,\text{nominal}} * K \tag{7.2}$$

Non-response probability, $P_{NR}(t)$, is obtained by using the Weibull distribution:

$$P_{NR}(t) = \exp -\{[(t/T_{1/2}) - C_i]/A_i\}^{Bi} \tag{7.3}$$

where t is the time available for completing a given task, A_i, B_i, and C_i represent the correlations obtained by the simulator experiments, and i indicates the skill, rule, and knowledge behavior.

7.1.3 SLIM

SLIM (success likelihood index methodology) is a structured expert-judgment-based method, also known as FLIM (failure likelihood index methodology) [10]. Basic steps for conducting HRA using SLIM are:

1. Select tasks that have the same task characteristics (*i.e.*, same set of PSFs) to form a single group
2. Assign the relative importance or weight (w_i) between PSFs.
3. Determine the rating or the current status (r_i) of PSFs for each of the tasks under evaluation.
4. Calculate the Success Likelihood Index (SLI) using the relative importance and the rating of PSFs for each of the tasks ($\mathrm{SLI} = \sum w_i r_i$).
5. Convert the SLI into the HEP by using the following equation, $\log(\mathrm{HEP}) = a * \mathrm{SLI} + b$, where a and b are calculated from the anchoring HEP values.

7.1.4 HEART

HEART (human error assessment and reduction technique) was suggested by Jeremy Williams [11]. HEART was used mostly in nuclear industries earlier in its development. The method has gradually extended to other industries, including aerospace, medical domains, and chemical industries, owing to its simplicity and ease of use.

HEART provides a relatively simple framework composed of generic task type (GTT) and a set of PSFs. Nominal error probabilities (NEPs) are given according to 9 GTTs, and 38 error-producing conditions (EPCs) or PSFs are used to increase the likelihood of error occurrence.

The general application steps are:

1. Selection of GTT (determination of the nominal error probability)
2. Selection of PSFs relevant to task situations
3. Assessment of the rating for selected PSFs
4. Calculation of final HEP

The final HEP is obtained by the equation:

$$\text{Final HEP} = \text{NEP} * \Pi\ [\mathrm{R(i)} * (\mathrm{W(i)} - 1) + 1] \tag{7.4}$$

where the NEP is given for a selected GTT, and W(i) and R(i) are the weight and rating of the *i*th PSF, respectively.

7.2 Second-generation HRA Methods

7.2.1 CREAM

CREAM (cognitive reliability and error analysis method) [12] has been developed on the basis of a socio-contextual model, the Contextual Control Model (COCOM) [17]. CREAM suggests a new framework for human error analysis by providing the same classification systems for both retrospective and prospective analyses (*i.e.*, genotypes and phenotypes). CREAM's major modules for identification and quantification of cognitive function failures, based on the assessment of common performance conditions, are introduced in this section.

7.2.1.1 Contextual Control Model (COCOM)

CREAM was developed on the basis of COCOM [17] which evolved from the SMoC (Simple Model of Cognition) model [18]. COCOM assumes that human cognition is the controlled use of competence, such as skill and knowledge, adapted to the requirements of the context. Human cognitive process in COCOM is expressed by four cognitive functions: observation, interpretation, planning, and execution. The functions are performed by causal links determined by the specific context at the time under consideration and not by predetermined sequential paths. The level of control is classified into four control modes from the lowest level to the highest level: scrambled control, opportunistic control, tactical control, and strategic control. The definitions of control modes are:

- Scrambled control mode – the choice of action is unpredictable or haphazard. Scrambled control characterizes a situation where little or no thinking is involved in choosing what to do.
- Opportunistic control mode – the action is determined by salient features of current context rather than more stable intentions or goals. The person does very little planning or anticipation, perhaps because the context is not clearly understood or because there is limited time available.
- Tactical control mode – situations where performance follows a known procedure or rule. The time horizon goes beyond the dominant needs of the present. Planning is of limited scope or range. Needs taken into account may sometimes be *ad hoc*.
- Strategic control mode – actions are chosen after full consideration of functional dependencies between task steps and interaction between multiple goals.

7.2.1.2 Classification Systems

Common performance conditions, cognitive activity types, and error modes corresponding to the cognitive stages are provided as the basic classification system. Nine contextual factors are defined as Common Performance Conditions (CPC). The definitions and brief explanations are listed in Table 7.1.

Table 7.1. Definitions or descriptions of the common performance conditions (CPCs) in CREAM

Name of CPCs	Definitions or descriptions
Adequacy of organization	The quality of the roles and responsibilities of team members, additional support, communication systems, safety management system, instructions and guidelines for externally oriented activities, role of external agencies, *etc*.
Descriptors	*Very efficient / Efficient/ Inefficient/ Deficient*
Working conditions	The nature of the physical working conditions such as ambient lighting, glare on screens, noise from alarms, interruptions from the task, *etc*.
Descriptors	*Advantageous / Compatible / Incompatible*
Adequacy of HMI and operational support	The human–machine interface in general, including the information available on control panels, computerized workstations, and operational support provided by specifically designed decision aids
Descriptors	*Supportive / Adequate / Tolerable / Inappropriate*
Availability of procedures/ plans	Procedures and plans include operating and emergency procedures, familiar patterns of response heuristics, routines, *etc*.
Descriptors	*Appropriate / Acceptable / Inappropriate*
Number of Simultaneous goals	The number of tasks a person is required to pursue or attend to at the same time (*i.e.*, evaluating the effects of actions, sampling new information, assessing multiple goals, *etc*.)
Descriptors	*Fewer than capacity / Matching current capacity / More than capacity*
Available time	The time available to carry out a task and corresponds to how well the task execution is synchronized to the process dynamics
Descriptors	*Adequate / Temporarily inadequate / Continuously inadequate*
Time of day (circadian rhythm)	The time of day (or night) describes the time at which the task is carried out, in particular whether or not the person is adjusted to the current time (circadian rhythm). Typical examples are the effects of shift work. The time of day has an effect on the quality of work, and performance is less efficient if the normal circadian rhythm is disrupted
Descriptors	*Day-time (adjusted) / Night-time (unadjusted)*
Adequacy of training and preparation	The level and quality of training provided to operators as familiarization to new technology, refreshing old skills, etc. It also refers to the level of operational experience
Descriptors	*Adequate, high experience / Adequate, limited experience / Inadequate*
Crew collaboration quality	The quality of the collaboration between crew members, including the overlap between the official and unofficial structure, the level of trust, and the general social climate among crew members
Descriptors	*Very efficient / Efficient / Inefficient / Deficient*

Fifteen cognitive activity types are defined. The categorization of the cognitive activity types are based on verbs for describing major tasks that are used in procedures, such as emergency operation procedures (EOPs) in nuclear power plants. The cognitive activities include “coordinate,” “communicate,” “compare,” diagnose," "evaluate," "execute," "identify," "maintain," "monitor," "observe," "plan," "record," "regulate," "scan," and "verify." Cognitive activities are associated with the cognitive functions (Table 7.2).

Cognitive error modes represent cognitive function failures for each of the cognitive functions. The classification of cognitive function failures and their nominal and upper- and lower- bound HEP values are shown in Table 7.3.

7.2.1.3 The Basic Method for Quantitative Error Prediction

The basic method is used for overall assessment of performance reliability of a task (*i.e.*, an estimation of the probability of performing an action incorrectly for the task as a whole). This stage provides a screening criterion for further detailed analysis which is done in the extended method. The basic method consists of three steps:

- Step 1: Construction of event sequence and task analysis
  Analysis of detailed information on accident scenarios and required tasks is conducted in this step. The hierarchical task analysis (HTA) [19, 20] or the goals–means task analysis (GMTA) [17] techniques are used as a task analysis method.

Table 7.2. The association matrix between the cognitive activities and the cognitive functions

Cognitive activity type	COCOM functions			
	Observation	Interpretation	Planning	Execution
Coordinate			◆	◆
Communicate				◆
Compare		◆		
Diagnose		◆	◆	
Evaluate		◆	◆	
Execute				◆
Identify		◆		
Maintain			◆	◆
Monitor	◆	◆		
Observe	◆			
Plan			◆	
Record		◆		◆
Regulate	◆			◆
Scan	◆			
Verify	◆	◆		

- Step 2: Assessment of CPCs
  Each CPC is evaluated, and the combined CPC score is calculated. Dependencies among CPCs are reflected in the calculation of the combined CPC score because CPCs have interdependent characteristics. The CPCs that have a dependency with other CPCs include working conditions, available time, number of goals, crew collaboration quality. The influence of other CPCs predefined in CREAM is considered in calculating the final combined CPC score, when the expected effect of these CPCs on the performance reliability is "not significant."
  Stages for evaluating CPCs are: (1) determine the expected level and evaluate the effect of CPCs on performance reliability (Table 7.3); (2) evaluate other influencing CPCs and perform an adjustment, if necessary, when the expected effects of the four CPCs that have a dependency with other CPCs are evaluated to be "not significant"; and (3) calculate the combined CPC score: $[\Sigma_{\text{reduced}}, \Sigma_{\text{not significant}}, \Sigma_{\text{improved}}]$.

- Step 3: Determination of probable control mode
  Determine probable control mode using the reduced or improved score of combined CPC score (Figure 7.1). Determine the range of action failure probability according to the control mode (Table 7.4).

7.2.1.4 The Extended Method for Quantitative Error Prediction

More detailed analysis of tasks screened by the basic method is performed in the extended method. Detailed analysis consists of three steps:

- Step 1: Development of a cognitive demands profile of the task
  An appropriate cognitive activity is determined from the list of cognitive activities for each of the task procedures or steps analyzed in the basic method. Cognitive activities are used to identify cognitive functions related to the task as well as to compose the cognitive profile. Major cognitive functions for performing task steps are determined by the relationship between a given cognitive activity and cognitive function(s) (Table 7.2).

- Step 2: Identify the likely cognitive function failure
  Identify the likely cognitive function failures that occur while performing corresponding task steps based on the information from the task analysis and assessment of CPCs (Table 7.3). The analysts determine the most probable failure mode among the candidates. Selection of likely cognitive function failures is skipped if the likelihood for occurrence of all the candidate failure modes is negligible.

- Step 3: Determination of specific action failure probability
  The nominal cognitive failure probability (CFP) is assigned for determined cognitive function failure (Table 7.3). The probability is adjusted to reflect a given context in which the task step is performed by multiplying an appropriate weighting factor that is determined by the assessment of CPCs.

Table 7.3. Types of cognitive function failures and nominal failure probability values

Cognitive function	Generic failure type	Lower bound (5%)	Basic value	Upper bound (95%)
Observation	O1. Wrong object observed	3.0E-4	1.0E-3	3.0E-3
	O2. Wrong identification	2.0E-2	7.0E-2	1.7E-2
	O3. Observation not made	2.0E-2	7.0E-2	1.7E-2
Interpretation	I1. Faulty diagnosis	9.0E-2	2.0E-1	6.0E-1
	I2. Decision error	1.0E-3	1.0E-2	1.0E-1
	I3. Delayed interpretation	1.0E-3	1.0E-2	1.0E-1
Planning	P1. Priority error	1.0E-3	1.0E-2	1.0E-1
	P2. Inadequate plan	1.0E-3	1.0E-2	1.0E-1
Execution	E1. Action of wrong type	1.0E-3	3.0E-3	9.0E-3
	E2. Action at wrong time	1.0E-3	3.0E-3	9.0E-3
	E3. Action at wrong object	5.0E-5	5.0E-4	5.0E-3
	E4. Action out of sequence	1.0E-3	3.0E-3	9.0E-3
	E5. Missed action	2.5E-2	3.0E-2	4.0E-2

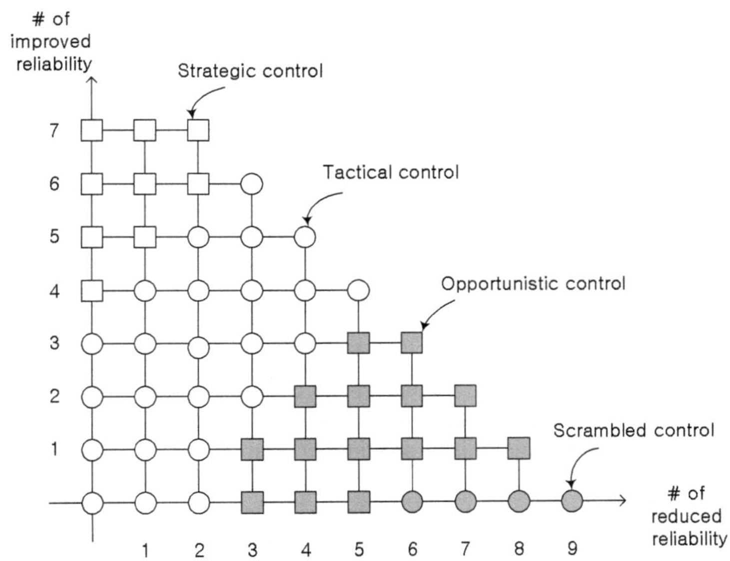

Figure 7.1. Relations between CPC score and control modes

Table 7.4. Control modes and probability intervals

Control mode	Reliability interval (probability of action failure)
Strategic	0.5E-5 < p < 1.0E-2
Tactical	1.0E-3 < p < 1.0E-1
Opportunistic	1.0E-2 < p < 0.5E-0
Scrambled	1.0E-1 < p < 1.0E-0

7.2.2 ATHEANA

ATHEANA (a technique for human event analysis) was developed under the auspices of US NRC, in order to overcome the limitations of first-generation HRA methods [13]. ATHEANA analyzes various human UAs including EOC and identifies the context or conditions that may lead to such UAs. EOCs are defined as inappropriate human interventions that may degrade plant safety condition. ATHEANA introduces "error-forcing context (EFC)" which denotes the context in which human erroneous actions are more likely to occur. EFC is composed of plant conditions and performance-shaping factors (PSFs). Determination of error-forcing context starts from the identification of deviations from the base-case scenario with which the operators are familiar, and then with other contributing factors, including instrumentation failures, support systems failures, and PSFs.

ATHEANA provides nine steps for identification and assessment of human failure events (*HFEs) for inclusion into the PSA framework:

- Step 1: Define the issue
- Step 2: Define the scope of analysis
- Step 3: Describe the base-case scenario
- Step 4: Define *HFE and UA
- Step 5: Identify potential vulnerabilities in the operators' knowledge base
- Step 6: Search for deviations from the base-case scenario
- Step 7: Identify and evaluate complicating factors and links to PSFs
- Step 8: Evaluate the potential for recovery
- Step 9: Quantify *HFE and UA

7.2.2.1 Step 1: Define the Issue

Analysts define the purpose of analysis by using the ATHEANA framework in the first step. ATHEANA is applied to various applications, such as for developing a new PRA model, upgrading a conventional PRA, or analyzing a specific issue/accident/scenario.

7.2.2.2 Step 2: Define the Scope of Analysis

The second step defines the scope of analysis on the basis of the purpose of analysis defined in the first step. Priorities of initiating events, accident sequences, and functions/systems/components are determined in the analysis.

7.2.2.3 Step 3: Describe the Base-case Scenario

The base-case scenario analysis implies description of the consensus operator mental model of plant responses and required human actions under a specific initiating event. The description of the base-case scenario is composed of the development of a consensus operator model (COM) and a reference analysis for a scenario, which includes neutronics and thermo-hydraulic analysis. This step is used as a reference for the deviation scenario analysis which is covered in Step 4.

*7.2.2.4 Step 4: Define *HFE(s) and/or UAs*

Candidate *HFEs and UAs are derived on the basis of the analysis on required function/functional failure mode/EOC or EOO/*HFE/UAs for the corresponding function/system/component. The *HFE/UA is defined not only in this step but also in the stage of deviation scenario analysis (Step 6), recovery analysis (Step 8), and quantification (Step 9) in a new or more detailed manner. ATHEANA provides detailed classifications of UAs and *HFEs (Tables 9.6–9.9 of ATHEANA [13]).

7.2.2.5 Step 5: Identify Potential Vulnerabilities in the Operators' Knowledge Base

Potential vulnerabilities in the operator knowledge base for a specific initiating event or scenario that results in the *HFEs or UAs is identified in this step. This identification supports the deviation analysis in Step 6 by:

- Investigation of potential vulnerabilities from operator expectation in a specific scenario. Differences of recency, frequency, and similarity along the scenarios is considered.
- Identification of a base-case scenario timeline and inherent difficulties associated with the required actions: (1) initial conditions or pre-trip scenario, (2) initiator and near simultaneous events, (3) early equipment initiation and operator response, (4) stabilization phase, (5) long-term equipment and operator response.
- Operator action tendency and informal rule: identification of operator action tendencies related to target HFEs/UAs and operating conditions which cause such tendency are performed. The identification of informal rules related to target HFEs/UAs is also performed.
- Analysis of expected formal rules and emergency operating procedures based on a given scenario: the points of decision making, movement to other procedure, important component control procedure, and reconfiguration of components are identified.

7.2.2.6 Step 6: Search for Deviations from the Base-case Scenario

UAs and *HFEs are identified in this step, based on the analysis of physical deviations from the base-case scenario.

- HAZOP guide words for identification of scenario deviation: No or Not/ More/Less/ Late/Never/Early/ Inadvertent/ Too quick or slow/ Too short or long/As well as/ Part of.
- Possible mismatches among timing and parameter values of physical deviations, and procedures or formal rules are investigated after the

identification of scenario deviations. Possible error types or inappropriate operator responses, in case of mismatch, are also investigated.

- Characteristics of deviation scenarios are considered in the identification of operator UAs and *HFEs by referring to: "Operator action tendencies (ATHEANA, Tables 9.12a, 9.12b)," "Scenario characteristics and description (ATHEANA, Table 9.15a)," "Scenario characteristics and associated error mechanisms, generic error types, and potential PSFs (ATHEANA, Table 9.15b)," "Questions to identify scenario relevant parameter characteristics (Table 9.16a)," and "Error mechanisms, generic error types, and potential PSFs as a function of parameter characteristics {Table 9.16b}."

7.2.2.7 Step 7: Identify and Evaluate Complicating Factors and Links to PSFs

Additional factors, such as physical conditions of (1) performance-shaping factors (PSFs), (2) hardware failures or indicator failures, are investigated, in addition to basic EFCs covered in Step 6.

7.2.2.8 Step 8: Evaluate the Potential for Recovery

Potential recovery possibilities are analyzed by identifying: (1) definition of possible recovery actions for *HFEs/UAs, (2) available time for recovery actions to prevent a severe consequence, such as reactor core damage, (3) availability and timing of cues to the operator for the requirement of the recovery actions, (4) availability and timing of additional resources to assist in recovery, (5) an assessment as to the strength of recovery cues with respect to initial EFCs and likelihood of successful recovery.

*7.2.2.9 Step 9: Quantify *HFE and UA*

The probability of an *HFE related to a specific UA for a specific scenario is defined as:

$$\mathrm{P(HFE} \mid S) = \sum_i \mathrm{P(EFC}_i \mid S) * \mathrm{P(UA} \mid \mathrm{EFC}_i, S) \tag{7.5}$$

where $\mathrm{P(EFC}_i \mid S)$ implies the probability of an EFC under a specific accident scenario, and $\mathrm{P(UA} \mid \mathrm{EFC}_i, S)$ implies the probability of a UA under a given EFC.

ATHEANA provides an overall analysis framework for the quantification of identified *HFEs. A summary description of the quantification framework provided by the current version of ATHEANA includes:

- Quantification of EFCs
  The EFC is defined as the combination of plant condition and PSFs which affect identified UAs. Quantification of EFCs implies the calculation of probability that a specific context occurs under a specific initiating event condition. Information required for quantification of plant condition depends on the EFC identified in the steps above. Information that is required may include:

- Frequencies of initiators
- Frequencies of certain plant conditions (*e.g.*, plant parameters, plant behavior) within a specific initiator type
- Frequencies of certain plant configurations
- Failure probabilities for equipment, instrumentation, indicators
- Dependent failure probabilities for multiple pieces of equipments, instrumentation, indicators
- Unavailabilities of equipments, instrumentation, indicators due to maintenance or testing

The following methods are used according to: (1) statistical analyses of operating experience, (2) engineering calculations, (3) quantitative judgments from experts, and (4) qualitative judgments from experts.

PSFs are grouped into two categories: (1) triggered PSFs that are activated by plant conditions for a specific deviation scenario, (2) non-triggered PSFs that are not specific to the context in the defined deviation scenario. Their quantification is performed on the basis of expert opinions from operator trainers and other knowledgeable plant staffs. Some parameters are calculated based on historical records.

- Quantification of UAs
  The current version of ATHEANA does not provide a clear technique or data for the quantification of UAs. Possible quantification methods that ATHEANA suggests are divided into: (1) the expert subjective estimation, (2) simulator experiment-based estimation, (3) estimation using other HRA methods, such as HEART and SLIM.

- Quantification of Recovery
  The probability of non-recovery for a UA is quantified in a subjective manner in consideration of: (1) the time available before severe core damage, (2) the availability of informative cues such as alarms and indications, and (3) the availability of support from other crew members or operating teams, such as the technical support center (TSC).

7.2.3 The MDTA-based Method

The MDTA (misdiagnosis tree analysis)-based method has been developed for assessing diagnosis failures and their effects on human actions and plant safety. The method starts from the assessment of potential for diagnosis failure for a given event by using a systematic MDTA framework [15].

The stages required for assessing *HFEs from diagnosis failures consist largely of:

- Stage 1: Assessment of the potential for diagnosis failures
- Stage 2: Identification of *HFEs that might be induced due to diagnosis failures
- Stage 3: Quantification of *HFEs and their modeling in a PRA model

7.2.3.1 Stage 1: Assessment of the Potential for Diagnosis Failures

The analysis of the potential for diagnosis failures (or misdiagnosis) is performed using the MDTA technique [15] (Figure 7.2). MDTA is constructed on the basis of two constituents (*i.e.*, diagnosis rules and misdiagnosis causes). The results of MDTA represent all the possible diagnosis results including misdiagnosis events.

Contributors to diagnosis failures are identified as: (1) plant dynamics (PD), (2) operator error (OE), and (3) instrumentation failure (IF), through the analyses of the NPP operating events that involved (the potential for) diagnosis failures, such as TMI-2 [21], Fort-Calhoun [22], and Palo Verde 2 [23]. Definitions of the three factors are: (1) plant dynamics (PD): mismatch between values of plant parameters and decision criteria of EOP diagnostic rules due to system dynamic characteristics, (2) operator error (OE): errors during information gathering or interpretation, and (3) instrumentation failure (IF): problems in the information systems.

Guidelines for a qualitative and quantitative consideration of misdiagnosis causes in the MDTA are provided for each cause. The qualitative and quantitative considerations of plant dynamics (PD) for decision rules of MDTA are made according to the following steps.

- Step 1: Classification of an event into sub-groups
  The contribution of the PD factor for an event at a decision rule is evaluated by estimating the fraction of an event spectrum where the behavior of the decision parameter does not match the established criteria

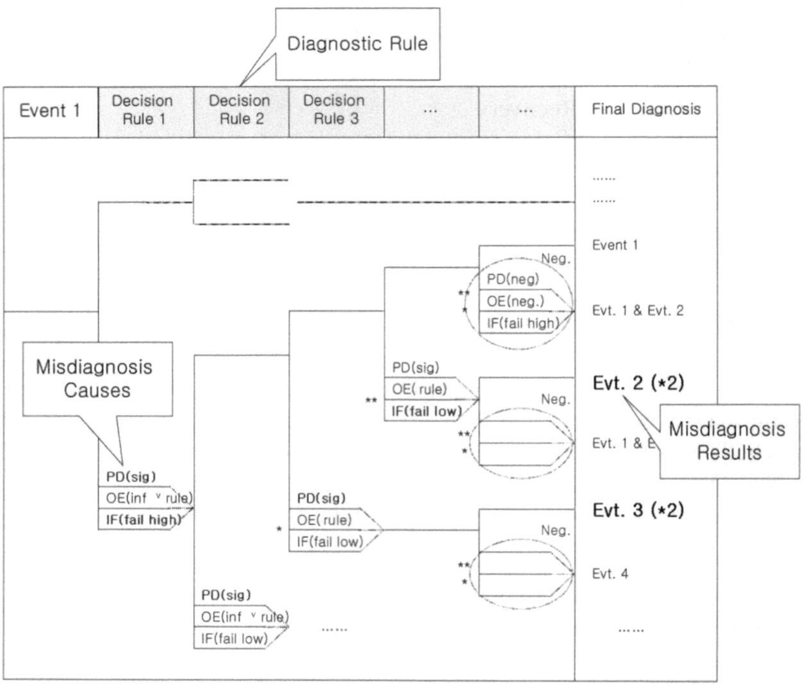

Figure 7.2. The basic structure of the MDTA

of the decision rule at the time of the operators' event diagnosis, to the overall spectrum of an event. The event under analysis is classified into sub-groups by considering plant dynamic behaviors from the viewpoint of operator event diagnosis, because plant behaviors are different according to break location or failure mode, even under the same event group. Each of the sub-groups becomes a set for thermal-hydraulic code analysis. Classification of an event is made according to event categorization and operative status of mitigative systems. An example of an event classification is found in Table 7.5.

Event categorization is done when the behavior of any decision parameter appears to be different according to break location, failure mode, or existence of linked systems. The status of mitigative systems means the combinatorial states of available trains of required mitigative systems, including those implemented by human operators. The frequency of each event group is later used for screening any event group of little importance in view of the likelihood.

- Step 2: Identification of suspicious decision rules
  Suspicious decision rules are defined as decision rules that have potential to be affected by the dynamics of an event progression in the way that the plant behavior mismatches the established decision criteria. Those suspicious decision rules are identified for each of the decision rules by each event group after categorizing the event groups.
  The representative group, in an event category, is defined as the most suspicious one with the highest likelihood by the judgment of analysts. The other event groups that show similar features in their dynamic progression to the representative one can be screened out for a further analysis by considering their importance by their relative likelihood. For example, all

Table 7.5. Composition of event groups for evaluating the contribution of plant dynamics to a diagnosis failure

Event category	Status of mitigative systems	Event group #	Frequency
E_Cat. 1 (f_1)	MSS. A (p_{1A})	1A	F_{1A} $(=f_1 * p_{1A})$
	MSS. B (p_{1B})	1B	F_{1B} $(=f_1 * p_{1B})$
	MSS. C (p_{1C})	1C	F_{1C} $(=f_1 * p_{1C})$
	...	...	...
E_Cat. 2 (f_2)	MSS. A (p_{2A})	2A	F_{2A} $(=f_2 * p_{2A})$
	MSS. B (p_{2B})	2B	F_{2B} $(=f_2 * p_{2A})$
	MSS. C (p_{2C})	2C	F_{2C} $(=f_2 * p_{2A})$
	...	...	...
...	...	...	...

the event groups belonging to the event category, E_Cat. 1, in Table 7.5 are assumed to show similar features for the identified decision parameters. Then, the groups such as 1B <E_Cat. 1 – MSS. B>, 1C <E_Cat. 1 – MSS. C> are screened out for a further analysis based on their relative likelihood when the 1A group, which is composed of <E_Cat. 1> and <MSS. A>, is defined as the representative one.

- Step 3: Qualitative assignment of the PD factor in the MDTA
  The contribution of the PD factor for taking a wrong path is acknowledged in the decision rule where the plant dynamics of the most suspicious group (Table 7.5) turns out to have a mismatch with established criteria. A more detailed thermal-hydraulic code analysis is performed for these event groups and decision parameters to assess the contribution of the PD factor quantitatively (*i.e.*, how much of an event spectrum contributes to the mismatch).

- Step 4: Quantitative assignment of the PD factor in the MDTA
  The purpose of this step is to establish the range of an event spectrum that mismatches with established criteria of a decision parameter. Further thermal-hydraulic analysis is performed to decide the range of the mismatch for the event group that showed the potential for a mismatch in Step 3. The fraction of an event spectrum in a mismatched condition at a decision rule is obtained by establishing the ranges of the mismatches for all potential event groups.

The contribution of operator errors (OE) for taking a wrong path at a decision point is assessed by assigning an appropriate probability to the selected items according to a cognitive function. The operator error probabilities for the selected items are provided in Table 7.6. These values were derived from expert judgment and cause-based decision tree methodology (CBDTM) [24].

The potential for a recovery via a checking of critical safety functions (CSFs) is considered, where applicable, for the decision rules with operator errors assigned, because the EOP system of the reference plant requires the shift technical advisor (STA) to conduct a check of the CSFs when the operators enter an EOP consequent upon a diagnosis. A non-recovery probability of 0.5 (assuming HIGH dependency with initial errors) is assigned to operator error probabilities for correspondent decision rules.

The contribution of instrumentation failures (IF) is assessed as follows. Factors affecting the unavailability of an instrumentation channel are classified into four categories: (1) an independent failure, (2) an unavailability due to a test and maintenance, (3) human miscalibration, and (4) a common-cause failure (CCF) [25]. The operators are assumed to be able to identify the failed state of an instrumentation when a single channel fails during a normal operation, since most of the instruments in large-scale digital control systems have 2 or 4 channels. The likelihood of functional failure during an accident progression is also considered to be negligible. The failure of multiple channels in a common mode during a normal operation is considered in this study. These common-mode failures are assumed not to be identified during both normal and abnormal operations.

Table 7.6. Operator error probabilities assigned to the selected items

Cognitive function	Detailed items	Basic HEP
Information gathering	Existence of other confusing information similar to the required information	BHEP = 1.0E-2
	Information on more than one object is required	BHEP = 1.0E-2
Rule interpretation	The logic of a decision rule	(Refer to CBDTM, $p_c g$)
	AND or OR	BHEP = 3.0E-4
	NOT	BHEP = 2.0E-3
	NOT & (AND or OR)	BHEP = 6.0E-3
	AND & OR	BHEP = 1.0E-2
	NOT & AND & OR	BHEP = 1.6E-2

The possibility of human miscalibration, where a dependency exists between channels, and the possibility of a CCF between instrumentation channels are considered as potential candidates. The possibility of human miscalibration is, however, also neglected because an initial miscalibration is expected to be identified during the process of functional testing or plant start-up operation after the calibration. Only the possibility of a CCF between the instrumentation channels for a diagnostic parameter is considered in the MDTA framework. Failure modes such as "Fail High" and "Fail Low" are considered in the MDTA framework since these modes are related to the actual phenomena of an instrumentation failure such as a "zero shift" or a "span error." These are expected to be relatively difficult for identification during normal operation, especially for a composite failure mechanism of both a "zero shift" and a "span error".

The probability of a CCF of instrumentation channels for a given diagnostic parameter is calculated by the β -factor model [26]:

$$Q_{CCF} = \beta * Q_T \tag{7.6}$$

where Q_{CCF} and Q_T denote the probability of a CCF and of a total failure, respectively, and β denotes the beta factor, which represents the portion of a CCF contributing to total failure probability.

The total failure probability, Q_T, is approximated to an independent failure probability, Q_I. The independent failure probability, Q_I, for the case where a fault is identified by a test, is calculated using Equation 7.7:

$$Q_I = \frac{1}{2}\lambda T \tag{7.7}$$

where λ denotes the failure rate of an instrumentation channel and T denotes the test interval.

*7.2.3.2 Stage 2: Identification of *HFEs*

A guideline for identifying *HFEs from diagnosis failures is provided in this section. This guideline is based on the following principles or assumptions:

- Operator tendency to maintain an initial diagnosis [27, 28]
- Operator tendency to focus on major (essential) systems or functions depending on the diagnosed event
- Operator actions through the guidance of a wrongly selected response procedure.

The identification of *HFEs starts from the definition of required functions according to an initiating event, on the basis of the above principles or assumptions. *HFEs result from UAs related to both required functions and unrequired or unnecessary functions. UAs in view of both functions are defined:

- UAs related to required functions
  - Failure to initiate required functions
  - Failure to maintain required functions (including an inappropriate termination of running systems or a failure to restart temporarily stopped systems)
- UAs related to unrequired or unnecessary functions
  - Manual initiation or operation of unrequired or unnecessary functions

*HFEs related to the above-mentioned two functions are identified using the following guidance, in the case that operators misdiagnose event A (the occurred event) as event B (the misdiagnosed event):

The required functions for the two events (*i.e.*, actual event and misdiagnosed event) are defined by referring to PRA event sequences and related EOPs. Two sources (*i.e.*, PRA event sequences and related EOPs) contain different information (*i.e.*, the PRA event sequences provide vitally requisite (essential) functions for placing the plant in a safe condition while the EOPs provide generally required functions to optimally control the plant in a safe and stable condition). The defined functions for the two events and two sources should be made as a table for comparison purposes. An example of required functions is given in Table 7.7 for a small loss of coolant accident (SLOCA) event and excessive steam demand event (ESDE) as a misdiagnosed event in an NPP.

UAs relevant to the two functions (*i.e.*, required function and unrequired function) are determined on the basis of a constructed table (Table 7.7).

UAs related to required functions are identified according to the following guidance:

- One or more essential functions may exist that become an essential function for an actual event, which are not for a misdiagnosed event, when the essential functions (requisite functions on the PRA event sequence) are compared between an actual event and a misdiagnosed event. The identified essential functions are assumed to have the potential for operators to commit UAs related to the required functions. UAs related to the failure to maintain the required functions are considered only when there are relevant stopping rules in the corresponding EOP.

UAs related to unrequired functions are identified according to the following guidance:

- The required functions that are not required by the actual event but are required by the misdiagnosed event are identified. Only the functions that may have an impact on plant safety are considered for a risk assessment.

The identified UAs by the above guidance are changed into appropriate HFEs to be modeled in a PRA model.

Table 7.7. An example of required functions for two events, SLOCA and ESDE

Required functions for SLOCA (the occurred event)		Required functions for ESDE (the misdiagnosed event)	
On the PSA event sequence	On the EOP (LOCA)	On the PSA event sequence	On the EOP (ESDE)
Reactor trip	Reactor trip	Reactor trip	Reactor trip
High-pressure safety injection (HPSI)	HPSI	(None)	HPSI
Low-pressure safety injection in case of HPSI failure	(None)	(None)	(None)
(None)	Isolation of the LOCA break location	(None)	Isolation of the faulted SG
RCS cooldown using the steam generators	RCS cooldown using the steam generators	RCS cooldown using the steam generators	RCS cooldown using the steam generators
RCS cooldown using the shutdown cooling system	RCS cooldown using the shutdown cooling system	RCS cooldown using the shutdown cooling system	RCS cooldown using the shutdown cooling system

*7.2.3.3 Stage 3: Quantification of the *HFEs and Their Modeling into a PRA*

A rough quantification method for identified *HFEs is dealt with in his section. The quantification scheme proposed in the MDTA framework is for a preliminary or rough assessment of the impact of diagnosis failures on plant risk. A theoretical or empirical basis may be deficient. The values provided in the proposed scheme appear to fall within a reasonable range of a human error probability.

The quantification of identified *HFEs is composed of "Estimation of the probability of a diagnosis failure," "Estimation of the probability of performing an UA under the diagnosis failure," and "Estimation of the probability of a non-recovery" (Equation 7.8). This is consistent with the ATHEANA quantification framework.

Probability of an *HFE = (Probability of a diagnosis failure) * (Probability of an UA under the diagnosis failure) * (Probability of a non-recovery) (7.8)

The selection of influencing factors and assigning appropriate values are based on expert judgments or by referring to existing HRA methods, such as CBDTM [24]. The availability of procedural rules for deciding to perform or not to perform actions related to identified UAs is selected as the key influencing factor affecting the likelihood of UAs. The probability of an UA under a diagnosis failure is assigned according to the availability of procedural rules as:

- When there is no procedural rule for the actions: 1.0
- When there are procedural rules for the actions:
  - When the plant conditions satisfy the procedural rules for committing UAs: 1.0
  - When the plant conditions do not satisfy the procedural rules for committing UAs (for UAs of an omission, this means that plant conditions satisfy the procedural rules for required actions): 0.1–0.05 (This probability represents the likelihood of operators committing UAs under a diagnosis failure even though plant conditions do not satisfy the procedural rules)

Table 7.8. The non-recovery probability assigned to two possible recovery paths (adapted from CBDTM [24])

Recovery Path (RP)	Available time	Probability of non-recovery
RP1: The procedural guidance on the recovery	$T_a > 30$ min	0.2
RP2: The independent checking of the status of the critical safety functions	30 min $< T_a <$ 1 h	0.2
	$T_a >$ 1 h	0.1

The following two paths as a potential way to recovery of committed UAs are considered:

- By procedural guidance for a recovery other than the procedural rules related to UAs
- By an independent checking of the status of CSFs by, for example, STA

The non-recovery probability for the two paths is assigned according to time available for operator recovery actions by adapting the values from the CDBTM (Table 7.8).

7.3 Concluding Remarks

Major HRA methods involving first-generation HRAs and second-generation HRAs, which have been mostly used and developed in the domain of the safety assessment of NPPs, are introduced in this chapter.

Major differences between first-generation HRAs and second-generation HRAs are summarized as: First-generation HRAs focus on developing quantification models for estimating the probability of an *HFE. Second-generation HRAs direct the focus on the identification of qualitative conditions or context under which UAs are likely to occur. Various kinds of human error modes have not been considered in first-generation HRAs (*i.e.*, a representative expression of error mode is "the failure to perform a required action within the time available"). Second-generation HRAs have the capability of dealing with various kinds of error modes or UAs, including EOCs. Another major change in second-generation HRAs is the major attention given operator cognitive errors that might occur during information gathering, situation assessment or diagnosis, or decision-making by using an appropriate model of human cognition or decision-making.

CREAM provides a new systematic framework and concrete classification systems for analyzing and predicting human erroneous actions through task analysis and assessment of overall task context on the basis of a human contextual control model. CREAM is viewed as a generic approach that is used irrespective of application field or industrial domain, rather than a specifically designed method for a specific industrial context. The method has some limitations for incorporating into the analysis the effect of dynamic features of an accident scenario of an NPP on human behavior.

ATHEANA provides a more specially designed framework for analyzing human performance problems under accident scenarios of NPPs. ATHEANA uses the term of "error-forcing context (EFC)" to represent a specific condition, which is composed of plant conditions and PSFs, in which the likelihood of occurrence for an expected UA is potentially significant. The ATHEANA method is very comprehensive in attempting to cover all time periods of event scenarios with consideration of various combinations of deviated scenarios and other PSFs and psychological error mechanisms. However, this comprehensiveness may induce a complexity or an inconsistency in using the method.

The MDTA-based method specifies contributing factors for diagnosis failure, such as plant dynamics (PD), operator errors (OE), and instrumentation failures

(IF). These factors are also considered as important contributors to a misdiagnosis in other methods, such as ATHEANA [13], Julius' method [29], and the confusion matrix [30]. However, the MDTA framework provides a more structured system with explicit guidelines for assessing the contribution of the three factors in evaluating the potential for a diagnosis failure. Analysts identify all combinations of misdiagnosis causes for all possible misdiagnosis results for a given event (or events). The MDTA framework, therefore, helps the analysts identify dominant contributors and decision paths leading to diagnosis failures. The MDTA-based method is utilized in evaluating the appropriateness of the diagnostic procedure in future large-scale digital control systems as well as in assessing the risk impact of diagnosis failures in existing ones.

All HRA methods, including both first- and second-generation HRAs, treat I&C systems or HMIs as important contributors to human reliability. However, the interactions between human operators and information systems are only partly modeled. The interdependency of human operators and the plant I&C systems, based on a reliability model integrating I&C systems and human operators, is dealt with in more detail in Part IV of this book.

References

[1] Bogner MS (1994) Human error in medicine. Lawrence Erlbaum Associates, Hillsdale, New Jersey.

[2] Reason J (1990) Human error. Cambridge University Press.

[3] Dougherty EM, Fragola JR (1998) Human reliability analysis: a systems engineering approach with nuclear power plant applications. John Wiley & Sons.

[4] Kirwan B (1994) A guide to practical human reliability assessment. Taylor & Francis.

[5] IAEA (1995) Human reliability analysis in probabilistic safety assessment for nuclear power plants. Safety series no.50, Vienna.

[6] Julius JA, Jorgenson EJ, Parry GW, Mosleh AM (1996) Procedure for the analysis of errors of commission during non-power modes of nuclear power plant operation. Reliability Engineering and System Safety 53: 139–154.

[7] Dougherty E (1992) Human reliability analysis – where shouldst thou turn? Reliability Engineering and System Safety 29: 283–299.

[8] Swain A, Guttmann HE (1983) Handbook of human reliability analysis with emphasis on nuclear power plant applications. NUREG/CR-1278, US NRC.

[9] Hannaman GW, Spurgin AJ, Lukic YD (1984) Human cognitive reliability model for PRA analysis. NUS- 4531, Electric Power Research Institute.

[10] Embrey DE, Humphreys P, Rosa EA, Kirwan B, Rea K (1984) SLIM-MAUD: an approach to assessing human error probabilities using structured expert judgment. NUREG/CR-3518, US NRC.

[11] Williams JC (1988) A data-based method for assessing and reducing human error to improve operational performance. Proceedings of the IEEE Fourth Conference on Human Factors and Power Plants, Monterey, California.

[12] Hollnagel E (1998) Cognitive reliability and error analysis method (CREAM). Elsevier, Amsterdam.

[13] Barriere M, Bley D, Cooper S, Forester J, Kolaczkowski A, Luckas W, Parry G, Ramey-Smith A, Thompson C, Whitehead D, Wreathall J (2000) Technical basis and implementation guideline for a technique for human event analysis (ATHEANA). NUREG-1624, Rev. 1, US NRC.

[14] Kim J, Jung W, Park J (2005) A systematic approach to analysing errors of commission from diagnosis failure in accident progression. Reliability Engineering and System Safety 89: 137–150.
[15] Kim J, Jung W, Son Y (2007) The MDTA-based method for assessing diagnosis failures and their risk impacts in nuclear power plants. Reliability Engineering and System Safety 93: 337–349.
[16] Rasmussen J (1983) Skills, rules, and knowledge; signals, signs, and symbols, and other distinctions in human performance models. IEEE Transactions on Systems, Man, and Cybernetics 13: 257–266.
[17] Hollnagel E (1993) Human reliability analysis: context and control. London, Academic Press.
[18] Hollnagel E, Cacciabue PC (1991) Cognitive modelling in system simulation. Proceedings of the Third European Conference on Cognitive Science Approaches to Process Control. Cardiff.
[19] Annett J, Duncan KD (1967) Task analysis and training design. Occupational Psychology 41: 211–221.
[20] Stanton NA (2006) Hierarchical task analysis: developments, applications and extensions, Applied Ergonomics 37: 55–79.
[21] Kemeny J (1979) The need for change: report of the President's commission on the accident at TMI. New York: Pergamon Press.
[22] Meyer OR, Hill SG, Steinke WF (1993) Studies of human performance during operating events: 1990–1992. NUREG/CR-5953, US NRC.
[23] MacDonald PE, Shah VN, Ward LW, Ellison PG (1996) Steam generator tube failures. NUREG/CR-6365, US NRC.
[24] Grobbelaar J, Julius J (2003) Guidelines for performing human reliability analyses. Draft Report.
[25] Min K, Chang SC (2002) Reliability study: KSNPP engineered safety feature actuation system. KAERI/TR-2165, KAERI.
[26] USNRC (1998) Guidelines on modeling common-cause failures in probabilistic risk assessment. NUREG/CR-5485, US NRC.
[27] Wickens C, Hollands J (2000) Engineering psychology and human performance. Prentice-Hall Inc.
[28] Mosneron-Dupin F, Reer B, Heslinga G, Straeter O, Gerdes V, Saliou G, Ullwer W (1997) Human-centered modeling in human reliability analysis: some trends based on case studies. Reliability Engineering and System Safety 58: 249–274.
[29] Julius J, Jorgenson E, Parry GW, Mosleh AM (1995) A procedure for the analysis of errors of commission in a probabilistic safety assessment of a nuclear power plant at full power. Reliability Engineering and System Safety 50: 189-201.
[30] Wakefield DJ (1988) Application of the human cognitive reliability model and confusion matrix approach in a probabilistic risk assessment. Reliability Engineering and System Safety 22: 295–312.

8

Human Factors Engineering in Large-scale Digital Control Systems

Jong Hyun Kim[1] and Poong Hyun Seong[2]

[1]MMIS Team, Nuclear Engineering and Technology Institute
Korea Hydro and Nuclear Power (KHNP) Co., Ltd.
25-1, Jang-dong, Yuseong-gu, Daejeon, 305-343, Korea
jh2@khnp.co.kr

[2]Department of Nuclear and Quantum Engineering
Korea Advanced Institute of Science and Technology
373-1 Guseong-dong, Yuseong-gu, Daejeon, 305-701, Korea
phseong@kaist.ac.kr

An approach to improve human reliability is introduced in this chapter. Major methods to analyze human reliability have been reviewed in Chapter 7. Chapter 8 presents human factors-related activities to design a human-machine interface (HMI), especially for nuclear power plant (NPP) applications. Human factors engineering (HFE) is strictly applied in the nuclear industry. Designing a good HMI enhances human reliability and prevents human errors, as well as helping with training and proceduralization. An HFE process to design an HMI for a safety critical system that requires high reliability for operators consists of three steps: analysis, design, and verification & validation (V&V).

The analysis step identifies what will be designed, who will use it, and how/when/where it will be used. A coupling of a system, tasks, and operators is shown in Figure 8.1. The analysis considers the system and its function as a work domain. Tasks performed by operators need to be identified. Finally, the cognitive characteristics of operators are taken into account as a user model of HMI.

The design step designs an HMI based on information from the analysis. Visual display unit (VDU)-based HMIs are the focus of this chapter, because main control rooms (MCRs) are being digitalized and computerized in newly constructed or modernized NPPs as technologies progress. Major trends of HMI design in a computerized MCR and human factors-related issues are presented and compared with conventional ones.

The V&V step ensures that the design conforms to HFE principles (verification) and supports operator performance in the real operation of NPPs through experiments (validation).

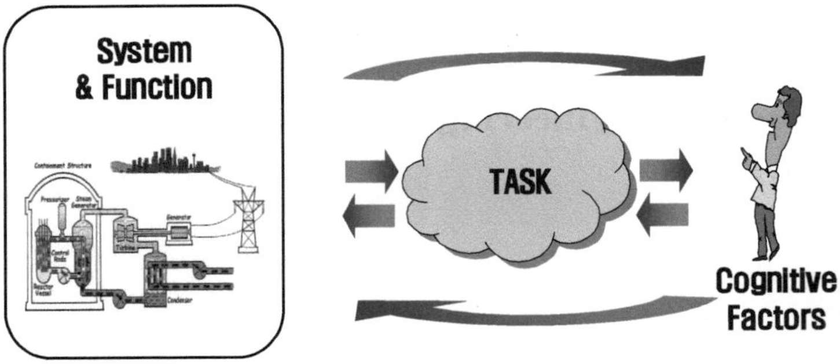

Figure 8.1. A coupling of a system, tasks, and operators

8.1 Analyses for HMI Design

8.1.1 Function Analysis

A function is a goal of system that operators should achieve by performing their tasks. The function analysis identifies the functions that must be performed to satisfy plant safety objectives in NPPs, prevent postulated accidents, or mitigate the consequences [1]. A function is usually classified into several sub-functions. The functional decomposition should start at "top-level" functions where a general picture of major functions is described, and continue to lower levels until a specific critical end-item requirement emerges (*e.g.*, a piece of equipment, software, or human action).

The central safety problem in the design of an NPP is to assure that radioactive fission products remain safely confined at all times during the operation of the NPP, refueling of the reactor, and the preparation and shipping of spent fuels [2]. Safety functions of NPPs are based on the concept of multiple barriers so that the escape of radioactive fission products to the public is prevented. The barriers consist of (1) fuel and cladding, (2) reactor coolant system (RCS), including reactor vessel, and (3) containment (Table 8.1). The fuel contains the fissile and fissionable materials within solid fuel elements. A layer of cladding surrounds the fuel to prevent the escape of the fission product gases and to confine fission products emitted near the surface of the fuel. The lower level of safety functions controls the reactivity of the reactor core by shutting the reactor down to reduce heat production in the reactor core for the integrity of the fuel and cladding. RCS also cools down the reactor core continuously.

The second barrier is the reactor coolant, which is typically water that comes in contact with the fuel and moves in one or more closed loops. The RCS removes heat from the reactor core and transfers it to boilers to generate steam. Fission products escaped from the fuel, neutrons, activated atoms in the coolant and picked up by the coolant are confined within the RCS. Pressure and inventory of RCS are controlled within a safe range to maintain the integrity of the RCS.

The third barrier, containment, which is made of thick reinforced concrete with a steel liner, contains radioactivity that is released either from the RCS or from the reactor vessel. All pipes and connections to the outside of containment are closed in situations where radioactivity may be released to the public. The pressure and temperature of containment are controlled within design limits to maintain the integrity of containment. The concentration of combustible gases (e.g., H_2 gas) should be controlled to prevent explosions.

Safety control functions are assigned to (1) personnel, (2) automatic control, or (3) combinations of personnel and automatic control; this is called function allocation. Function allocation has traditionally been based on a few simple principles. These include the left-over principle, the compensatory principle, or complementarity principle [3]. Function allocation in accordance with the left-over principle means that people are left with the functions that have not been automated or that could not be automated due to technical or economical reasons. The compensatory principle uses a list or table of the strong and weak features of humans and machines as a basis for assigning functions and responsibilities to various system components. A famous list is Fitts' list (Table 8.2 in Section 8.2.2.2). The complementarity principles allocate functions to maintain operator control of the situation and to support the retaining of operator skills.

The operator roles in executing safety functions are assigned as a supervisory role, manual controller, and backup of automation. The supervisory role monitors the plant to verify that the safety functions are accomplished. The manual controller carries out manual tasks that the operator is expected to perform. The backup of automation carries out a backup role to automation or machine control.

Table 8.1. Multiple barriers for the NPP safety

Barriers	Safety functions
1 : fuel and cladding	reactivity control
	RCS and core heat removal
2 : reactor coolant system	RCS inventory control
	RCS pressure control
3 : containment	containment isolation
	containment temperature and pressure control
	containment combustible gas control

8.1.2 Task Analysis

The task analysis defines what an operator is required to do [4]. A task is a group of related activities to meet the function assigned to operators as a result of the function allocation activity. The task analysis is the most influential activity in the HMI design process. The results of task analysis are used as inputs in almost all HFE activities.

The task analysis defines the requirements of information needed to understand the current system status for monitoring and the characteristics of control tasks needed for operators to meet safety functions. Information requirements related to monitoring are alarms, alerts, parameters, and feedback needed for action. Characteristics of control tasks include (1) types of action to be taken, (2) task frequency, tolerance and accuracy, and (3) time variable and temporal constraints. Task analysis provides those requirements and characteristics for the design step. The design step decides what is needed to do the task and how it is provided. The HMI is designed to meet the information requirement and reflect the characteristics of control tasks.

Task analysis considers operator cognitive processes, which is called cognitive task analysis. Cognitive task analysis addresses knowledge, thought processes, and goal structures that underlie observable task performances [5]. This analysis is more applicable to supervisory tasks in modern computerized systems, where cognitive aspects are emphasized more than physical ones. The control task analysis and the information flow model are the examples of cognitive task analyses.

The results of task analysis are used as an input for various HFE activities as well as for HMI design. The task analysis addresses personnel response time, workload, and task skills, which are used to determine the number of operators and their qualifications. The appropriate number of operators is determined to avoid operator overload or underload (*e.g.*, boredom). Skill and knowledge that are needed for a certain task are used to recruit operational personnel and develop a training program to provide necessary skill and system knowledge.

The task analysis is also used to identify relevant human task elements and the potential for human error in HRA. The quality of HRA depends to a large extent on analyst understanding of personnel tasks, the information related to those tasks, and factors that influence human performance of those tasks. Detail of HRA methods are found in Chapter 7.

8.1.2.1 Task Analysis Techniques

Each task analysis technique has particular strengths and weaknesses. A number of tools and techniques are available for task analysis [6, 7]. An appropriate technique for a specific purpose or mix of two or more techniques need(s) to be selectively applied for a successful analysis. This chapter introduces three useful techniques relevant to the HMI design in NPP applications.

(A) Hierarchical Task Analysis

Hierarchical task analysis (HTA) is widely used in a variety of contexts, including interface design and error analysis in both individual and team tasks, and in a variety of areas, including NPPs and command/control systems [8, 9]. The process of the HTA is to decompose tasks into sub-tasks to any desired level of detail. Each task, that is, operation, consists of a goal, input conditions, actions, and feedback. Input conditions are circumstances in which the goal is activated. An action is a kind of instruction to do something under specified conditions. The relationship between a set of sub-tasks and the superordinate task is defined as a plan. HTA identifies actual or possible sources of performance failure and to propose suitable remedies, which may include modifying the task design and/or providing appropriate training. The HTA is a systematic search strategy that is adaptable for use in a variety of different contexts and purposes within the HFE [9].

The HTA is a useful tool for NPP application because task descriptions are derived directly from operating procedures. Most tasks are performed through well-established written procedures in NPPs. The procedures contain goals, operations, and information requirements to perform the tasks. A part of the HTA derived from the procedure to mitigate a steam generator tube rupture (STGR) accident is shown in Figure 8.2.

(B) Control Task Analysis: Decision Ladder

A framework of the task analysis to represent various states of knowledge and information processes was introduced by Rasmussen [10, 11]. This "decision ladder" is expressed in terms independent of the specific system and its immediate control requirements.

The basic structure of the ladder is illustrated in Figure 8.3. The boxes correspond to information-processing activities, whereas the circles correspond to states of knowledge. This sequence, which has been developed from the analysis of decision making in a power plant control room, includes the following phases. The operator detects the need for intervention and starts to pay attention to the situation. The present state of the system through analyzing the information available is

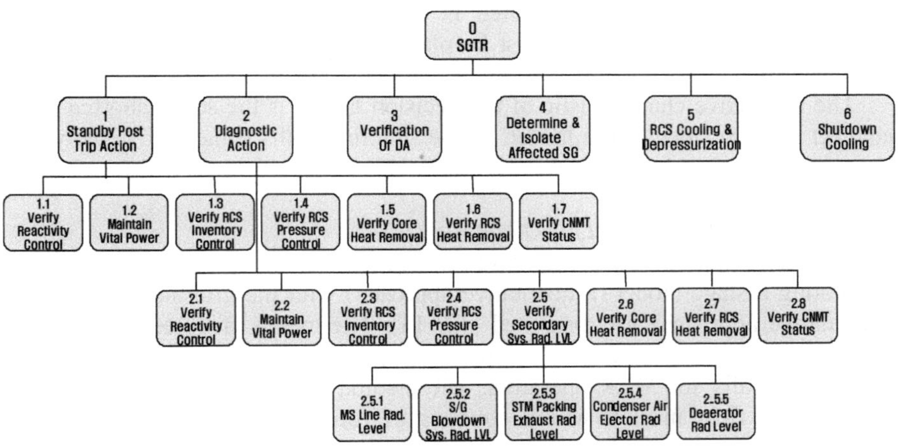

Figure 8.2. A part of HTA for SGTR accident

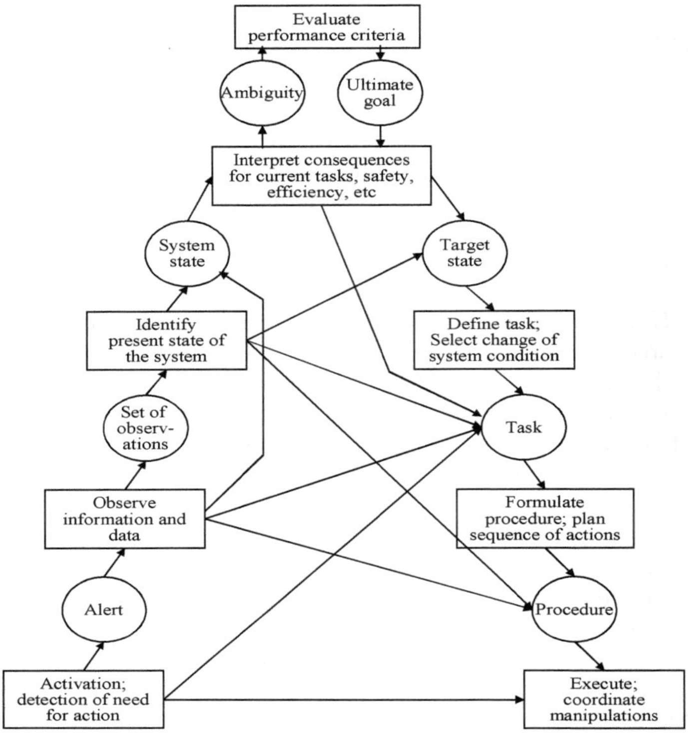

Figure 8.3. Typical form of decision ladder [11]

identified. The operator predicts the consequences in terms of goals of the system (or operation) and constraints based on the identified state. The operator evaluates the options and chooses the most relevant goal if there are two or more options available. The task to be performed is selected to attain the goal. A proper procedure (*i.e.*, how to do it), must be planned and executed when the task has been identified.

The distinctive characteristic of the decision ladder is the set of shortcuts that connect the two sides of the ladder. These shunting paths consist of stereotypical processes frequently adopted by experts [10, 11].

(C) Information Flow Model for Diagnosis Tasks

A method to quantify the cognitive information flow of diagnosis tasks by integrating a stage model (a qualitative approach) with the information theory (a quantitative approach) was proposed [12, 13]. The method focuses on diagnosis tasks which are one of the most complex and mental resource-demanding tasks in NPPs, especially for MCR operators. The method includes: (1) constructing the information flow model, which consists of four stages based on operating procedures of NPPs; and (2) quantifying the information flow using Conant's model, a kind of information theory. An information flow model for NPP diagnosis tasks is illustrated in Figure 8.4. The circles and the numbers in the circles represent the state of information and the sequence of processing, respectively.

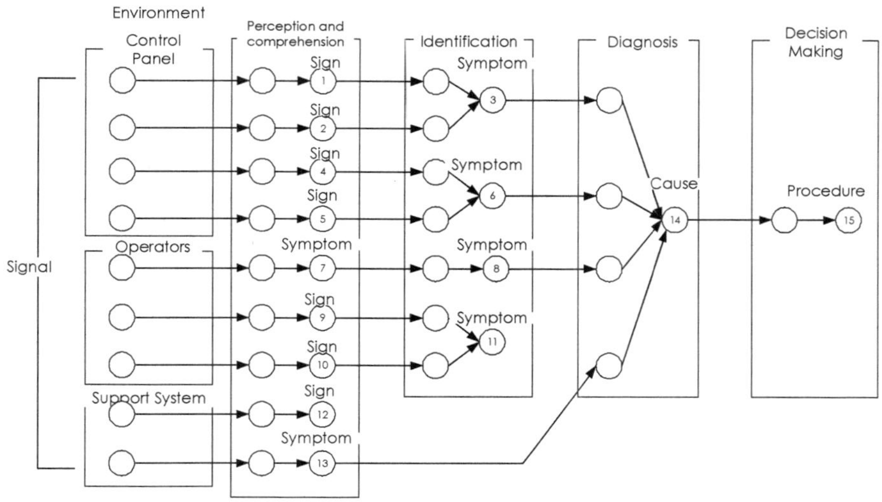

Figure 8.4. A typical form of information flow model [12]

The model represents operator dynamic behaviors between stages, such as moving back to the previous stage or skipping a stage by numbering the information. The information flow model consists of four stages: perception and comprehension, identification, diagnosis, and decision making. The stages perform the function of information transformation by mapping. Five types of state of information are defined according to knowledge and abstraction that it contains: signal, sign, symptom, cause, and procedure. This model assumes that information processing in the stages is carried out through mapping (e.g., many-to-one and one-to-one), transferring to the next stages, or blocking. Readers are referred to references for detail of the information flow model. The relationship between the method and operator performances, that is, time-to-completion and workload, has also been shown by laboratory and field studies [13, 14].

8.1.3 Cognitive Factors

A general understanding of operator cognitive factors is needed to understand the background against which the design of the HMI takes place. Important cognitive factors related to NPP operators are information-processing model, mental model, problem-solving strategy, mental workload, and situation awareness. Cognitive factors help HMI designers understand operator error mechanisms and an inherent human limitations which are crucial factors for ensuring safe operation of NPPs. The designer model of the system should also be consistent with that of the user to minimize error affordances [15]. Important cognitive factors are illustrated in this section, that deserve consideration in the HMI design for NPPs.

8.1.3.1 Cognitive Models

(A) Information-processing Model

A model of information processing helps designers understand different psychological processes used in interacting with systems and operator limitations with cognitive resources. A number of models for human information processing have been proposed. The model frequently used in human engineering was proposed by Wickens [16]. The information-processing model is illustrated in Figure 8.5. The model is conveniently represented by different states at which information gets transformed: (1) sensory processing, (2) perception of information, (3) situation awareness, (4) response planning, and (5) action execution. Sensory processing is an activity in which information and events in the environment must gain access to the brain. Raw sensory data are interpreted or given their meaning, through the stage of perception. The interpretation may be driven both by sensory input (which is called the bottom-up processing) or by inputs from long-term memory about what events are expected (which is called the top-down processing). The situation awareness, achieved through perception and cognitive operations, triggers an action–the selection of a response. The response selected is executed through actions, (*e.g.*, "start a pump"). The feedback loop at the bottom of the model indicates that actions are directly sensed by the operator.

Working memory and attention are limited resources of human operators in the information-processing model. Working memory has limited capacity and without sustained attentional resources, information decays rapidly. Information can be lost due to: (1) loss of attentional resources to keep it active, (2) overload of working memory limited capacity, or (3) interference from other information in working memory. The upper limit or capacity of working memory is known to be around 7 ± 2 chunks of information for optimistic situations [17]. A chunk is the unit of working memory space, defined jointly by the physical and cognitive properties that bind items together. Tasks that may overload operator working memory need to be avoided. HMI design considers some aids to reduce the burden, when exceeding the working memory capacity. The aid may be processed information to reduce computational or inferential load. Information is continuously displayed at any place in the HMI to remove the necessity for memorization.

The limitation of human attention represents one of the most formidable bottlenecks in human information processing. Attention is currently viewed as a finite limited resource that is assigned across the elements of cognition associated with perceptual mechanisms, working memory, and actions execution that compete for this limited resource (Figure 8.5) [16]. A multiple-resource model with attentional processing resources was proposed [18], by dividing into two dimensions: (1) processing stage (*i.e.*, perceptual and central processes require different resources than response processes); and (2) input modality (*i.e.*, spatial and verbal mental representation require different resources from linguistic information). Attentional limitation is difficult to reflect in the HMI design, because attention allocation is largely dependent upon situation and operator. The situation that requires an operator to perform more than one task simultaneously, that is, multi-tasking, should be avoided to reduce attentional problems. Attention needs to be divided into two dimensions of modalilty and not be concentrated in one, when multi-tasking is required.

(B) Mental Model

Mental models are operator internal models or their understandings on actual, physical systems. Operators formulate mental models of systems through interaction with a complex system, training, or education [19]. Operators learn system dynamics, its physical appearance and layout, and causal relations among its components [20]. Models are functional and always less complex than the original system.

Mental models may also be similar across large groups of people. Those mental models are defined as a population stereotype [16]. For example, consider the relationship between the display lighting of a room and the movement of a light switch. Flipping the switch up turns the light on in North America, while the opposite (up is off) occurs in Europe. HMI is expected to be compatible with both mental models and real systems. The compatibility needs to be considered from two viewpoints, that is, information display and control. The compatibility of information display is achieved in both static aspects (*i.e.*, the contents and the organization of the information), and dynamic aspects (*i.e.*, movements of information) [19]. The compatibility of control considers primary and secondary controls performed by operators. A primary control is the main control activity, like "actuates a component or a system." In a computerized HMI, to perform a primary task, operators also have to perform interface management tasks with a mouse or a keyboard, such as navigation, which are called secondary tasks. The HMI for primary controls needs to be compatible with the operator mental model of system and population stereotypes. The interface design for secondary controls needs to be compatible with operator expectations that have been obtained through the use of ordinary personal computers.

8.1.3.2 Strategy

Strategies in human decision-making are defined as a sequence of mental and effector (action on the environment) operations used to transform an initial state of knowledge into a final goal state of knowledge [21]. Strategies are defined as the

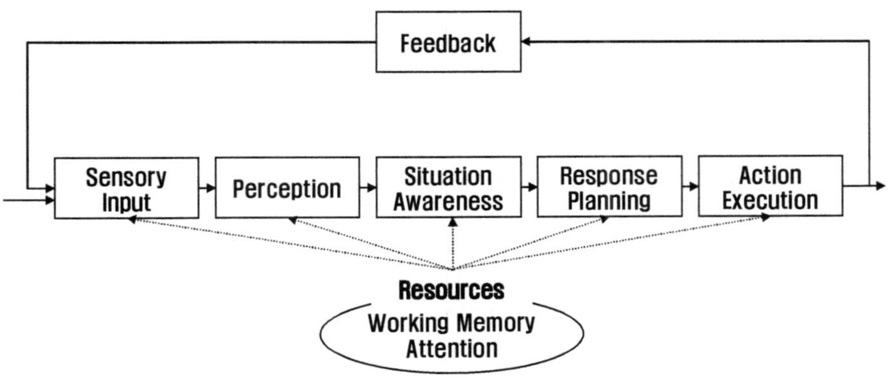

Figure 8.5. A general information-processing model

generative mechanisms by which a particular task can be achieved, if tasks or control tasks are the goals that need to be achieved [4, 10]. Strategies are adaptively selectable to cope with the limitation of human cognitive resources under complex task environments. For example, technicians switch strategies to avoid exceeding their resource constraints [22, 23], spontaneously switching to another strategy to meet the task demands in a more economic fashion, when one strategy becomes too effortful. Three major classes of factors influence which strategy is used to solve a particular decision problem [21]: characteristics of the person, characteristics of the decision problem, and characteristics of the social context. The characteristics of the person include cognitive ability and prior knowledge. Prior knowledge, obtained either through experience or training, will determine which strategies are available to a decision maker in his or her memory. Experience in a decision domain may also impact the frequency and recency with which available strategies have been used. Characteristics of the problem, such as how information is displayed, can affect how much cognitive effort is needed to implement various strategies. Characteristics of the social context influence the relative importance of such factors as the justifiability of a decision in determining strategy selection.

Computerized operator support systems, which is one of trends in computerized MCRs (Section 8.2.3), potentially have negative consequences without sufficient consideration of the strategies and, furthermore, may become a new burden on the operators. Application of an advanced alarm system, as a shift from tile annunciator alarm systems to computer-based alarm systems eventually collapsed, necessitated a return to older technology, because strategies to meet the cognitive demands of fault management that were implicitly supported by the old representation were undermined in the new representation [24]. Computerized operator support systems need to be consistent in content and format with the cognitive strategies and mental models employed by the operator [25].

8.1.3.3 Situation Awareness

A general definition of situation awareness (SA) describes SA as "the perception of the elements in the environment within a volume of time and space (Level 1 SA), the comprehension of their meaning (Level 2 SA) and the projection of their status in the near future (Level 3 SA)" [26]. Perception of cues, Level 1 SA, is fundamental. The correct awareness of a situation is hardly constructed without a basic perception of important information. The 76% of SA errors in pilots are traced to problems in the perception of needed information [27].

SA encompasses how people combine, interpret, store, and retain information, Level 2 SA. Operators integrate multiple pieces of information and determine their relevance to a person's goal at this level. Twenty percent of SA errors were found to involve problems with Level 2 SA [27].

The highest level of SA is to forecast future situation events and dynamics, Level 3 SA. This ability to project from current events and dynamics to anticipate future events (and their implications) allows for timely decision making.

An HMI design may largely influence SA by determining how much information can be acquired, how accurately it can be acquired, and to what degree it is compatible with operator SA needs [28]. Several advantages of the SA concept in the design process are:

- A means of designing for dynamic, goal-oriented behavior, with its constant shifting of goals
- A means of moving from a focus on providing operators with data to providing operators with information
- A means of incorporating into the design a consideration of the interplay of elements, wherein more attention to some elements may come at the expense of others
- A means for assessing the efficacy of a particular design concept that an examination of underlying constructs (attention, working memory) does not provide

8.1.3.4 Mental Workload

Operator mental workload is also an important consideration when designing an HMI. Excessive high levels of mental workload can lead to errors and system failure, whereas underload can lead to boredom and eventual error. The term workload refers to that portion of operator limited capacity actually required to perform a particular task [29]. The theoretical assumption underlying this definition is that human error results from limited information-processing capacity or processing resource. Greater task difficulty increases the requirement for mental -processing resources. Performance decrements result, if the processing demands of a task or tasks exceed available capacity.

The measure of mental workload is used for a variety of purposes in the HFE. The measurements play an important role in (1) allocating functions and tasks between humans and machines based on predicted mental workload; (2) comparing alternative equipment and task designs in terms of the workloads imposed; (3) monitoring operators of complex equipment to adapt to the difficulty of the allocation of functions in response to increases and decreases in mental workload; and (4) choosing operators who have higher mental workload capacity for demanding tasks [30]. A number of approaches to measuring mental workload have been proposed. The approaches are summarized [32, 33].

8.2 HMI Design

The HMI design determines what should be displayed and how it should be displayed ("when and where it should be displayed" may be included in "how"). The HMI design step selects the contents of display ("what"), based on the results of the task analysis that provide the information requirements for monitoring and control. The way the information is displayed may result from investigation of available display formats as well as function, task, and cognitive analyses. The design addresses HFE principles, guidelines, and standards that are relevant to design elements, such as text, symbol, and color. Useful guidelines for the NPP applications are NUREG-0700 [25] and 5908 [33], MIL-STD-1472F [34], and EPRI-3701 [35].

This chapter focuses on computerized HMIs. Modern computer techniques are available and proven for the application to the design of MCRs of NPPs. The Three Mile Island unit 2 (TMI-2) accident demonstrated that various and voluminous information from conventional alarm tiles, indicators, and control devices imposed a great burden on operators during emergency control situations. Modern technologies have been applied to MCR design in newly constructed or modernized plants to make for simpler and easier operation.

There are three important trends in the evolution of advanced MCRs [36]. The first is a trend toward the development of computer-based information display systems. Computer-based information display provides the capability to process data of plants and use various representation methods, such as graphics and integrated displays. Plant data are also presented in an integrated form into a more abstract level of information. Another trend is toward increased automation. An enhanced ability to automate tasks traditionally performed by an operator becomes possible with increased application of the digital control technology. Computerized operator support systems are developed as the third trend, based on expert systems and other artificial intelligence-based technologies. These applications include aids such as alarm processing, diagnostics, accident management, plant monitoring, and procedure tracking. The three trends and related issues are reviewed in this section in more detail.

8.2.1 Computer-based Information Display

8.2.1.1 Design Considerations

Graphic capability and computational power of computer systems produce a variety of ways in which information is displayed. Information in conventional MCRs is displayed through analog/digital indicator or status light at a fixed location. Computer-based information systems provide richer diversities in selecting the display formats than conventional types of control rooms. Designers determine the display format, that is, how the information is presented or how the tasks are supported by the information display system due to the richness of representation. Approaches to the display format are introduced in Section 8.2.1.2.

Operators view only a small amount of information at a time because of the limited viewing area of VDUs, even if the storage capacity of modern computer systems makes it possible to store unlimited data about an NPP. Guideline documents for the HMI design specify that the total amount of information on each screen is minimized by presenting only what is necessary to the operator [37]. It is recommended that the display loading (the percent of active screen area) should not exceed 25 % of whole screen [38]. The display density generally should not exceed 60% of the available character spaces [39]. Empirical evidence consistently addresses that as long as the information necessary to perform the task is presented, that human performance tends to deteriorate with increasing display density.

Information display in computerized HMIs is organized into multiple pages due to this spatial constraint. Where information must necessarily be spread over several VDU pages, careful decisions have to be made about the division of such information into pages. A means of browsing and navigating between these pages is designed in a consistent manner so that the interface management does not add significantly to the task load of the operator, namely, a secondary workload.

The following aspects need to be considered [25], when designing an interface into multiple pages:

- The organization of a display network reflects an obvious logic based on task requirement and be readily understood by operators.
- The display system provides information to support the user in understanding the display network structure.
- A display is provided to show an overview of the structure of an information space, such as a display network or a large display page.
- Easily discernable features appear in successive views and provide a frame of reference for establishing relationship across views.
- Cues are provided to help the user retain in a sense of location within the information structure.

8.2.1.2 Approaches to Information Display

Representative approaches to the information display for NPPs are introduced. Those become available due to the rich representation capability of computerized display.

(A) Graph

Graphs, which are classical, are also best suited in advanced displays for providing approximate values, such as an indication of deviation from normal, a comparison of operating parameter to operating limits, a snapshot of present conditions, or an indication of the rate of change [40]. This taxonomy involves bar graph, X-Y plot, line graph, and trend plot.

Two interesting psychological factors are concerned in designing graphs: population stereotypes and emergent features. Population stereotypes (Section 8.1.3.1) define mappings that are more directly related to experience [18], or expectancy that certain groups of people have for certain modes of control expectation or display presentation. Any design that violates a strong population stereotype means that the operator must learn to inhibit his/her expectancies [41]. People tend to resort to population stereotypes under high stress levels, despite being trained to the contrary, and become error-prone.

An emergent feature is a property of the configuration of individual variables that emerges on the display to signal a significant, tasks-relevant, and integrated variable [16]. An example of bar graphs to indicate pressurizer variables is shown in Figure 8.6. The emergent feature in Figure 8.6(b) is the horizontally dashed line. The emergent feature provides a signal for the occurrence of an abnormal situation in the pressurizer at a glance when the normal state, that is, straight line is broken.

(B) Configural and Integral Display

A configural display is a display that arranges low level of data into a meaningful form which is an emergent feature. The polygonal display is an example of a configural display, as shown in Figure 8.7 [42]. The display is adopted in the safety

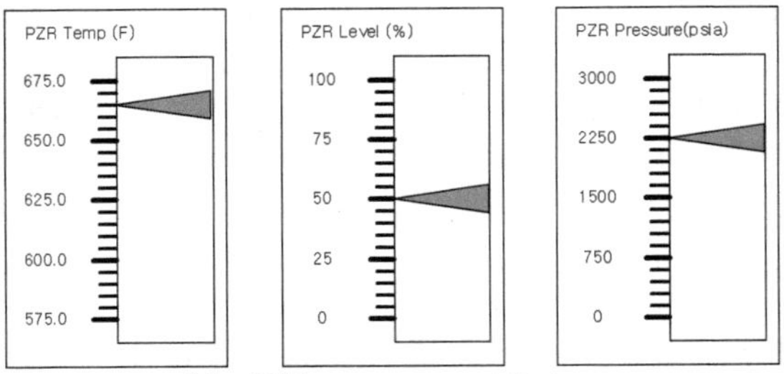

(a) Without emergent feature

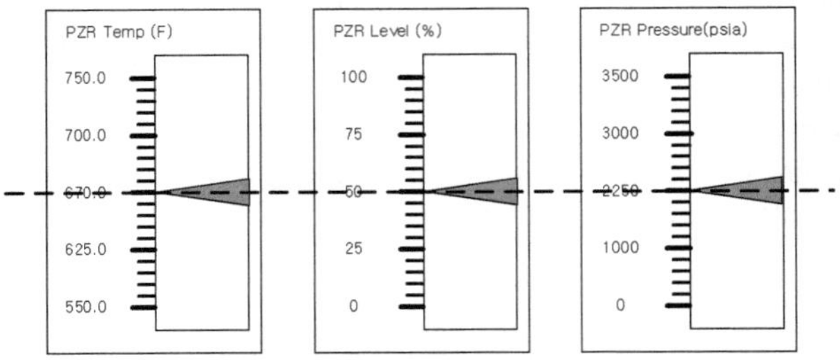

(b) With emergent feature

Figure 8.6. Bar graphs for pressurizer variables

parameter display system (SPDS) of NPPs. The operator can readily see whether the plant is in a safe or unsafe mode by glancing at the shape of the polygon.

An integral display is a display in which many process variables are mapped into a single display feature, such as, an icon. The integral display provides the information about the overall status of a system with a single feature, whereas an individual parameter is available in the configural display. An example of an integral display is shown in Figure 8.8. The symbol indicates characteristics of wind in the weather map. The symbol contains the information about the direction and speed of wind and cloudiness in an icon. Another example of integral displays is a single alarm that contains warnings of two or more parameters.

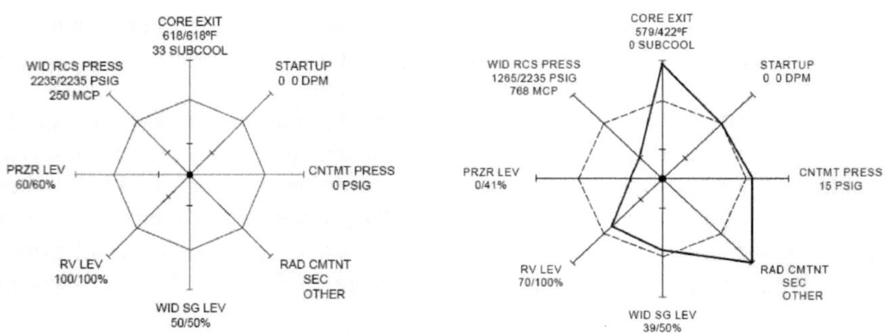

Figure 8.7. Polygonal display [42]

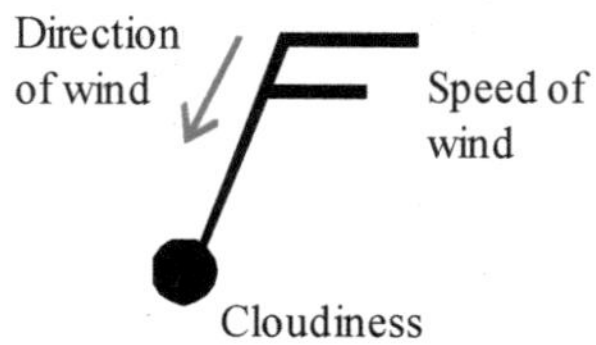

Figure 8.8. Integral display (a symbol for indicating wind)

(C) Ecological Interface Design

Ecological interface design (EID) is a theoretical framework for designing HMIs for complex systems such as NPPs [43]. EID is theoretically based on two concepts proposed by Rassmussen [44]: abstraction hierarchy and taxonomy of skills, rules, and knowledge (SRK). Abstraction hierarchy is a multi-level knowledge representation framework for describing functional structure of work domains. Abstraction hierarchy is defined by means–ends relations between levels, with higher levels containing functional information and lower levels containing physical information [45]. Abstraction hierarchy is used as a basis for selection of information to be represented in a display. The SRK taxonomy provides a basis about how information is displayed in the interface. EID recommends that information should be presented in such a way as to promote skill- and rule-based behavior, allowing operators to deal with task demands in a relatively efficient and reliable manner. Knowledge-based behavior is also supported by embedding an abstraction hierarchy representation in the interface. The usefulness of EID in SRK-based behaviors was shown experimentally in several studies [46, 47].

The concept seems to be immature for NPP applications that require proven and reliable technologies, although the EID is apparently beneficial for diagnosing novel situations. Some issues have been addressed for EIDs [42], including the lack of real applications in NPPs. EIDs needs to be compatible and integrated with other activities in human factors engineering of NPPs, such as analysis, training, procedure development, and evaluation. Standard designs of NPP systems, such as RCS and steam generators, also need to be developed for the promotion of real EID applications.

(D) Information-rich Display

Information-rich display, proposed by the OECD Halden Reactor Project for petroleum application [48], is an alternative to resolve the problem of limited viewing area of VDU-based displays. The concept presents information in a condensed form, presenting just more data on each display. Problems related to the keyhole effect are expected to be diminished by reducing the area of process information displays. The process value is presented by both a trend line and actual value. Normal and abnormal regions are represented with light and darker grays, respectively. The detail of description is shown in Figure 8.9.

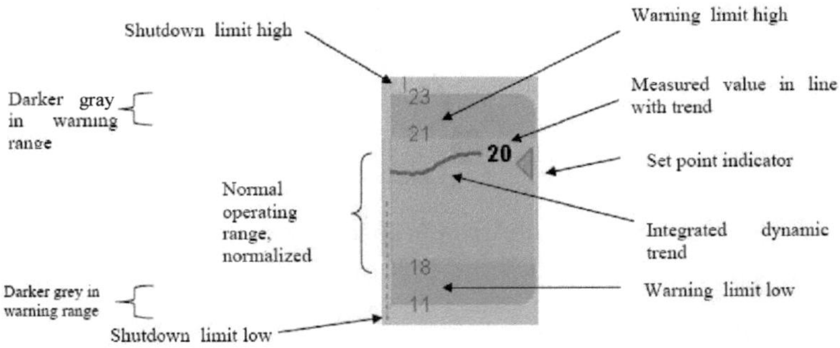

Figure 8.9. Information-rich display [48]

8.2.1.3 Issues in the Computer-based Information Display

Poor implementation of information systems creates human performance problems [42]. Modern computer-based information systems have many advantages, such as flexibility of display types, unlimited data storage, and computational power. However, designer should take into account several issues that may deteriorate operator performance. Three important issues and possible resolutions are introduced in this section.

(A) Information Overload

Some highly troublesome situations occur during conditions of information overload in complex, dynamic processes like NPPs [49]. Computer-generated displays are not limited by physical space, unlike conventional indicators on panels, but can present arbitrary amounts of information by means of scrolling, overlapping windows, and hierarchies of displays. Information is presented faster than the eye or the brain can comfortably deal with it.

Operators develop a number of strategies to cope with conditions of overload imposed by a vast quantity of information. The strategies are omission, reducing precision, queuing, filtering, cutting categories, and escape [49, 50]. An HMI, if properly designed, supports some of the strategies, like queuing, filtering, and categorizing information. For example, filtering is supported if the interface categorizes or prioritizes plant data and measurements according to their urgency. Different forms of codings (*e.g.*, color, size, shape, and patterns) are used to categorize or prioritize plant data. An alarm system may use color coding to distinguish the importance of alarms; the first priority of alarms is represented by red, the second by yellow, and the third by green.

(B) Interface Management Tasks

Operators must perform interface management tasks, or so-called secondary tasks, in a computer-based display system, such as navigating, configuring, and arranging the interface. A broad survey of the effects of interface management on the plant safety and operator performance has been performed by Brookhaven National Laboratory [51]. There are two forms of negative effects of interface management tasks: (1) primary task performance declines because operator attention is directed toward the interface management task, and (2) under high workload, operators minimize their performance of interface management tasks, thus failing to retrieve potentially important information for their primary tasks. These effects were found to have potential negative effects on safety.

There are three trade-offs related to navigation with respect to design. The first is a trade-off between distributing information over many display pages that require a lot of navigation and packing displays with data potentially resulting in a crowded appearance that requires less navigation. Initially crowded displays may become well liked and effective in supporting performance as operators gain experience with them [51]. The second is a trade-off between depth and breadth in hierarchical structure of display pages. Depth increases as breadth is decreased, when multiple pages are organized into a hierachical structure for navigation. Performance is best when depth is avoided. Empirical studies show that greater breadth is always better than introduction of depth. The third trade-off is related to the number of VDUs [51]. Fewer VDUs means smaller control rooms, more simplicity in that there are fewer HMI to integrate, less cost, and a lower maintenance burden. The demand of secondary tasks, on the contrary, is reduced by increasing the number of VDUs, because operators can view more information at a time.

Interface management tasks are relieved by introducing design concepts [52]:

- Improving HSI predictability
- Enhancing navigation functions
- Automatic interface management features
- Interface management training

(C) Keyhole Effect

The limited viewing area of VDUs brings about a new issue which is referred to as the "keyhole effect" [52]. Operators are required to navigate repeatedly and get focused on a small area of the interface without recognizing the overall state of the plant, just like the view from outside of a door through a keyhole. The keyhole effect interferes with operator situation awareness about the overall state of the plant.

The keyhole effect becomes significant in a computerized procedure system (discussed in Section 8.2.3.2) in NPPs when operators are required to perform multiple procedures [53]. Operators may lose a sense of where they are within the total set of active procedures, because only a portion of the procedures are observed at one time. The display space may be inadequate to allow simultaneous viewing of multiple procedures and associated plant data.

There are a few approaches to prevent keyhole effects in advanced control rooms. One is the introduction of a large display panel (LDP) which allows the operator to assess the overall plant process performance by providing information to allow a quick assessment of plant status. An LDP is legible from the workstations as well as from probable locations of observers or support personnel in an MCR for easy access to information. Information about the overall plant status helps operators to avoid performing an incorrect, focused task. Another is that all resources need to be integrated to permit operators to view the plant situation and recover any situation in an efficient way. For example, a computerized procedure system provides all the required information using normal resources and displays as much as possible, rather than dedicated and specific displays for every step of the procedure.

8.2.2 Automation

8.2.2.1 Automation in NPPs

The most appropriate definition of "automation" for NPP applications is the execution by a machine agent of a function which was previously carried out by a human operator [54], although the term can be used in many different ways. Systems or operations tend to be automated to make an NPP safer and more reliable, as information technology becomes mature.

There are some fundamental reasons for applying automation to NPPs. One is to ensure speed, accuracy, or reliability of a required action or operation which is beyond the capability of a human operator. This includes consideration of whether the operator can respond to an event with sufficient speed, or can control a process variable to the required degree of accuracy. In NPPs, for example, reactor trip or core protection systems are automated so that these systems can be quickly activated to control a core chain reaction and remove heat from the core when the integrity of the reactor is threatened. Another example in NPPs is the control of steam generator water level. Controlling the level of steam generator is a tough task for turbine operators because of complex thermodynamics. Human operators usually take over the control as a backup when the automatic control fails.

The second reason is that automation is applied to tasks which must be carried out in an unacceptably hostile environment, inaccessible to personnel. The task in containments needs to be automated because containment comprises high levels of radiation and temperature.

A third reason is a reduction of total workload that may exceed the capability of the available operators or a reduction in number of operators required to operate a station by replacing operators with automation systems.

Methods for system automation in NPPs are classified into several categories [55]. The first is computer-based operator aids which process and summarize plant information in a succinct form, including information analysis systems, information reduction and management systems, equipment monitoring systems, diagnostic systems, and procedure support systems. These systems are called computerized operator support systems (COSSs), which will be dealt with in the next section. The next category, which is a typical automation system, is automatic functions which aid or supplement the operator control over a sequence or process, such as plant sequence control and closed loop control. These functions typically are placed under manual control when desired. The systems automate the whole or part of functions which are difficult for operators to perform, usually, for the purpose of the reduction of operator workload. The third is the function of quick initiation for ensuring plant safety. The automatic features detect variables which exceed safety limits and initiate appropriate safety actions, such as reactor trip and/or initiation of safeguard equipment. This category includes systems which prevent unsafe conditions such as interlocks. For example, the isolation valves of a pump are automatically closed in the NPP (*i.e.*, interlocked) to protect the integrity of the pump when the pump is suddenly unavailable.

8.2.2.2 Issues in Automation

The development and introduction of automation systems has improved the accuracy, speed, and reliability in a variety of domains, including NPPs. Automation does not simply replace operator activities, but changes operator roles. The issues of automation are related to a large extent to breakdowns in interaction and balance between human operators and automated systems.

(A) Balance Between Automation Systems and Human Operators

A balance between operators and automation systems is achieved by properly allocating functions to operators, machines, or to a cooperation of operators and machines. Functions are assigned to human or automation systems through function allocation [55, 56]. Function allocation is a part of function analysis (Section 8.1.1). Detailed individual functions and operations for safe and effective running of systems are clearly defined. Functions are then allocated based on knowledge of the capabilities and limitations of human operator and available technology for design, manufacturing, and implementation of the system. A human operator is known to be more capable in reasoning, pattern recognition, and responding to unexpected situations than is a machine agent. On the contrary, the strong points of machines are speed, accuracy, multi-tasking, and consistency within the boundary of a design. Fitts' list [57], which describes the advantages of human and machines, is shown in Table 8.2. Function allocation in NPPs should also consider (1) operating experience of former plants, (2) regulatory requirements, and (3) existing practices [1]. The criteria are defined to determine how functions are assigned between operators and automation systems.

Identification of the level of automation is another indispensable activity for balance between operators and automation. The level of automation is a key element in the design of automation systems. The roles of operators, automation

Table 8.2. Fitts' list

Human	Machine
• Ability to detect small amount of visual or acoustic energy • Ability to perceive patterns of light or sound • Ability to improvise and use flexible procedures • Ability to store very large amounts of information for long periods and to recall relevant facts at the appropriate time • Ability to reason inductively • Ability to exercise judgment	• Ability to respond quickly to control signals and to apply great forces smoothly and precisely • Ability to perform repetitive, routine tasks • Ability to store information briefly and then to erase it completely • Ability to reason deductively, including computational ability • Ability to handle highly complex operations, *i.e.*, to do many different things at once

systems, and the interaction between them are defined after the level of automation is clearly defined. Other important design factors related to the level of automation are the authority (*i.e.*, ultimate decision-maker) on the control function and the feedback from the automation system. The level of automation incorporates issues of authority and feedback (issues of interaction between automation systems and operators), as well as relative sharing of functions for determining options, selecting options and implementing [58].

A classification proposed by Sheridan is used to determine the level of automation [59]:

1. Human does the whole job up to the point of turning it to the machine to implement
2. Machine helps by determining the options
3. Machine helps to determine options and suggests one, which human need not follow
4. Machine selects action and human may or may not do it
5. Machine selects action and implements it if human approve
6. Machine selects action, informs human in plenty of time to stop it
7. Machine does whole job and necessarily tells human what it did
8. Machine does whole job and tells human what it did only if human explicitly asks
9. Machine does whole job and decides what the human should be told
10. Machine does the whole job if it decides it should be done, and if so, tells human, if it decides that the human should be told

(B) Interaction Between Automation Systems and Human Operators

Breakdown in the interaction between automation systems and human operators brings about unanticipated problems and failures like inadvertent reactor trips in NPPs. Factors that cause breakdown are inappropriate feedback from automation systems, inappropriate transparency, operator overreliance of underreliance on automatic systems, and authority between automation systems and operators.

Inappropriate feedback from automation systems moves the operator out of the control loop. Poor feedback becomes crucial, especially when the operator takes over the control from an automation system. Operators cannot properly handle the situation transferred from the automatic mode because of bad situation awareness, if the operator is out of the loop and the automation system fails. Some cases of unexpected reactor trips during the transition between the manual mode and the automatic mode are reported in NPPs [60]. Automation systems provide the following feedback information for better operator situation awareness about automation systems:

- Informing what tasks are performed by the automation system.
- Providing information which operators want to know to understand a given situation (*e.g.*, the reason why the operation by the automation system is actuated).
- Alerting operators to the state in which automatic controls are, or will be soon activated, especially when the operation is under manual control.

Sudden activations by automatic systems which operators are not aware of can cause them to be confused or embarrassed.

An automation system needs to be transparent. The system needs to provide information which operators want to know, giving a better understanding of a given situation, especially, related to automatic operation. Information which operators can query during an operation includes the condition that initiates the automatic operation and which (or how the) automation system is now operating.

Operator reliance on automation systems is most related to operator trust in automation. Seven possible characteristics of human trust in human–machine systems have been suggested by Sheridan [59]: (1) reliability, (2) competence or robustness, (3) familiarity, (4) understandability, (5) explication of intent, (6) usefulness, and (7) dependency. Operators will trust more reliable, robust, familiar, and predictable systems. A system needs to provide a way operators can see the system through the interface to the underlying system to gain operator understandability. Operators will rely on a system which responds in a useful way to create something valuable. Operator trust is also related to the willingness that they depend on automation system. Over-reliance causes the misuse of automation, resulting in several forms of human error, including decision biases and failures of monitoring [54]. On the contrary, an operator who has under-reliance on automation tends to ignore the information from the automation system, even when the information is correct and useful. This situation may lead to a serious problem if the operator does not recognize the occurrence of an abnormal situation.

The ultimate authority for operations that directly or indirectly influence the safety of NPPs should be given to operators. The problem of assigning authority is related to the level of automation and feedback from automation. Operators are in command only if they are provided with or have access to all information necessary to assess the status and behavior of the automation systems and to make decisions about future courses of actions based on this assessment [56]. The operator has the authority to instruct, redirect, and override the automation system if there is a need to escape from automation when deemed necessary. Therefore, the designer should select a proper level of automation for the operator to keep authority and should design an interface to easily obtain information referenced and expert-judgment referenced comparisons.

8.2.3 Computerized Operator Support Systems

8.2.3.1 Computerized Operator Support Systems in NPPs

COSS is a computer system which supports operator cognitive activities such as diagnosis, situation awareness, and response planning in adverse situations. COSS is also categorized as a type of automation system according to the definition of automation (Section 8.2.2).

A need for COSS in NPP operations was firstly issued after the Three Mile Island unit 2 (TMI-2) accident. A lesson learned from the experience of TMI-2 is that a shortage of significant information during an accident at NPPs can have catastrophic results. The accident pointed out the need to develop a systematic approach to manage plant emergency responses, to identify a better decision-making process, and to implement real-time information support for safety monitoring and decision-making [62]. The alarm system of the plant presented so many nuisance alarms that alarms were not helpful for operators to diagnose plant status. The safety parameter display system (SPDS), an example of COSSs, has been suggested as a result of a research on the TMI-2 accident [63]. The system has proved helpful to operators and has been successfully implemented for commercial plants. The SPDS for on-line information display in the control room has been developed into a licensing requirement in the USA.

COSS has operator needs in NPPs. Difficulties often arise as a result of the inability to identify the nature of the problem in abnormal situations [64]. Operator responses to plant states are well described in operation procedures, if plant status is correctly evaluated. The operator needs timely and accurate analysis of actual plant conditions.

COSSs are based on expert systems or knowledge-based systems. Expert systems are interactive computer programs whose objective is to reproduce the capabilities of an exceptionally talented human experts [65, 66]. An expert system generally consists of knowledge bases, inference engines, and user interfaces. The underlying idea is to design the expert system so that the experience of the human experts and the information on the plant structure (knowledge base) are kept separate from the method (inference engine) by which that experience and information are accessed. The knowledge base represents both the thinking process of a human expert and more general knowledge about the application. Knowledge is usually based on "IF-THEN" rules; expert systems may be called rule-based systems. The inference engine consists of logical procedures to select rules. The inference engine chooses which rules contained in the knowledge base are applicable to the problem at hand. The user interface provides the user access window with powerful knowledge residing within the expert system.

8.2.3.2 Applications in NPPs

A large number of COSSs have been developed for application in NPPs, although not all were successfully implemented. Surveys of COSS applications for NPPs have been presented [65, 67]. A brief introduction to those systems is provided in this chapter.

COSSs support the elements of human information processing, although the boundaries between the systems are becoming more ambiguous as systems are integrated. Operator support systems for real-time operation assist on-line management of adverse events by assisting the detection, diagnosis, response planning, and response implementation of human operators. The relationship of real-time operator support systems and cognitive activities is shown in Figure 8.10. Alarm systems usually support monitoring and detection tasks. Alarm information includes the deviation of parameters, the state of equipment, and the parametric cause of a reactor trip. Conventional alarm systems, like tile-style alarm systems, suffer from several common problems, including too many nuisance alarms and the annunciation of too many conditions that should not be a part of an integrated warning system. Advanced alarm systems have general alarm-processing functions such as categorization, filtering, suppression, and prioritization in order to cope with these problems [25].

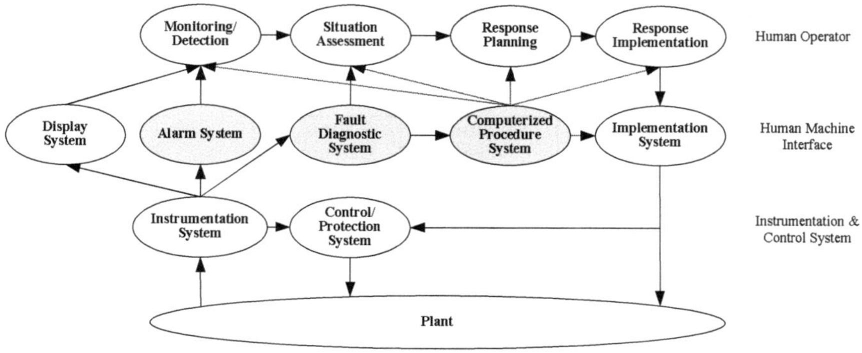

Figure 8.10. COSS and cognitive activities

Fault diagnosis systems have been developed to support operators in cognitive activities, such as fault detection and diagnosis. Real-time diagnosis support is done at several different levels (*e.g.*, component, subsystem, function or event). A diagnosis is also made at an event level such as a loss of coolant accident (LOCA) or a system level like a turbine generator. Various demonstrative diagnostic systems have also been developed, mostly in academic and research institutes. Advanced alarm systems are regarded as a representative system in advanced MCRs. Most fault diagnostic systems have seen relatively little success in terms of the application to real plants for technical and/or practical reasons. The fault detection function of a fault diagnostic system tends to be implicitly implemented in early fault detection of alarm systems and in automatic procedure identification of computerized procedure systems. Computerized procedure systems (CPSs) are also another representative system of advanced MCRs [68, 69]. CPSs were developed to assist human operators by computerizing paper-based procedures, whose purpose is to guide operator actions in performing their tasks in order to increase the likelihood that the goals of the tasks are safely achieved [25]. CPSs may support cognitive functions, such as monitoring and detection, and situation assessment according to the level of automation, while they were originally developed to support response planning.

8.2.3.3 Issues of COSS

(A) Instrument Paradigm

A COSS is an instrument rather than prosthesis for operators [70]. The prosthesis approach focuses on operator passive role, such as gathering data for a COSS and accepting COSS solutions (Figure 8.11(a)). A COSS solves a problem, on the assumption that the problem was beyond the skill of the original agent in the prosthesis concept, while the instrument approach emphasizes the more active role of human operators. The instrumental perspective defines a COSS as a consultant to provide a reference or a source of information for the problem solver (Figure 8.11(b)). The operator is in charge of decision making and control. The prosthesis approach to decision aiding may show critical flaws in the face of unexpected

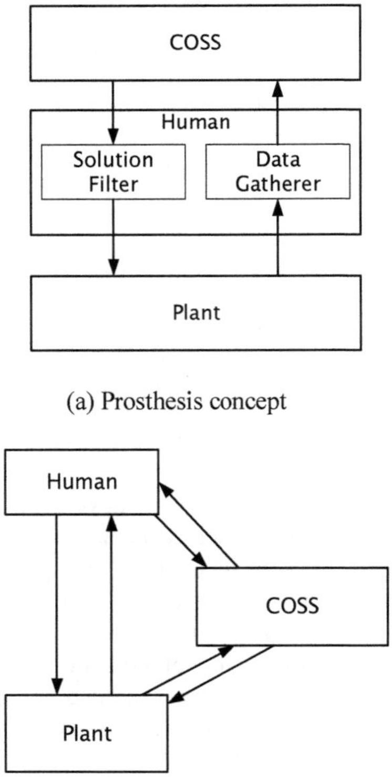

(a) Prosthesis concept

(b) Instrument concept

Figure 8.11. COSS paradigms

situations [71, 72]. A COSS needs to be an instrument that the operator can use from necessity rather than a prosthesis that restricts operator behavior.

(B) Compatibility with Strategy
A COSS potentially has negative consequences without sufficient consideration of strategies and may become a new burden on operators. A shift in an MCR, in which the alarm system is changed from the tile annunciator alarm systems to a computer-based alarm system, eventually collapsed, because strategies to meet the cognitive demands of fault management that were implicitly supported by the old representation were undermined in the new representation [71]. The effects of information aid types on diagnosis performance can differ according to the strategies that operators employ [73, 74].

Three elements need to be considered in designing a control room feature for NPPs [74]: (1) operational tasks that must be performed; (2) a model of human performance for these tasks; and (3) a model of how control room features are intended to support performance. Operational tasks define classes of performance that must be considered. A model of human performance makes more explicit the requirements for accurate and efficient performance and reveals potential sources of error. The model of support allows the generation of specific hypotheses about how performance is facilitated in the control room. The model of support systems needs to be developed based on the performance model. Diagnostic strategies serve as a performance model to design an information aiding system for fault identification. Operator support systems, therefore, need to support the strategies employed by operators rather than provide the evaluation results of the system about the plant status.

(C) Verification and Validation

V&V are indispensable activities for the application of COSSs to NPPs. A COSS needs to be verified and validated by a knowledge base V&V and a software V&V. The knowledge base should be faultless and accurate for a COSS to provide correct solutions [76]. COSSs, especially for safety-critical applications, need to be carefully verified and validated through all phases of the software life cycle to ensure that the software complies with a variety of requirements and standards (Chapters 4 and 5) [77].

(D) System Complexity

A COSS may cause an increase in system complexity [78]. COSSs were originally intended to reduce operator workload by providing useful information for operation. On the contrary, these systems may increase the workload because the addition of a system increases the complexity of systems that operators need to handle. A COSS can be a new burden on operators. Operator benefits from the COSS should be larger than the inconvenience from the system complexity, although it is important that the system is designed as simple as possible, to hide the effect from the increased system complexity. To do so, the development of a COSS should start with the operator necessity and a prudent task analysis.

8.3 Human Factors Engineering Verification and Validation

HFE verification ensures not only that the HMI design contains all the minimal inventories, that is, information and controls, but also that the design conforms to HFE principles. The former is the availability verification and the latter is the suitability verification. HFE validation is a performance-based evaluation which determines if the design in terms of actual operator, the full control room, the actual procedure, and the real plant dynamics, can actually work together as a whole.

8.3.1 Verification

The objective of the availability verification is to verify that HMI design accurately describes all HMI components, inventories, and characteristics that are within the scope of the HMI design. The activity reviews the design in terms of the following aspects:

- If there are unavailable HMI components which are needed for task performance (*e.g.*, information or control)

- If HMI characteristics do not match the task requirements, *e.g.*, range or precision of indication
- If there are unnecessary HMI components that are not required for any task

Suitability verification determines if the design conforms to HFE design principles. HFE principles for HMI components are found in the psychological and HCI-related handbooks. Some guidelines that compile the available principles into a book are available for NPP applications [25] or other applications [34]. This activity is usually performed by HFE experts using a checklist. The checklists provide the review guidelines about displayed elements (*e.g.*, size, color, information density).

8.3.2 Validation

Validation is a performance-based evaluation, while verification is performed by using checklists. Validation determines if the operators can reach the system operational goal (*e.g.*, safe shutdown), with the HMI. In this test, trained operators perform assigned scenarios using the HMI under simulated situations. A wide range of measures are used to assess human performance and system performance. Some examples of measured performances that are used for an NPP application are introduced in this section.

8.3.2.1 System Performance

System performance measures the discrepancy between ideal values and values obtained during the simulation on a predefined set of system parameters. The criterion for choosing a parameter is that the final state of the parameter reflects the cumulative effect of operator performance. The parameters differ according to scenarios. Examples of parameters that are selected in NPPs are subcooling margin and pressure/temperature of the primary system, which represent the integrity or stability of the primary system and fuel. The amplitude, frequency, and trends during a period of time are also alternate choices.

8.3.2.2 Task Performance

Primary tasks (*i.e.*, monitoring and control) and secondary tasks (*i.e.*, interface management) performed by operators are identified and assessed. Popular objective measures are reaction time, duration, and accuracy (or conversely, error) [79]. Reaction time is the time between the occurrence of an event requiring an action on the part of operators or team and the start of the action demanded by the event. Duration is the time from the stimulus for task initiation to the time of task completion. Accuracy is the most useful measure of human performance, that is applied to the system or situation where time aspect is not important, but the number of errors is critical.

8.3.2.3 Workload

Many approaches to measuring operator workload have been suggested [29, 31, 32]. Techniques for measuring mental workload are divided into two broad types: predictive and empirical [80]. Predictive techniques are usually based on mathematical modeling, task analysis, simulation modeling, and expert opinions. These techniques do not require operators to participate in simulation exercises. Predictive techniques are typically used in the early stages of design process and therefore, are thought not to be suitable for validation [81]. Empirical techniques are divided into three types: performance-based technique, subjective rating-based technique, and physiological measure-based technique [80]. A description on the merits and demerits of empirical measures are given in Table 8.3. A basic assumption of the performance-based approach is that deteriorated or erratic performance may indicate an unacceptable level of workload. Performance-based techniques are categorized into primary task measures and secondary task measures. Primary task measures monitor the changes of performance in actual tasks as the demands of the task vary. Operators are required to perform concurrently a simple task like counting numbers as well as primary tasks (*i.e.*, actual tasks) in the secondary task measure. This measure assumes that the secondary task performance may be affected by the load of primary tasks. Primary task measures are not suitable for measurement of cognitive workload associated with monitoring or decision-making tasks like in NPPs. Secondary task measures have the drawback that they can contaminate human performance by interfering with primary tasks [18].

Subjective methods involve a self-report or a questionnaire that asks the operators to rate their level of mental effort during the tasks. This method has strong merits, such as operator acceptance and ease of use. The methods have several inherent weaknesses such as susceptibility to operator memory problem and bias, as well as to the operator experience and degree of familiarity with the tasks. Representative subjective measures are overall workload (OW), modified Cooper–Harper scale (MCH), subjective workload assessment technique (SWAT), and national aeronautic and space administration task load index (NASA-TLX).

Table 8.3. Comparison of empirical measures for workload

Method	Description	Merits	Demerits
Subjective	Operators rate the level of mental effort that	Operator acceptance Ease of use	Susceptible to operator's memory problems, bias, and experience & degree of familiarity
Performance-based	Use operator's performance to determine workload	Direct and objective measure (primary task measure) Sensitive to low workload level, diagnostic (secondary task measure)	Only available when the primary tasks are specified Intrusive
Physiological	Measure changes in operator's physiology that are associated with cognitive task demands	Not intrusive Objective and continuous measure	Implicit measure

The NASA-TLX is superior in validity and NASA-TLX and OW are superior in usability [82].

Physiological techniques measure the physiological change of the autonomic or central nervous systems associated with cognitive workload [18]. Electroencephalogram (EEG), evoked potential, heart-rate-related measures, and eye-movement-related measures are representative tools for cognitive workload evaluation based on the physiological measurements.

8.3.2.4 Situation Awareness

A relatively small number of methods, compared to the methods for measuring workload, have been proposed for measuring situation awareness. Measurement techniques of SA for the validation are categorized into three groups: (1) direct query and questionnaire; (2) subjective rating; and (3) physiological measurement techniques [81, 83]. Direct query and questionnaire techniques are categorized into post-test, on-line-test, and freeze techniques according to the evaluation point over time [84]. These kinds of techniques are based on questions and answers regarding the SA. The most well-known method is the situational awareness global assessment technique (SAGAT) [26]. The experiment session is stopped and operators are asked to answer questions to assess operator situation awareness, and the operator answers are compared with correct answers, when SAGAT is used. However, SAGAT is intrusive in that it contaminates any performance measures. This is related to the concern that the questions may suggest some details of the scenario to participants (*e.g.*, operators), setting up expectancy for certain types of questions [81, 85, 86].

Subjective rating techniques typically involve assigning a numerical value to the quality of SA during a particular period of an event [87]. Subjective ratings techniques are popular because these techniques are fairly inexpensive, easy to administer, and non-intrusive [87, 88]. Subjective measures that use questionnaires can be evaluated by operators [89] or subject matter experts [90].

Physiological measurement techniques have been used to study complex cognitive domains such as mental workload and fatigue. Very few experiments have been conducted to study SA [91]. Eye fixation measurement, which is called VISA (visual indicator of situation awareness) has been used as an indicator for SA [92]. Time spent on eye fixation has been proposed as a visual indicator of SA, as a result of a study of VISA. The results of the VISA study also showed that a subjective measure named SACRI (situation awareness control room inventory) scores were correlated with VISA.

8.4 Summary and Concluding Remarks

Three activities, analysis, design, and V&V, are performed in an iterative way for HMI design. V&V identifies design discrepancies that need to be modified. The discrepancies may arise because the design is not consistent with HFE guidelines, any HMI feature is not available for operation, or the design causes serious operator inconvenience which can deteriorate performance. The discrepancy may be resolved through a simple modification of design or may require both re-

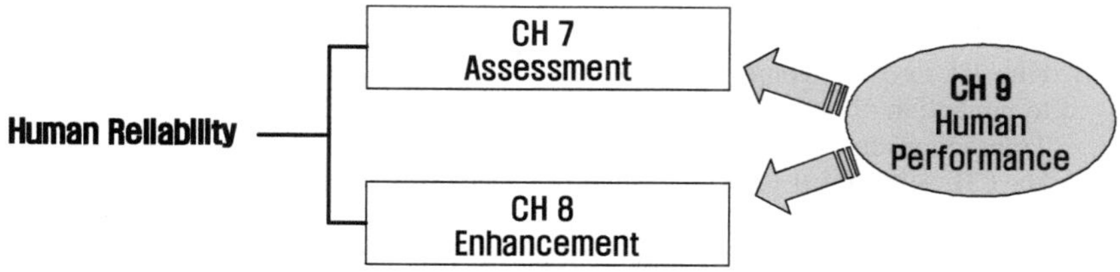

Figure 8.12. Relations among the chapters in Part III

analysis of function/task and re-design. Design modification is also verified and validated again, although it needs not to be over an entire range of V&V.

A support system to measure human performance is introduced in Chapter 9. Human performance measurement plays an important role in both human reliability assessment and enhancement (Figure 8.12). Fundamental human errors in terms of accuracy and time are detected in the operational situation for human reliability assessment. Other performances, like workload and mental stress, also need to be measured to obtain performance-shaping factors. A wide range of human performance is measured in the HFE validation to ensure that HMI can support safe and efficient operation for human reliability enhancement (Section 8.3). Human performance measured is translated to meaningful information for human reliability assessment and enhancement.

References

[1] US NRC (2002) Human Factors Engineering Program Review Model. NUREG-0711, Rev. 2

[2] Lamarsh JR (1983) Introduction to Nuclear Engineering, Addison Wesley

[3] Bye A, Hollnagel E, Brendeford TS (1999) Human-machine function allocation: a functional modelling approach. Reliability Engineering and System Safety: 64, 291–300

[4] Vicente KJ (1999) Cognitive Work Analysis. Lawrence Erlbaum Associates

[5] Schraagen JM, Chipman S F, Shalin V L (2000) Cognitive Tasks Analysis. Lawrence Erlbaum Associates

[6] Kirwan B, Ainsworth LK (1992) A Guide to Task Analysis. Taylor & Francis

[7] Luczak H (1997) Task analysis. Handbook of Human Factors and Ergonomics, Ed. Salvendy G. John Wiley & Sons

[8] Shepherd A (2001) Hierarchical Task Analysis. Taylor & Francis

[9] Annett J (2003) Hierarchical Task Analysis. Handbook of Cognitive Task Design, Ed. E. Hollnagel, Ch. 2, Lawrence Erlbaum Associates

[10] Rasmussen J, Pejtersen A M, Goodstein LP (1994) Cognitive Systems Engineering. Wiley Interscience

[11] Rasmussen J (1986) Information Processing and Human-Machine Interaction, North-Holland

[12] Kim JH, Seong PH (2003) A quantitative approach to modeling the information flow of diagnosis tasks in nuclear power plants. Reliability Engineering and System Safety 80: 81–94

[13] Kim JH, Lee SJ, Seong PH (2003) Investigation on applicability of information theory to prediction of operator performance in diagnosis tasks at nuclear power plants. IEEE Transactions on Nuclear Science 50: 1238–1252

[14] Ha CH, Kim JH, Lee SJ, Seong PH (2006) Investigation on relationship between information flow rate and mental workload of accident diagnosis tasks in NPPs. IEEE Transactions on Nuclear Science 53: 1450–1459

[15] Reason J (1990) Human Error. Cambridge University Press

[16] Wickens CD, Lee J, Liu Y, Becker SG (2004) An Introduction to Human Factors Engineering. Prentice-Hall

[17] Miller GA (1956) The magical number seven plus or minus two: Some limits on our capacity for processing information. Psychological Review 63: 81–97

[18] Wickens CD, Hollands JG (1999) Engineering Psychology and Human Performance. Prentice-Hall

[19] Gentner D, Stevens AL (1983) Mental Models. Lawrence Erlbaum Associates

[20] Moray N (1997) Human factors in process control. Ch. 58, Handbook of Human Factors and Ergonomics, Ed., G. Salvendy, A Wiley-Interscience Publication

[21] Payne JW, Bettman JR, Eric JJ (1993) The Adaptive Decision Maker. Cambridge University Press

[22] Rasmussen J, Jensen A (1974) Mental procedures in real-life tasks: A case study of electronic trouble shooting. Ergonomics 17: 293–307

[23] Rasmussen J (1981) Models of mental strategies in process plant diagnosis. In: Rasmussen J, Rouse WB, Ed., Human Detection and Diagnosis of System Failures. New York: Plenum Press

[24] Woods DD, Roth EM (1988) Cognitive Systems Engineering. Handbook of Human-Computer Interaction. Ed. M. Helander. Elsevier Science Publishers

[25] US NRC (2002) Human-System Interface Design Review Guidelines. NUREG-0700

[26] Endsley MR (1988) Design and evaluation for situation awareness enhancement. Proceedings of the Human Factors Society 32nd Annual Meeting: 97–101

[27] Jones DG, Endsley MR (1996) Sources of situation awareness errors in aviation. Aviation, Space and Environmental Medicine 67: 507–512

[28] Endsley MR (1995) Toward a theory of situation awareness in dynamic systems. Human Factors 37: 32–64

[29] O'Donnell RD, Eggenmeier FT (1986) Workload assessment methodology. Ch. 42, Handbook of Perception and Human Performance, Ed. Boff KR, et al., Wiley-Interscience Publications

[30] Sanders MS, McCormick EJ (1993) Human Factors in Engineering and Design. McGraw-Hill

[31] Tsang P, Wilson GF (1997) Mental workload. Ch. 13, Handbook of Human Factors and Ergonomics, Ed. Salvendy G, Wiley-Interscience Publications

[32] Gawron VJ (2000) Human Performance Measures Handbook. Lawrence Erlbaum Associates

[33] US NRC (1994) Advanced Human-System Interface Design Review Guidelines. NUREG/CR-5908

[34] Department of Defense (1999) MIL-STD-1472F, Design Criteria Standard

[35] EPRI (1984) Computer-generated display system guidelines. EPRI NP-3701

[36] O'Hara JM, Hall MW (1992) Advanced control rooms and crew performance issues: Implications for human reliability. IEEE transactions on Nuclear Science 39(4): 919–923

[37] Tullis TS (1988) Screen design. Handbook of Human-Computer Interaction, Ed. M. Helander, Elsevier Science Publishers

[38] Danchak MM (1976) CRT displays for power plants. Instrumentation Technology 23: 29–36

[39] NASA (1980) Spacelab Display Design and Command Usage Guidelines, MSFC-PROC-711A, George C. Marshall Space Flight Center

[40] IEEE (1998) IEEE Guide for the Application of Human Factors Engineering in the Design of Computer-Based Monitoring and Control Displays for Nuclear Power Generating Stations. IEEE-Std 1289

[41] US NRC (1983) Handbook of Human Reliability Analysis with Emphasis on Nuclear Power Plant Applications. NUREG/CR-1278

[42] US NRC (2000) Advanced Information Systems Design: Technical Basis and Human Factors Review Guidance. NUREG/CR-6633

[43] Vicente KJ, Rasmussen J (1992) Ecological interface design: theoretical foundations. IEEE Transactions on System, Man, and Cybernetics 22: 589–606

[44] Rasmussen J (1983) Skills, rules, and knowledge; signals, signs, and symbols, and other distinctions in human performance models. IEEE Transactions on System, Man, and Cybernetics 13: 257–266

[45] Rasmussen J (1985) The role of hierarchical knowledge representation in decision making and system management. IEEE Transactions on System, Man, and Cybernetics 15: 234–243

[46] Vicente KJ (1995) Supporting operator problem solving through ecological interface design. IEEE Transactions on System, Man, and Cybernetics 25: 529–545

[47] Ham DH, Yoon WC (2001) The effects of presenting functionally abstracted information in fault diagnosis tasks. Reliability Engineering and System Safety 73: 103–119

[48] Braseth AO, A building block for information rich dispays. IFEA Conference on Alarmhandtering on Gardermoen

[49] Hollnagel E, Bye A, Hoffmann M (2000) Coping with complexity – strategies for information input overload. Proceedings of CSEPC 2000: 264–268

[50] Hoffmann M, Bye A, Hollnagel E (2000) Responding to input information overload in process control – a simulation of operator behavior. Proceedings of CSEPC 2000: 103–108

[51] US NRC (2002) The Effects of Interface Management Tasks on Crew Performance and Safety in Complex, Computer Based Systems. NUREG/CR-6690

[52] Woods DD (1990) Navigating through large display networks in dynamic control applications. Proceedings of the Human Factors Society 34th Annual Meeting

[53] US NRC (2000) Computer-Based Procedure Systems: Technical Basis and Human Factors Review Guidance. NUREG/CR-6634

[54] Parasuraman P (1997) Humans and automation: use, misuse, disuse, abuse. Human Factors 39: 230–253

[55] IAEA (1992) The Role of Automation and Humans in Nuclear Power Plants. IAEA-TECDOC-668

[56] Sarter NB, Woods DD, Billings CE (1997) Automation surprise. Ch. 57, Handbook of Human Factors and Ergonomics, Ed. Salvendy G, Wiley-Interscience Publications

[57] Fitts PM (1951) Human Engineering for an Effective Air Navigation and Traffic Control System. Washington DC, National Research Council

[58] Endsley MR, Kaber DB (1999) Level of automation effects on performance, situation awareness and workload in a dynamic control task. Ergonomics 42: 462–492

[59] Sheridan T (1980) Computer control and human alienation. Technology Review 10: 61–73

[60] http://opis.kins.re.kr, Operational Performance Information System for Nuclear Power Plant
[61] Niwa Y, Takahashi M, Kitamura M (2001) The design of human-machine interface for accident support in nuclear power plants. Cognition, Technology & Work 3: 161–176
[62] Sun BK, Cain DG (1991) Computer application for control room operator support in nuclear power plants. Reliability Engineering and System Safety 33: 331–340
[63] Woods DD, Wise J, Hanes L (1982) Evaluation of safety parameter display concept. Proceedings of the Human Factors Society 25th Annual Meeting
[64] Bernard JA (1992) Issues regarding the design and acceptance of intelligent support systems for reactor operators. IEEE Transactions on Nuclear Science 39: 1549–1558
[65] Bernard JA, Washio T (1989) Expert System Application within the Nuclear Industry. American Nuclear Society
[66] Adelman L (1992) Evaluating Decision Support and Expert Systems. John Wiley & Sons
[67] Kim IS (1994) Computerized systems for on-line management of failures: a state-of-the-art discussion of alarm systems and diagnostic systems applied in the nuclear industry. Reliability Engineering and Safety System 44: 279–295
[68] Niwa Y, Hollnagel E, Green M (1996) Guidelines for computerized presentation of emergency operating procedures. Nuclear Engineering and Design 167: 113–127
[69] Pirus D, Chambon Y (1997) The computerized procedures for the French N4 series. IEEE Sixth Annual Human Factors Meeting
[70] Woods DD, Roth EM (1988) Cognitive systems engineering. Handbook of Human-Computer Interaction, Ch. 1, Ed. Helander M, Elsevier Science Publishers
[71] Hollnagel E, Mancini G, Woods DD (1988) Cognitive Engineering in Complex Dynamic Worlds. Academic Press
[72] Woods DD (1986) Paradigms for intelligent decision support. Intelligent Decision Support in Process Environments, Ed. Hollnagel E, Mancini G, Woods DD, New York: Springer-Verlag
[73] Kim JH, Seong PH (2007) The effect of information types on diagnostic strategies in the information aid. Reliability Engineering and System Safety 92: 171–186
[74] Yoon WC, Hammer JM (1988) Deep-reasoning fault diagnosis: an aid and a model. IEEE Transactions on System, Man, and Cybernetics 18: 659–676
[75] Roth EM, Mumaw RJ, Stubler WF (1992) Human factors evaluation issues for advanced control rooms: a research agenda. IEEE Fifth Conference on Human Factors and Power Plants: 254–259
[76] Kim JH. Seong PH (2000) A methodology for the quantitative evaluation of NPP diagnostic systems' dynamic aspects. Annals of Nuclear Energy 27: 1459–1481
[77] IEEE (1998) IEEE Standard for Software Verification and Validation. IEEE-Std 1012
[78] Wieringa PA, Wawoe DP (1998) The operator support system dilemma: balancing a reduction in task complexity vs. an increase in system complexity. IEEE International Conference on Systems, Man, and Cybernetics: 993–997
[79] Meister D (1986) Human Factors Testing and Evaluation. Elsevier
[80] Williges R, Wierwille WW (1979) Behavioral measures of aircrew mental workload. Human Factors 21: 549–574
[81] O'Hara JM, Stubler WF, Higgins JC, Brown WS (1997) Integrated System Validation: Methodology and Review Criteria. NUREG/CR-6393, US NRC
[82] Hill SG, Iavecchia HP, Byers JC, Bittier AC, Zaklad AL, Christ RE (1992) Comparison of four subjective workload rating scales. Human Factors 34: 429–440
[83] Endsley MR, Garland DJ (2001) Situation Awareness: Analysis and Measurement. Erlbaum, Mahwah, NJ

[84] Lee DH, Lee HC (2000) A review on measurement and applications of situation awareness for an evaluation of Korea next generation reactor operator performance. IE Interface 13: 751–758
[85] Sarter NB. Woods DD (1991) Situation awareness: a critical but ill-defined phenomenon. The International Journal of Aviation Psychology 1: 45–57
[86] Pew RW (2000) The state of situation awareness measurement: heading toward the next century. Situation Awareness Analysis and Measurement, Ed. Endsley MR, Garland DJ. Mahwah, NJ: Lawrence Erlbaum Associates
[87] Fracker ML, Vidulich MA (1991) Measurement of situation awareness: A brief review. Proceedings of the 11th Congress of the international Ergonomics Association: 795–797
[88] Endsley MR (1996) Situation awareness measurement in test and evaluation. Handbook of Human Factors Testing and Evaluation, Ed. O'Brien TG, Charlton SG. Mahwah, NJ: Lawrence Erlbaum Associates
[89] Taylor RM (1990) Situational Awareness: Aircrew Constructs for Subject Estimation, IAM-R-670
[90] Moister KL, Chidester TR (1991) Situation assessment and situation awareness in a team setting. Situation Awareness in Dynamic Systems, Ed. Taylor RM, IAM Report 708, Farnborough, UK, Royal Air Force Institute of Aviation Medicine
[91] Wilson GF (2000) Strategies for psychophysiological assessment of situation awareness. Situation Awareness Analysis and Measurement, Ed. Endsley MR, Garland DJ. Mahwah, NJ: Lawrence Erlbaum Associates
[92] Drøivoldsmo A, Skraaning G, Sverrbo M, Dalen J, Grimstad T, Andresen G (1988) Continuous Measure of Situation Awareness and Workload. HWR-539, OECD Halden Reactor Project

9

HUPESS: Human Performance Evaluation Support System

Jun Su Ha[1] and Poong Hyun Seong[2]

[1]Center for Advanced Reactor Research
Korea Advanced Institute of Science and Technology
373-1 Guseong-dong, Yuseong-gu, Daejeon, 305-701, Korea
hajunsu@kaist.ac.kr

[2]Department of Nuclear and Quantum Engineering
Korea Advanced Institute of Science and Technology
373-1 Guseong-dong, Yuseong-gu, Daejeon, 305-701, Korea
phseong@kaist.ac.kr

Research and development for enhancing reliability and safety in NPPs have been mainly focused on areas such as automation of facilities, securing safety margin of safety systems, and improvement of main process systems. Studies of TMI-2, Chernobyl, and other NPP events have revealed that deficiencies in human factors, such as poor control room design, procedure, and training, are significant contributing factors to NPPs incidents and accidents [1–5]. Greater attention has been focused on the human factors study. Modern computer techniques have been gradually introduced into the design of advanced control rooms (ACRs) of NPPs as processing and information presentation capabilities of modern computers are increased [6, 7]. The design of instrumentation and control (I&C) systems for various plant systems is also rapidly moving toward fully digital I&C [8, 9]. For example, CRT- (or LCD-) based displays, large display panels (LDP), soft controls, a CPS, and an advanced alarm system were applied to APR-1400 (Advanced Power Reactor-1400) [10]. The role of operators in advanced NPPs shifts from a manual controller to a supervisor or a decision-maker [11] and the operator tasks have become more cognitive works. As a result, HFE became more important in designing an ACR. The human factors engineering program review model (HFE PRM) was developed with the support of U.S. NRC in order to support advanced reactor design certification reviews [4]. The Integrated System Validation (ISV) is part of this review activity. An integrated system design is evaluated through performance-based tests to determine whether it acceptably supports safe operation of the plant [12]. NUREG-0711 and NUREG/CR-6393 provide general guidelines for the ISV. Appropriate measures are developed in consideration of the actual application environment in order to validate a real system. Many techniques for the evaluation of human performance have been developed in a variety of industrial area. The OECD Halden Reactor Project (HRP) has been conducting numerous

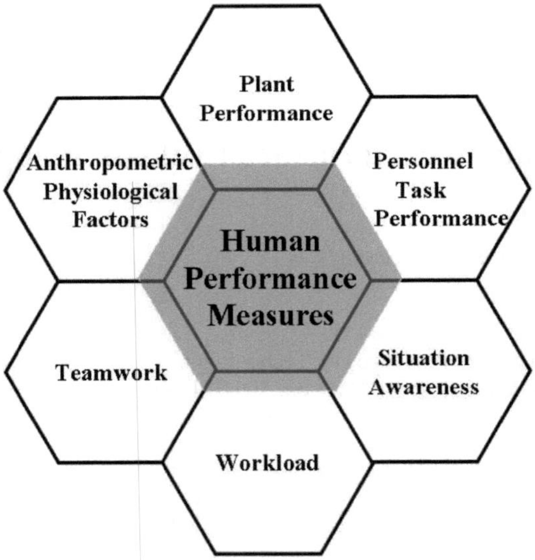

Figure 9.1. Factors for human performance evaluation

studies regarding human factors in the nuclear industry [13–18]. R&D projects concerning human performance evaluation in NPPs have also been performed in South Korea [10, 19]. These studies provide not only valuable background but also human performance measures helpful for the ISV. A computerized system based on appropriate measures and methods for the evaluation of human performance is very helpful in validating the design of ACRs.

A computerized system developed at KAIST, called HUPESS (human performance evaluation support system), is introduced in this chapter [14]. HUPESS supports evaluators and experimenters to effectively measure, evaluate, and analyze human performance. Plant performance, personnel task performance, situation awareness, workload, teamwork, and anthropometric and physiological factors are considered as factors for human performance evaluation in HUPESS (Figure 9.1).

Empirically proven measures used in various industries for the evaluation of human performance have been adopted with some modifications. This measure is called the main measure. Complementary measures are developed in order to overcome some of the limitations associated with main measures (Figure 9.1). The development of measures is based on regulatory guidelines for the ISV, such as NUREG-0711 and NUREG/CR-6393. Attention is paid to considerations and constraints for the development of measures in each of the factors, which are addressed in Section 9.1. The development of the human performance measures adopted in HUPESS is explained in Section 9.2. System configuration, including hardware and software, and methods, such as integrated measurement, evaluation, and analysis, are shown in Section 9.3. Issues related to HRA in ACRs are introduced and the role of human performance evaluation for HRA is briefly discussed in Section 9.4. Conclusions are provided in Section 9.5.

9.1 Human Performance Evaluation with HUPESS

9.1.1 Needs for the Human Performance Evaluation

The objective of the ISV is to provide evidence that the integrated system adequately supports plant personnel in the safe operation of the relevant NPP [12]. The safety of an NPP is a concept which is not directly observed but is inferred from available evidence. The evidence is obtained through a series of performance-based tests. The integrated system is considered to support plant personnel in the safe operation if the integrated system is assured to be operated within acceptable performance ranges. Operator tasks are generally performed through a series of cognitive activities such as monitoring the environment, detecting changes, understanding and assessing the situation, diagnosing the symptoms, decision-making, planning responses, and implementing the responses [5]. The HMI design of an ACR is able to support the operators in performing these cognitive activities by providing sufficient and timely data and information in an appropriate format. Effective means for system control are provided in an integrated manner. The suitability of the HMI design of an ACR is validated by evaluating human (operator) performance resulting from cognitive activities, which is effectively conducted with HUPESS.

9.1.2 Considerations and Constraints in Development of HUPESS

HUPESS is based on considerations and constraints (Figure 9.2). Plant performance, personnel task performance, situation awareness, workload, teamwork, and anthropometric and physiological factors are considered for the human performance evaluation (Figure 9.1), as recommended in regulatory guidelines. The evaluation of human performance with HUPESS provides regulatory support, when the ISV is conducted to get the operation license of an advanced NPP.

The operating environment in an ACR changes from a conventional analog-based HMI to a digitalized one. Increased automation, development of compact and computer-based workstations, and development of intelligent operator aids are three important trends in the evolution of ACRs [20]. Increased automation results in a shift of operator roles from a manual controller to a supervisor or a decision-maker. The role change is typically viewed as positive from a reliability standpoint, since unpredictable human actions are removed or reduced. The operator can better concentrate on supervising overall performance and safety of the system by automating routine, tedious, physically demanding, or difficult tasks. Inappropriate allocation of functions between automated systems and the operator may result in adverse consequences, such as poor task performance and out-of-loop control coupled with poor situation awareness [12]. The shift in the operator's role may lead to a shift from high physical to high cognitive workload, even though the overall workload is reduced. Computer-based workstation of ACRs, which has much flexibility offered by software-driven interface such as various display formats (*e.g.*, lists, tables, flow charts, graphs) and diverse soft controls (*e.g.*, touchscreen, mice, joysticks), also affects the operator performance. Information is presented in pre-processed or integrated forms rather than raw data of parameters, condensing information on a small screen. In addition, the operator has to manage the display in order to obtain data and information which he or she wants to check. Poorly designed displays may mislead and/or confuse the operator and excessively increase cognitive workload, which can lead to human errors. Operator tasks in an ACR are conducted in a different way from the conventional one due to these changes in operating environment. More attention should be paid to operator task performance and cognitive measures, such as situation awareness and workload.

The evaluation of human performance is practical and economic. Evaluation techniques are able in practice to provide technical bases in order to get an operation license since the aim of the performance evaluation is eventually to provide an effective tool for the validation of the HMI design of an ACR. The ISV is performed through a series of tests which require considerable resources (*e.g.*, time, labor, or money) from preparation to execution. Economic methods which are able to save resources are required. Measures generally used and empirically proven to be useful in various industries are adopted as main measures with some modifications. Complementary measures are developed to overcome some of the limitations associated with main measures in order to consider these constraints. Both the main measure and the complementary measure are used for evaluation of plant performance, personnel task performance, situation awareness, and workload. Teamwork and anthropometric and physiological factors are evaluated with only main measure. In addition, all the measures are developed for simultaneous evaluation without interfering with each other. For example, if simulator-freezing techniques, such as SAGAT or SACRI, are adopted for the evaluation of situation awareness, the evaluation of workload might be interfered with the simultaneous evaluation of situation awareness.

Evaluation criteria for performance measures should be clear. The criteria should, at least, be reasonable if it is not applicable to provide clear criteria. Performance measures represent only the extent of performance in relevant measures. Some scores, such as 4 or 6, represent the extent of workload induced by relevant tasks if NASA-TLX uses 7-point scale for the evaluation of workload for operator tasks in NPPs. Performance acceptability in each of the measures is evaluated on the basis of performance criteria. Approaches to establishing performance criteria are based on types of comparisons, such as requirement-referenced, benchmark-referenced, normative-referenced, and expert-judgment-referenced [12]. The requirement-referenced is a comparison of performance in the integrated system considered with an accepted and quantified performance requirement based on engineering analysis, technical specification, operating procedures, safety analysis reports, and/or design documents. Specific values in plant parameters required by technical specification and time requirements for critical operator actions are used as criteria for the requirement-referenced comparison. The other approaches are typically employed when the requirement-referenced comparison is not applicable. The benchmark-referenced is a comparison of performance in the integrated system considering a benchmark system which is predefined as acceptable under the same or equivalent conditions. A project for the ISV of a modernized NPP control room (CR) is based on the benchmark-referenced comparison [21]. The CR of the 30-year-operated NPP was renewed with modernization of the major part of the CR HMI. The human performance level in the existing CR is used as an acceptance criterion for the human performance in the modernized CR. The modernized CR is considered as acceptable if the human performance in the modernized CR is evaluated as better than or at least equal to that in the existing CR. This approach is also applicable to a totally new CR (*i.e.*, an ACR). For example, the operator workload in an ACR is compared with that in a reference CR (conventional one) which has been considered as acceptable. The normative-referenced comparison is based on norms established for performance measures through its use in many system evaluations. The performance in the integrated system is compared to the norms established under the same or equivalent conditions. The use of the Cooper–Harper scale and NASA-TLX for workload assessment are examples of this approach in the aerospace industry [12]. The expert-judgment-referenced comparison is based on the criteria established through the judgment of subject matter experts (SMEs). Measures generally used and empirically proven to be useful in various industries are adopted as main measures in order to provide clear or reasonable criteria. Attention has been paid to techniques which have been used in the nuclear industry so that the results of the studies are utilized as reference criteria. Main measures are used to determine whether the performance is acceptable or not, whereas complementary measures are used to compare and then scrutinize the performance among operators or shifts or supplement the limitation of the main measures. Human performance measures are described one by one with the performance criteria in the following section.

New technologies which are thought to be very helpful in the evaluation of the human performance are considered for the development of HUPESS. Techniques based on eye movement measurements provide a novel approach for the evaluation of human performance in ACRs. The changed environment in ACRs is coupled with several issues related to the human performance. Primary means of information input to the operator are through the visual channel in the majority of cases. An analysis of operator eye movement and fixation gives insights regarding several issues. One of the critical issues is configuration change of the HMI in ACRs. Difficulty in navigating through and finding important information fixed on a dedicated area and loss of the ability to utilize well-learned rapid eye-scanning patterns and pattern recognition from spatially fixed parameter displays in conventional control rooms are critical issues in the HMI in ACRs. Analysis of information-searching patterns of operators is a promising approach to deal with these issues. Some measurements of eye movement are effectively used for the evaluation of personnel task performance, situation awareness, and workload. Problems coupled with application of an eye-tracking system (ETS) to the study of human factors in NPPs are the intrusiveness for operator tasks and measurement quality of eye movement. An ETS of head-mounted type was cumbersome for operators to freely perform their tasks in NPPs. HUPESS is equipped with a state-of-the-art ETS using five measurement cameras (non-head-mounted type), which is not intrusive and has a high quality of measurement.

9.2 Human Performance Measures

9.2.1 Plant Performance

The principal objective of operators in an NPP CR is to operate the NPP safely. Operator performance is evaluated by observing whether the plant system is operated within an acceptable safety range which is specified by process parameters of the NPP. Operator performance measured by observing, analyzing, and then evaluating process parameters of an NPP, is referred to as plant performance. Plant performance is considered as crew performance rather than individual performance since an NPP is usually operated by a crew as a team. Plant performance is a result of operator activities, including individual tasks, cognitive activities, and teamwork. Measures of plant performance are considered as product measures, whereas other measures for personnel task performance, situation awareness, workload, teamwork, and anthropometric and physiological factors are considered as process measures. Product measures provide an assessment of results while process measures provide an assessment of how that result was achieved [17]. The achievement of safety and/or operational goals in NPPs is generally determined by values of process parameters. Values, such as setpoints, are required to assure the safety of NPPs (or the sub-systems of an NPP) in each of process parameters. Objective evaluation of plant performance is conducted because explicit data are obtainable. For example, an important goal is to maintain the pressurizer level in a LOCA, which is evaluated by examining the plant performance measure regarding the pressurizer level. However, information on how the pressurizer level is maintained in the required level is not provided by the plant performance measures. The plant performance measure in isolation may not be informative about human performance [21]. Plant performance is considered as

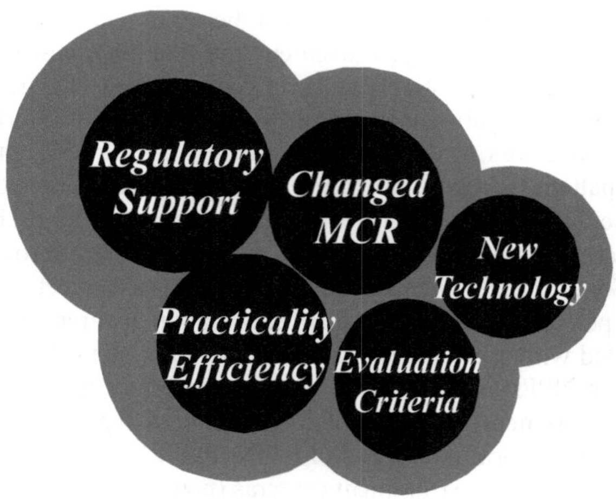

Figure 9.2. Key considerations and constraints in development of HUPESS

global performance of a crew's control or a product measure [17]. Human performance accounting for the process is evaluated by other measures for personnel task performance, situation awareness, workload, teamwork, and anthropometric and physiological factors. Another challenging case is where the plant is operated within acceptable ranges, even though design faults in human factors exist. For example, a highly experienced crew operates a plant system within acceptable range, even though the HMI is poorly designed. Plant performance is supplemented by other performance measures [12]. Attention is deliberately paid to preparation of test scenarios, selection of important process parameters, and integrated analysis with other measures in order to make plant performance more informative. Test scenarios are designed so that the effects of HMI design (*e.g.*, a new design or design upgrade) are manifested in operator performance. This is expected to improve the quality of evaluations with other performance measures. Process parameters sensitive to and representative of operator performance are selected as important process parameters. Plant performance is analyzed with other measures in an integrated manner (Section 9.3.3).

Operational achievement in important process parameters is used for evaluation of plant performance in HUPESS. Several important process parameters are selected by SMEs (process experts). Whether the values of the selected process parameters are maintained within upper and lower operational limits (within acceptable range) or not is used as a main measure for evaluation of plant performance. The discrepancy between operationally suitable values and observed values in the selected process parameters is utilized in order to score plant performance as a complementary measure. The process parameters should be within a range of values, called the target range, to achieve plant safety at the end of test scenarios. The elapsed time from an event (*e.g.*, transient or accident) to the target range in each of the selected process parameters is calculated with simulator logging data.

9.2.1.1 Main Measure: Checking Operational Limits

SMEs (process experts) select important process parameters (empirically 5 to 7) for each test scenario. Upper and lower operational limits for the safe operation of NPPs are determined by SMEs after reviewing operating procedures, technical specifications, safety analysis reports, and design documents. Whether the values of the selected parameters exceed those of the upper and lower limits or not is confirmed during validation tests. Plant performance is evaluated as acceptable if the values do not exceed the limits. The evaluation criterion of this measure is based on requirement referenced comparison. The values of parameters are obtained from logging data of a simulator.

9.2.1.2 Complementary Measure: Discrepancy Score and Elapsed Time from Event to Target Range

Discrepancies between operationally suitable values and observed values in selected process parameters are calculated during the test. This evaluation technique was applied to PPAS (Plant Performance Aassessment System) and effectively utilized for evaluation of plant performance [13, 17]. The operationally suitable value is assessed as a range and not a point value by SMEs, because of difficulty in assessing the operationally suitable value as a specific point value. The range value represents acceptable performance expected for a specific scenario (*e.g.*, LOCA or transient scenario). The assessment of an operationally suitable value is based on operating procedures, technical specifications, safety analysis reports, and design documents. The discrepancy is used for the calculation of the complementary measure if the value of a process parameter is beyond the range (*e.g.*, upper bound) or under the range (*e.g.*, lower bound). The discrepancy in each parameter is obtained as:

$$D_{d,i}(t) = \begin{cases} \dfrac{X_i(t) - S_{U,i}}{M_i}, & \text{if } X_i(t) > S_{U,i} \\ 0, & \text{if } S_{L,i} \le X_i(t) \le S_{U,i} \\ \dfrac{S_{L,i} - X_i(t)}{M_i}, & \text{if } X_i(t) < S_{L,i} \end{cases} \tag{9.1}$$

where :

$D_{d,i}(t)$ = discrepancy of parameter i at time t during the test

$X_i(t)$ = value of the parameter i at time t during the test

$S_{U,i}$ = upper bound value of the operationally suitable value

$S_{L,i}$ = lower bound value of the operationally suitable value

M_i = mean value of the parameter i during initial steady-state

t = simulation time after an event occurs

The discrepancy between observed and operationally suitable values in each parameter is normalized by dividing with the mean value of parameter i obtained during the initial steady state. All discrepancy in parameters is eventually integrated into a measure, giving a kind of total discrepancy. The normalized discrepancy of parameter i is summed up over test time T:

$$D_{d,i}^{avg} = \frac{\sum_{t=1}^{T} D_{d,i}(t)}{T} \tag{9.2}$$

where:

$D_{d,i}^{avg}$ = averaged sum of the normalized discrepancy of parameter i over the test time, T.

The next step is to obtain weights for selected process parameters. The analytic hierarchy process (AHP) is used to evaluate the weights. The AHP is useful in hierarchically structuring a decision problem and quantitatively obtaining weighting values. The AHP serves as a framework to structure complex decision problems and provide judgments based on expert knowledge and experience to derive a set of weighting values by using pair-wise comparisons [22]. The averaged sums of parameters are multiplied by weights of relevant parameters and multiplied values are summed as:

$$D_d = \sum_{i=1}^{N} \left(\omega_i \times D_{d,i}^{avg} \right) \tag{9.3}$$

where:

D_d = total discrepancy during the test

N = total number of the selected parameters

ω_i = weighing value of parameter i

Another measure of discrepancy is calculated at the end of the test; this represents the ability of a crew to complete an operational goal:

$$D_{e,i} = \begin{cases} \dfrac{X_i - S_{U,i}}{M_i}, & \text{if } X_i > S_{U,i} \\ 0, & \text{if } S_{L,i} \le X_i \le S_{U,i} \\ \dfrac{S_{L,i} - X_i}{M_i}, & \text{if } X_i < S_{L,i} \end{cases} \tag{9.4}$$

where:

$D_{e,i}$ = discrepancy of parameter i at the end of the test

X_i = value of parameter i at the end of the test

$S_{U,i}$ = upper bound value of the operationally suitable value

$S_{L,i}$ = lower bound value of the operationally suitable value

M_i = mean value of the parameter i during initial steady - state

The normalized discrepancy of parameter i is multiplied by the weight of each parameter i and the multiplied values are summed as:

$$D_e = \sum_{i=1}^{N} \left(\omega_i \times D_{e,i} \right) \tag{9.5}$$

where:

D_e = total discrepancy at the end of the test

A low total discrepancy means better plant performance. The total discrepancy is used for comparing performance among crews or test scenarios rather than for determining if it is acceptable or not.

The elapsed time from an event to the target range in each of the selected process parameters is based on the fact that a shorter time spent in accomplishing a task goal represents good performance. The elapsed time is calculated at the end of a test. The time to parameter stabilization is used as a measure of fluctuation in a parameter. The evaluation criteria of these measures are based on both requirement-referenced and expert-judgment-referenced comparisons.

9.2.2 Personnel Task Performance

Design faults result in unnecessary work being placed on operators, even though plant performance is maintained within acceptable ranges. Personnel task measures provide complementary data to plant performance measures. Personnel task measures reveal potential human performance problems, which are not found in the evaluation of plant performance [12]. Personnel tasks in the control room are summarized as a series of cognitive activities. The operator task is evaluated by observing whether relevant information about the situation is monitored or detected, whether correct responses are performed, and whether the sequence of operator activities is appropriate [18].

9.2.2.1 Main Measure: Confirming Indispensable Tasks and Completion Time

Whether the cognitive activities are performed correctly or not is evaluated by observing a series of tasks. Some elements of cognitive activities are observable, even though others are not observable, but inferable. Activities related to detection or monitoring and execution are considered as observable activities, whereas other cognitive activities are inferred from the observable activities [23]. Personnel task performance is evaluated by observing whether operators monitor and detect appropriate data and information, whether they perform appropriate responses, and whether the sequence of processes is appropriate. Primary task and secondary task are evaluated for the personnel task evaluation. A test scenario for the ISV is hierarchically analyzed and then an optimal solution of the scenario is developed for an analytic and logical measurement (Figure 9.3). The operating procedure provides a guide for the development of an optimal solution, since an operator task in NPPs is generally based on a goal-oriented procedure. The main goal refers to a goal to be accomplished in a scenario. The main goal is located at the highest rank and is divided into sub-goals; sub-goals are also divided, if needed. Detections, operations, and sequences are used to achieve the relevant sub-goal in the next rank. Detections and operations break down into detailed tasks to achieve the relevant detections and operations, respectively. Tasks located in the bottom rank comprise crew tasks required for completion of the main goal (Figure 9.3). Top-down and bottom-up approaches are utilized for the development of an optimal solution. Indispensable tasks required for safe NPP operation are determined by SMEs. SMEs observe operator activities, collect data, such as operator speech, behavior, cognitive process, and logging data, and then evaluate whether the tasks located in the bottom rank are appropriately performed or not during the test. Personnel task

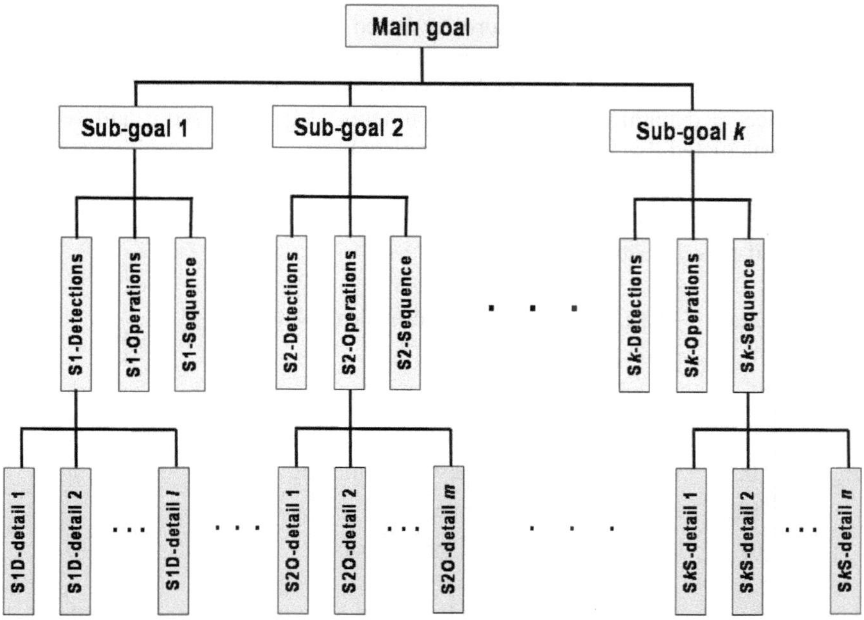

Figure 9.3. Optimal solution of a scenario in hierarchical form

performance is considered as acceptable if all indispensable tasks are satisfied. The evaluation criterion of this measure is based on both requirement referenced and expert-judgment-referenced comparisons. Operators may implement tasks in different ways from the optimal solution according to their strategy, which is not considered by the SMEs in advance. SMEs check and record operator activities during the test in this case. Some parts of the optimal solution are revised based on observed activities after the test. Task performance is re-evaluated with the revised solution and collected data.

Task completion time is also evaluated. Time to complete each of the tasks located in the bottom rank is evaluated based on experience and expertise of SMEs. Summation of evaluated times is interpreted as a required time to complete a goal. The completion time of the personnel task is considered as acceptable if the real time spent for the completion of a goal in a test is less than or equal to the required time.

9.2.2.2 Complementary Measure: Scoring Task Performance

The main measure, a kind of descriptive measure, is complemented by scoring the task performance, which is used for analyzing and comparing performance among crews or test scenarios. The weights of the elements in the optimal solution are calculated using the AHP. Operator activities are observed and evaluated during a test. Whether the respective tasks are satisfied in an appropriate sequence is evaluated by SMEs. Task performance is scored with observed and evaluated data and weights of the tasks. A higher score means a higher task performance. The evaluation criterion of this measure is based on expert-judgment referenced comparison. This technique was used in OPAS (Operator Performance Assessment System) and reported to be a reliable, valid, and sensitive indicator of human performance in dynamic operating environments [18]. Task score and sequence score are calculated. Each task score is calculated as:

$$T_j = \begin{cases} 0, & \text{if task-}j\text{ is not satisfied} \\ 1, & \text{if task-}j\text{ is satisfied} \end{cases} \tag{9.6}$$

where, T_j = task-*j* score. Each sequence score is calculated, as:

$$SEQ_k = \begin{cases} 1, & \text{if sequence - }k\text{ is very appropriate} \\ 0.75, & \text{if sequence - }k\text{ is appropriate} \\ 0.5, & \text{if sequence - }k\text{ is somewhat confusing} \\ 0.25, & \text{if sequence - }k\text{ is inappropriate} \\ 0, & \text{if sequence - }k\text{ is very inappropriate} \end{cases} \tag{9.7}$$

where SEQ_k = sequence-*k* score. Finally, personnel task score is calculated by summing up the weighted task scores and sequence scores:

$$S_{PT} = \sum_{j=1}^{M} (\omega_j \times T_j) + \sum_{k=1}^{L} (\omega_k \times SEQ_k) \tag{9.8}$$

where:

S_{PT} = the personnel task score

M = total number of the tasks in the bottom rank

L = total number of the sequences considered

ω_j = weighting value of task-*j*

ω_k = weighting value of sequence-*k*

9.2.3 Situation Awareness (SA)

Operator actions are always based on identification of the operational state of the system in NPPs. Incorrect SA contributes to the propagation or occurrence of accidents, as shown in the TMI-2 accident [24]. SA is frequently considered as a crucial key to improve performance and reduce error [25–27]. Definitions of SA have been discussed [28–31]. An influential perspective of SA has been put forth by Endsley, who notes that SA concerns “knowing what is going on” [31]. SA is defined more precisely as “situation awareness is the perception of the elements in the environment within a volume of time and space, the comprehension of their meaning, and the projection of their status in the near future” [31]. Operator tasks are significantly influenced by operator SA. Tasks in NPPs are summarized as a series of cognitive activities, such as monitoring, detecting, understanding, diagnosing, decision-making, planning, and implementation. Correct SA is one of the most critical contributions to safe operation in NPPs. The ACR in APR-1400 adopts new technologies, such as CRT- (or LCD-) based displays, LDP, soft-controls, a CPS, and an advanced alarm system. The changed operational environment can deteriorate SA of operators, even though operators are expected to be more aware of situations with new technologies. Difficulty in navigating and finding important information through computerized systems, loss of operator vigilance due to automated systems, and loss of the ability to utilize well-learned and rapid eye-scanning patterns and pattern recognition from spatially fixed parameter displays are considered as potential challenges [20]. A new ACR design is validated throughout ISV tests; that is, performance-based tests.

Measurement techniques which were developed for SA measurement can be categorized into four groups, such as performance-based, direct query and questionnaire, subjective rating, and physiological measurement techniques [12, 27]. Performance-based techniques are not suitable for ISV tests because they have both logical ambiguities in their interpretation and practical problems in their administration [12]. Direct query and questionnaire techniques are categorized into post-test, on-line-test, and freeze techniques according to the evaluation point over time [32]. These techniques are based on questions and answers regarding the SA. The post-test technique takes up much time to complete detailed questions and answers, which can lead to incorrect memory problems of operators. The operator has a tendency to overgeneralize or rationalize their answers [33]. The on-line-test techniques require questions and answers during the test to overcome the memory problem. Questions and answers are considered as another task, which may distort operator performance [12]. Freeze techniques require questions and answers by randomly freezing the simulation to overcome the demerits of the post-test and on-line-test techniques. A representative technique is SAGAT (Situation Awareness Global Assessment Technique) which has been employed across a wide range of dynamic tasks, including air traffic control, driving, and NPP control [34]. The SAGAT has advantages of being easy to use (in a simulator environment), possessing good external indices of information accuracy, and possessing well-accepted face validity [35]. A criticism of the SAGAT has been that periodic interruptions are too intrusive, contaminating any performance measures, which is related to the concern that the questions may cue participants (*e.g.*, operators) to some details of the scenario, setting up an expectancy for certain types of questions [12, 36, 37]. Performance measures (*e.g.*, kills and losses in an air-to-air fighter sweep mission) are not significantly affected by conditions of simulation freeze or non-freeze [34], question point in time, question duration, and question frequency for the SAGAT measurement [38, 39]. It is impossible to "prove" that SAGAT does not influence performance, even though all the studies indicate that it does not appear to significantly influence performance as long as the stops (or freezes) are unpredictable to subjects [34]. The SACRI, which was adapted after SAGAT for use in NPPs, have been studied [40, 41]. The SACRI has been developed for use in the NORS simulator in HRP. Subjective rating techniques typically involve assigning a numerical value to the quality of SA during a particular period of event

[42]. Subjective rating techniques are popular because they are fairly inexpensive, easy to administer, and non-intrusive [35, 42]. However, there have been criticisms. Participants (or operators) knowledge may not be correct and the reality of the situation may be quite different from what they believe [43]. SA may be highly influenced by self-assessments of performance [35]. Operators may rationalize or overgeneralize about their SA [43]. In addition, some measures such as SART and SA-SWORD include workload factors rather than limiting the techniques to SA measurement itself [12]. Physiological measurement techniques have been used to study complex cognitive domains, such as mental workload and fatigue. Very few experiments have been conducted to study SA [44]. Physiological measures have unique properties considered attractive to researchers in the SA field, even though the high cost of collecting, analyzing, and interpreting the measures is required, compared with the subjective rating and performance-based measurement techniques. Intrusive interference such as freezing the simulation is not required. Continuous measurement of the SA can be provided. It is possible to go back and assess the situation, because physiological data are continuously recorded. Eye fixation measurement called VISA has been used as an indicator for SA in the nuclear industry [16]. Time spent on eye fixation has been proposed as a visual indicator of SA in an experimental study of VISA. The results of the VISA study showed that SACRI scores correlated with VISA, which was somewhat inconsistent between two experiments in the study. Physiological techniques are expected to provide potentially helpful and useful indicators regarding SA, even though these techniques cannot clearly provide how much information is retained in memory, whether the information is registered correctly, or what comprehension the subject has of those elements [33, 44].

A subjective rating measure is used as the main measure for SA evaluation in HUPESS, even though it has the drawbacks mentioned above. Eye fixation measurement is also used as complementary measures.

9.2.3.1 Main Measure: KSAX

KSAX [10] is a subjective ratings technique adapted from the SART [45]. Operators subjectively assess their own SA on a rating scale and provide the description or the reason why they give the rating after completion of a test. One of the crucial problems in the use of SART was that workload factors are not separated from SA evaluation. Endsley's SA model has been applied to the evaluation regime of SART in KSAX. KSAX has been successfully utilized in the evaluation of HMI design in the ACR of APR1400 [10]. Operators are not inconvenienced by evaluation activities since SA is evaluated based on a questionnaire after a test. The evaluations of the other performance measures, especially cognitive workload, are not affected by the evaluation of SA, which leads to economic evaluation of human performance for the ISV. All measures considered in HUPESS are evaluated in one test. KSAX results from an antecedent study for APR-1400 [10] are utilized as a criterion based on the benchmark-referenced comparison, which is considered as an important merit [12]. The KSAX questionnaire consists of several questions regarding level 1, 2, and 3 SAs defined by Endsley. Usually, a 7-point scale is used for the measurement. The rating scale is not fixed but the use of a 7-point scale is recommended, because the antecedent study used a 7-point scale. Questions used in KSAX are asked such that SA in an advanced NPP is compared with that of already licensed NPPs. Operators who have been working in licensed NPPs are selected as participants for validation tests. The result of the SA evaluation is considered as acceptable if the result of SA evaluation in an advanced NPP is evaluated as better than or equal to that in the licensed NPP. The evaluation criterion of this measure is based on the benchmark-referenced comparison.

9.2.3.2 Complementary Measure: Continuous Measure Based on Eye Fixation Measurement

The subjective measure of SA is complemented by a continuous measure based on eye fixation data, which is a kind of physiological measurement. It is not possible to continuously measure operator SA and secure objectivity since KSAX is subjectively evaluated after a test. A physiological method generally involves the measurement and data processing of one or more variables related to human physiological processes. Physiological measures are known as being objective and providing continuous information on activities of subjects. Eye-tracking systems which have the capability to measure subject eye movement without direct contact have been developed. The measurement of eye movement is not intrusive for subject activities. The primary means of information input to the operator are through the visual channel in the majority of cases. An analysis of the manner in which operator eyes move and fixate gives an indication of information input. The analysis of the eye movement is used as a complementary indicator for the SA evaluation, even though it cannot exactly tell the operator SA. There are many information sources to be monitored in NPPs. However, operators have only limited capacity of attention and memory. Operators continuously decide where to allocate their attentional resources, because it is impossible to monitor all information sources. This kind of cognitive skill is called selective attention. Operators use this cognitive skill to overcome the limitations of human attention. The stages of information processing depend on mental or cognitive resources, a sort of pool of attention or mental effort that is of limited availability and is allocated to processes as required [46]. Operators try to understand what is going on in the NPP when an abnormal situation occurs. Operators receive information from the environment (*e.g.*, indicators or other operators) and process the information to establish a situation model based on their mental model. A situation model is an operator understanding of the specific situation. The model is constantly updated as new information is received [47]. The mental model refers to general knowledge governing the performance of highly experienced operators. The mental model includes expectancies on how NPPs will behave in abnormal situations. For example, when a LOCA occurs, the pressurizer pressure, temperature, and level will decrease, and the containment radiation will increase. These expectancies form rules for the dynamics of NPPs. The mental model is based on these rules [48]. Operators usually first recognize an abnormal or accident situation by the onset of salience, such as alarm or deviation in process parameters from the normal condition. Then, they develop situation awareness or establish their situation model by selectively paying attention to important information sources. Maintenance of situation awareness or confirmation of their situation model is accomplished by iterating selective attention. The operator allocates their attentional resources not only to the salient information sources but also to valuable information sources in order to effectively monitor, detect, and understand the state of a system. Eye fixation on area of interest (AOI), which is important in solving problems, is considered as an index of monitoring and detection, which then can be interpreted by perception of the elements (level 1 SA). An action is delayed or not executed at all, as perceived information is thought about or manipulated in working memory [46]. Time spent on the AOIs by operators is understood as an index for comprehension of their meaning (level 2 SA). Selective attention is associated with the expectancy of the near future. The projection of their status in the near future (level 3 SA) is inferred from the sequence of eye fixations.

Eye fixation on AOIs, time spent on AOIs, and the sequence of fixations are used for SA evaluation in HUPESS. SMEs analyze eye fixation data after the completion of a test. The analysis is recommended for specific periods representing the task steps in the optimal solution of personnel task performance. The times spent for achieving sub-goals in the optimal solution are used as specific periods for the analysis. Attention is paid to finding out deficiencies of HMI design or operator incompetence leading to inappropriate ways of eye fixation. SMEs analyze eye fixation data and evaluate the SA as one of three grades, excellent, appropriate, or not appropriate, for each of the periods.

The evaluation criterion of this measure is based on expert-judgment-referenced comparisons. This technique has the drawback that eye fixation data are analyzed by SMEs, which requires more effort and time. SMEs can provide meaningful evaluation from the eye fixation data, because SMEs have usually most knowledge and experience about the system and the operation. An experimental study has been performed with a simplified NPP simulator [49]. Eye fixation data during complex diagnostic tasks were analyzed in the experiments. The results showed that eye fixation patterns of subjects with high, medium, or low expertise were different in the same operation conditions. A high-expertise subject fixated various information sources with a shorter fixation time. Important information sources were iteratively fixated and the situation was reported with high confidence at the end of simulation. On the other hand, a low-expertise subject spent more time on salient information sources. Various information sources important to solving the problem were not fixated and the situation was reported with low confidence (it seemed just a guess). A computerized system in HUPESS for the eye fixation analysis facilitates the SA evaluation (Figure 9.4). The number centered in the circle represents the order of the fixation. The area of the circle is proportional to the fixation time.

9.2.4 Workload

Workload has an important relationship to human performance and error [12]. A generally accepted definition of cognitive workload is not available, despite its importance [50–52]. One definition of the workload is "the portion of operator's limited capacity actually required to perform a particular task" [53]. More mental resources are required as cognitive workload is increased. Human errors may occur

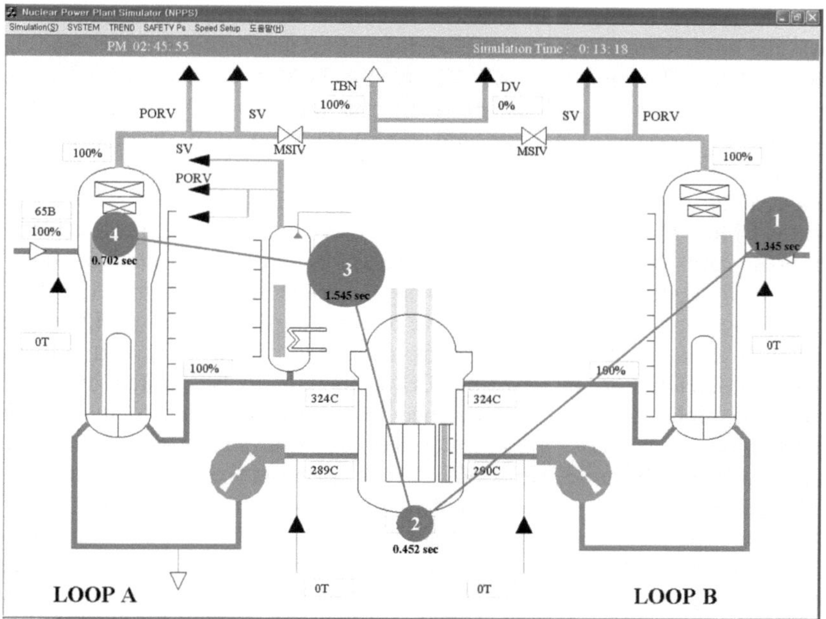

Figure 9.4. A computerized system for the eye fixation analysis

causing deterioration of human performance if the cognitive workload exceeds the limit of operator capacity [54]. Advanced information technologies are applied to ACRs. Operators are required to play the role of supervisor or decision-maker rather than manual controller. Operator tasks are expected to require increased mental activities rather than physical activities. The evaluation of cognitive workload has been considered as one of the most important factors to be evaluated for the ISV. Techniques for measuring cognitive workload are divided into two broad types: predictive and empirical [12]. Predictive techniques are usually based on mathematical modeling, task analysis, simulation modeling, and expert opinions. These techniques do not require operators to participate in simulation exercises. They are typically used in the early stages of design process and thought not to be suitable for the ISV stage [12]. Empirical techniques are divided into performance-based, subjective ratings, and physiological measures [55]. Performance-based techniques are categorized into primary task measures and secondary task measures. Primary task measures are not suitable for the measurement of cognitive workload associated with monitoring or decision-making tasks like in NPPs. Secondary task measures have the drawback that the measurement itself contaminates human performance by interfering with primary tasks [46]. Subjective ratings techniques measure the cognitive workload experienced by a subject (or an operator) through a questionnaire and an interview. Subjective measures have been most frequently used in a variety of domains since they have been found to be reliable, sensitive to changes in workload level, minimally intrusive, diagnostic, easy to administer, independent of tasks (or relevant to a wide variety of tasks) and possessive of a high degree of operator acceptance [56–62]. There are representative subjective measures, such as overall workload (OW), modified Cooper–Harper scale (MCH), subjective workload assessment technique (SWAT), and National Aeronautic and Space Administration task load index (NASA-TLX). The models of SWAT, NASA-TLX, OW, and MCH have been verified by examining the reliability of the methods [62]. NASA-TLX was evaluated as superior in validity and NASA-TLX and OW were evaluated as superior in usability [62]. Physiological techniques measure the physiological change of the autonomic or central nervous systems associated with cognitive workload [46]. Electroencephalogram (EEG), evoked potential, heart-rate-related measures, and eye-movement-related measures are representative tools for cognitive workload evaluation based on physiological measurements [63, 64].

EEG measures have proven sensitive to variations of mental workload during tasks such as in-flight mission [65, 66], air traffic control [67], and automobile driving [68]. However, the use of EEG is thought to be limited for the ISV, because multiple electrodes are attached to an operator head to measure EEG signals, which may restrict the operator activities and thus interfere with operator performance in dynamic situations. Wave patterns regarding latencies and amplitudes of each peak in evoked potential (EP) or event-related-potential (ERP) analysis are analyzed after providing specific stimulations. The EP is thought not to be applicable to the study of complex cognitive activities in the ISV, because events evoking the EP are simple and iterated many times [69]. Measures of heart rate (HR) and heart rate variability (HRV) have proven sensitive to variations in the difficulty of tasks such as flight maneuvers and phases of flight (*e.g.*, straight and level, takeoffs, landings) [70, 71], automobile driving [68], air traffic control [67], and electroenergy process control [72]. However, the heart-rate-related measures do not always produce the same pattern of effects with regard to their sensitivity to mental workload and task difficulty since they are likely to be influenced by the physical or psychological state of a subject [73–75]. Eye-movement-related measures are generally based on blinking, fixation, and pupillary response. Many studies suggest that eye-movement-related measures are effective tools for the evaluation of cognitive workload [76–80]. Cumbersome equipment, such as a head-mounted ETS, is used to obtain eye movement data and is thought to be intrusive to the operator tasks. Recently, ETSs which can measure eye movement data without direct contact (non-intrusively) have been developed [81, 82].

NASA-TLX, a widely used subjective ratings technique, is used as the main measure for evaluation of cognitive workload. Blinking- and fixation-related measures are used as the complementary measures in HUPESS.

9.2.4.1 Main Measure: NASA-TLX

A subjective measure is considered an indicator related to participant internal experience. Subjective rating techniques have been widely used for the evaluation of workload in various fields. NASA-TLX has been extensively used in multi-task contexts, such as real and simulated flight tasks [83–86], air combat [87, 88], remote control of vehicles [89] and simulator-based NPP operation [10, 15, 16, 80, 90, 91]. NASA-TLX is a recommended instrument for assessing cognitive workload by U.S. NRC [92]. An important merit is that NASA-TLX results from antecedent studies for APR-1400 [10, 91] are utilized as reference criteria for the

ISV. NASA-TLX divides the workload experience into the six components: mental demand, physical demand, temporal demand, performance, effort, and frustration [93]. Operators subjectively assess their own workload on a rating scale and provide the description or reason why they give a rating after completion of a test. In HUPESS, the six questions used in NASA-TLX are made such that workload in an advanced NPP is compared with that in already licensed NPPs. The result of NASA-TLX evaluation is considered as acceptable if the result of NASA-TLX in an advanced NPP is evaluated as lower than or equal to that in the licensed NPP. A 7-point scale is used for the measurement. The rating scale is not fixed but the use of a 7-point scale is recommended, because the antecedent studies used a 7-point scale. The evaluation criterion of this measure is based on the benchmark referenced comparison.

9.2.4.2 Complementary Measure: Continuous Measures Based on Eye Movement Measurement

The subjective measure of cognitive workload is complemented by continuous measures based on eye movement data in a similar way as the evaluation of SA. It is not possible to continuously measure operator workload and secure objectivity, because NASA-TLX is evaluated subjectively after the completion of a test. Continuous measures based on eye movement data are utilized as complementary measures for the evaluation of cognitive workload. Blink rate, blink duration, number of fixations, and fixation dwell time are used as indices representing the cognitive workload. Blinking refers to a complete or partial closure of the eye. A reduced blink rate helps to maintain continuous visual input since visual input is disabled during eye closure. The duration and number of eye blinks decreases when the cognitive demands of the task increase [77, 78]. A recent study showed that blink rates and duration during the diagnostic tasks in simulated NPP operation correlated with NASA-TLX and MCH scores [80]. Some studies have shown that a higher level of arousal or attention increased the blink rate [94, 95]. Considering that the operator tasks in NPPs are a series of cognitive activities, the increased blink rate is used as a clue indicating the point where a higher level of concentration or attention is required. Eye fixation parameters include the number of fixations on AOI and the duration of the fixation, also called dwell time. The more eye fixations are made for problem-solving, the more information processing is required. A longer fixation duration means that more time is required to correctly understand the relevant situation or object. The number of fixations and fixation dwell time are increased if an operator experiences a higher cognitive workload. The number of fixations and fixation dwell time were found to be sensitive to the measurement of mental workload [76, 96]. The dwell time serves as an index of the resources required for information extraction from a single source [46]. Dwells are largest on the most information-rich flight instrument, and dwells are much longer for novice than for expert pilots, reflecting the novice greater workload [97]. The low-expertise subject spends more time for fixation on a single component than the high-expertise subject during complex diagnostic tasks in simulated NPP operations [49]. Eye fixation pattern (or visual scanning) is used as a diagnostic index of the workload source within a multi-element display environment [46]. More frequent and extended dwells were made for fixation on more important instruments during diagnostic tasks [49]. Long novice dwells were coupled with more frequent visits and served as a major "sink" for visual attention [44]. Little time was left for novices to monitor other instruments, and as a result, their performance declined on tasks using those other instruments. Eye fixation parameters are effectively used for evaluating the strategic aspects of resource allocation. The evaluation of these measures is performed by SMEs to find valuable aspects. These measures are based on expert-judgment-referenced comparison. The eye-movement-related measures, such as blink rate, blink duration, number of fixations, and fixation dwell time correlate with NASA-TLX and MCH scores in an experimental study with an NPP simulator [80]. Continuous measures based on eye movement data are very useful tools for complementing the subjective rating measure.

9.2.5 Teamwork

An NPP is operated by a crew (not an individual operator). There are individual tasks which are performed by individual operators and there are some tasks which require the cooperation of the crew. The cooperative tasks are appropriately divided and allocated to operators to achieve an operational goal. The ACR of APR-1400 is equipped with LDP designed to support team performance by providing a common reference display for discussions. The ACR design also allows operators to be located nearer to one another than conventional CRs and to access the plant information from workstations allocated to operators for exclusive use. These interface changes are expected to improve operator performance by facilitating verbal and visual communication among operators [90, 98] and thus improve team work. BARS (behaviorally anchored rating scale) is used for the evaluation of team work in HUPESS [90]. BARS includes evaluation components, such as task focus/decision-making, coordination as a crew, communication, openness, and team spirit. Several example behaviors (positive or negative) and anchors (or critical behaviors), indicating good/bad team interactions in each of the components, are identified by SMEs during a test. Identified example behaviors and anchors are used as criteria for final (or overall) rating of team work by the SMEs after the test. Usually, a 7-point scale (1–7) is used for BARS ratings with 7 being the best team interaction. The rating scale is not fixed. The use of a 7-point scale is recommended, because BARS results with a 7-point scale from an antecedent study for APR-1400 [10] are utilized as reference criteria. Attention is focused on findings which influence team work. SMEs determine whether the teamwork is acceptable or not based on experience and knowledge. Thus, evaluation criterion of this measure is based on expert-judgment-referenced comparisons.

9.2.6 Anthropometric and Physiological Factors

Anthropometric and physiological factors include visibility and audibility of indication, accessibility of control devices to operator reach and manipulation, and the design and arrangement of equipment [12]. Generally many of the concerns are evaluated earlier in the design process with HFE V&V checklists. Attention is focused on those anthropometric and physiological factors that can only be addressed in real or almost real (simulation with high fidelity) operating conditions (*e.g.*, the ability of the operators to effectively use or manipulate various controls, displays, workstations, or consoles in an integrated manner), since the ISV is a kind of feedback step for design validation and improvement [12]. Items related to these factors in HFE V&V checklists are selected before the validation test and then reconfirmed during the validation test by SMEs. Attention is paid to anthropometric and physiological problems caused by unexpected design faults. The evaluation criterion of this measure is based on both requirement-referenced (HFE V&V checklist) and expert-judgment-referenced comparisons.

9.3 Human Performance Evaluation Support System (HUPESS)

9.3.1 Introduction

The measures for plant performance, personnel task performance, situation awareness, workload, team work, and anthropometric/physiological factors are used for human performance evaluation. Human performance evaluation is required throughout the development process of a system. Experiment-based evaluation, called ISV, is generally required in the validation step of the design of an integrated system. HUPESS has been developed for the ISV of ACRs, specifically for the ISV of the ACR of APR-1400. HUPESS consists of hardware systems and software systems. HUPESS supports evaluators (or experimenters) to effectively measure and analyze human performance for the ISV. Human performance measures considered in HUPESS provide multi-lateral information for the ISV. The evaluation of human performance is performed in an integrated manner to produce results with the most information. HUPESS was designed to provide an integrated analysis environment.

9.3.2 Configuration of HUPESS

9.3.2.1 Hardware Configuration

The hardware system of HUPESS consists of two HUPESS Servers (HSs), two stationary evaluation stations (SESs), four mobile evaluation stations (MESs), a set of multi-channel AV (audio/video) systems, an ETS with five measurement cameras, a high-capacity storage station (HCSS), and a TCP/IP network system (Figure 9.5). The HSs are connected to the simulator of APR-1400 (another type of simulator can be connected to HUPESS). The primary HS between the two HSs is ordinarily used. The secondary HS is used in the case of failure of the primary HS. Data representing the plant state (*e.g.*, process parameters and alarms) and control activities performed by operators are logged in the simulator. The logging data are then transferred to HUPESS through the HS for evaluation of human performance. The SESs are fixed to the evaluator desk, whereas the MESs are used at any place where a wireless network service is available. The evaluator moves around to observe operator activities or HMI displays and immediately evaluate the observed items with the MES. The AV system provides sounds and scenes to the evaluator which cannot be heard and seen at the evaluator desk. The AV system also records the sounds and the scenes regarding the operation. All the activities related to the operation may not be observed and evaluated by the evaluator during a test. Activities which were missed or not processed by the evaluator during a test are evaluated with the recorded AV data after the test. The ETS measures eye movement of a moving operator on a wheeled chair with five measurement cameras (Figure 9.6). Coverage of eye movement measurement is about 2 meters from right-hand side to left-hand side. All the data and information related to the evaluation of human performance and the plant system are stored in the HCSS.

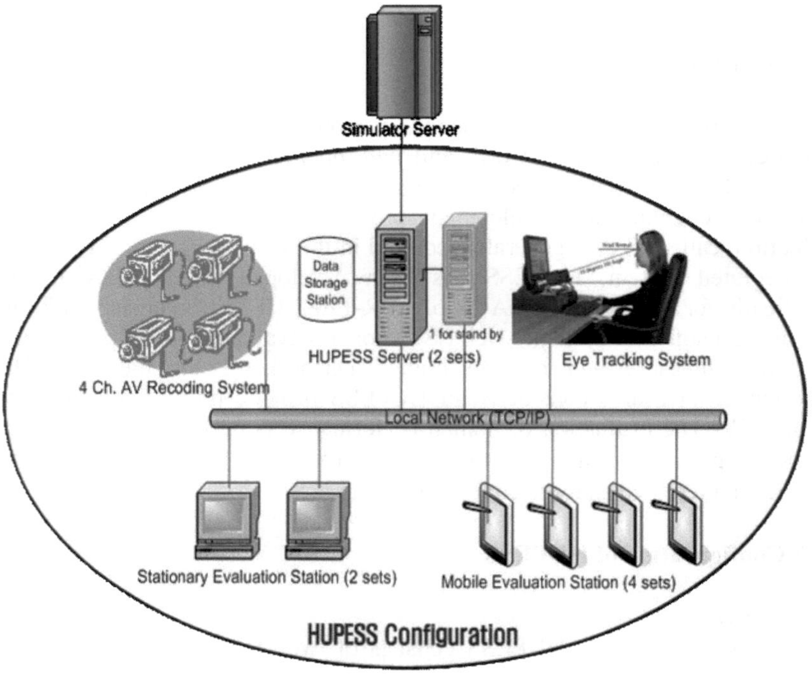

Figure 9.5. HUEPSS H/W configuration

Figure 9.6. Eye-tracking system with five measurement cameras

9.3.2.2 Software Configuration

The software system installed in HUPESS includes HS application software, SES application software, MES application software, and COTS (commercial-off-the-shelf) application software such as ETS application software and AV system application software (Figure 9.7).

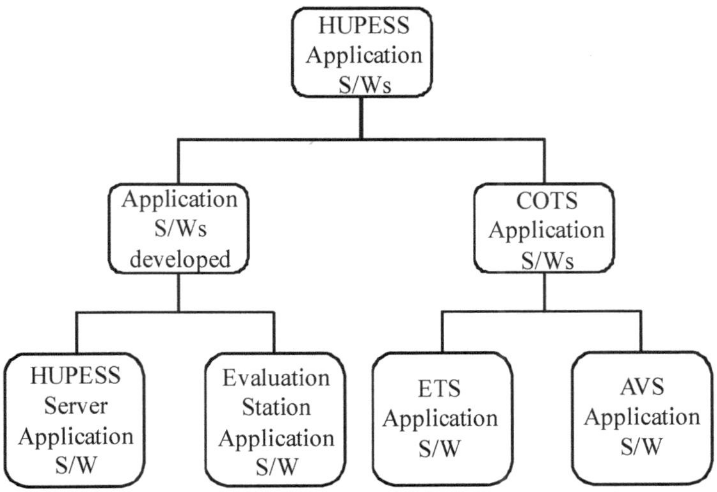

Figure 9.7. HUPESS software configuration

9.3.3 Integrated Measurement, Evaluation, and Analysis with HUPESS

The evaluation of human performance is conducted in a step-by-step process (Figure 9.8). Various test scenarios representing a wide spectrum of plant operating conditions are generally used for the ISV. Each test scenario is analyzed by SMEs in the scenario analysis step. Important process parameters are selected and then weighted for the evaluation of plant performance. The weighting values of the important parameters are calculated with the AHP. The optimal solution is developed for evaluation of personnel task performance. The tasks to be performed by operators are also weighted with the AHP. All the information and settings performed in this step are stored in a computer file. This procedure is computerized in HUPESS for later convenient use.

Information and settings regarding the evaluation are managed in the experimental management step. Information about evaluators and operators who will participate in tests is inputted, stored, retrieved, and revised according to user requests. Measures for evaluation of human performance are selected based on the purpose of each test. The computer file generated in the scenario analysis step is loaded for evaluation with HUPESS. Options related to ETS and KSAX, NASA-TLX, and BARS questionnaires are set up. Preparation for evaluation of human performance is done when the scenario analysis step and experiment management step are completed.

Measures for the evaluation of human performance are evaluated in real-time and post-test steps (Figure 9.9). The times of operator activity are recorded in order to effectively evaluate human performance. Operator activity includes bottom-rank tasks considered in the evaluation of personnel task performance, example behaviors and critical behaviors in teamwork evaluation, and activities belonging to anthropometric and physiological factors. Time-tagging is easily conducted with

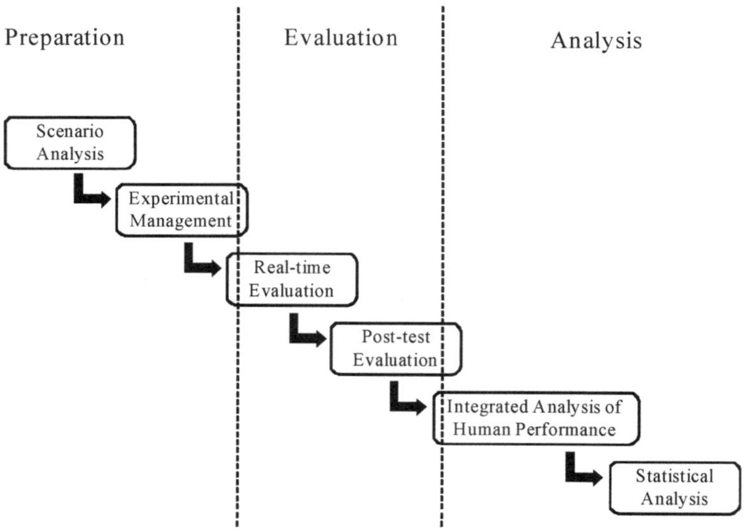

Figure 9.8. Evaluation procedure with HUEPSS

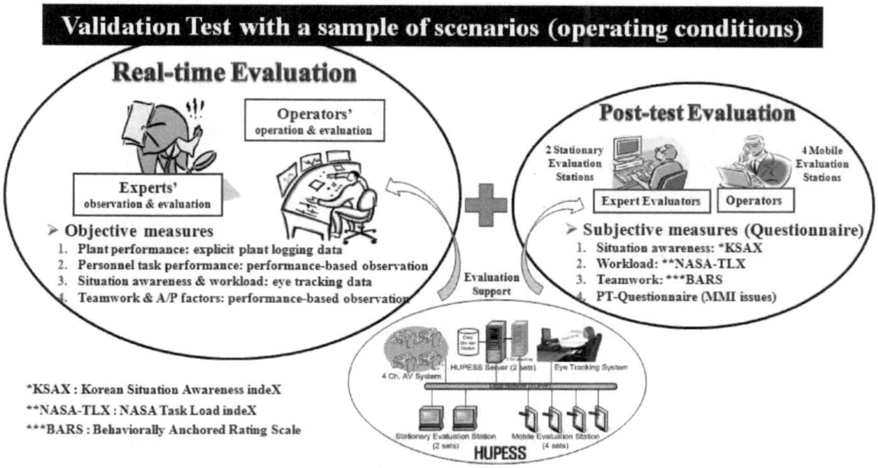

Figure 9.9. Overall scheme for the evaluation with HUEPSS

HUPESS. All that SMEs (as evaluators) have to do are to check items listed in HUPESS based on their observation. HUPESS records automatically the checked items and the relevant times. Time-tagged information facilitates the integrated evaluation of human performance in the analysis steps. Plant performance is connected to personnel task performance by time-tagged information. HUPESS is connected to a simulator of the plant system to acquire logging data representing the plant state (*e.g.*, process parameters and alarms) and control activities performed by operators. Process parameters are observed and evaluated to determine how the plant system is operating. Design faults or shortcomings may require unnecessary work or an inappropriate manner of operation, even though plant performance is maintained within acceptable ranges. This problem is solved by analyzing plant performance (or process parameters) with operator activity. Inappropriate or unnecessary activities performed by operators are compared with logging data representing the plant state if operator activity is time-tagged. This analysis provides diagnostic information on operator activity. For example, if the operators should navigate the workstation or move around in a scrambled way in order to operate the plant system within acceptable ranges, the HMI design of the ACR is considered inappropriate. As a result, some revisions are followed, even though the plant performance is maintained within acceptable ranges. Eye-tracking measures for the SA and workload evaluation are connected to personnel task performance with time-tagged information. Eye-tracking measures are analyzed for each of the tasks defined in the optimal solution. SA and workload are evaluated in each task step by considering the cognitive aspects specified by the task attribute, which is expected to increase the level of detail for the measurement. Eye fixation data are used for determining if the operators are correctly monitoring and detecting the environment. This information is used for evaluation of personnel task performance. The evaluations of personnel task performance, the teamwork, and the anthropometric/physiological factors are analyzed in an integrated manner with time-tagged information, which provides diagnostic information for human performance evaluation. Teamwork is required for operator tasks. Example behaviors and critical behaviors attributable to teamwork are investigated in the series of operator tasks with time-line analysis. Behaviors attributable to teamwork are evaluated whether they contribute to good or poor performance of the operator tasks. On the other hand, overloaded operator tasks are evaluated whether they inhibit teamwork or not. Unexpected anthropometric/physiological problems observed during a test are analyzed in the context of operator tasks, which are useful for analyzing the cause of anthropometric/physiological problems. AV recording data are effectively utilized with real-time evaluation data. The AV recording data provides information which may be missed or not processed by SMEs during a test. Scenes in ACRs, including the operator activities and HMI displays during specific time periods are replayed with AV recording data. The time-tagged information is compared and analyzed with the AV recording data.

Several questionnaires are evaluated by evaluators and operators who have participated in the tests after the completion of the real-time evaluation. The questionnaire-evaluations include KSAX for the SA evaluation, NASA-TLX for the workload evaluation, BARS for the teamwork evaluation and the PT (post-test) questionnaire for the evaluation of issues which cannot be covered by human performance measures adopted in HUPESS. All the questionnaires are provided in computerized form in HUPESS. SMEs as evaluators use the SES for evaluation of BARS and operators use the MES for evaluation of KSAX, NASA-TLX, and PT questionnaire, respectively. Evaluators and operators simultaneously evaluate relevant questionnaires after running a scenario.

An integrated analysis for a test and statistical analyses for several tests of interest are performed in HUPESS. All the items evaluated during and after a test are investigated through time-line analysis in the integrated analysis for a test. However, the integrated analysis for a test provides only insights regarding a test. The integration of the insights from the tests representing various operating conditions is conducted by statistical analyses. The results of the statistical analyses are considered to be important criteria for the ISV, because the design of ACRs must support safe operation of the plant system regardless of shifts, scenarios, and other operating conditions. An acceptable performance level is assured from the evaluation results of a series of tests, which is done by statistical analyses. HUPESS provides statistical analyses, such as descriptive statistics, linear regression analysis, *t*-test, *z*-test, ANOVA, and correlation analysis.

Operator tasks in NPPs are generally based on a goal-oriented procedure. Operator tasks are analyzed and then constructed into an optimal solution in a hierarchical form. The optimal solution consists of the main goal, sub-goals, observable cognitive tasks, and sub-tasks. The relative importance (or weight value) of the elements in the optimal solution is obtained with the AHP. Operator tasks are ranked with weight values for tasks. Analysis resources are allocated according to the relative importance of the tasks. An important task is analyzed with more resources (*e.g.*, more time or more additional consideration is allocated to the analysis). Much more time is required for the analysis of human performance in a test. Many tests covering a sufficient spectrum of operational situations in an NPP are performed to validate the HMI design. Consequently, the importance-based approach is thought to be an efficient strategy.

9.4 Implications for HRA in ACRs

9.4.1 Issues Related to HRA

Little study has been conducted on HRA in ACRs [20]. One controversial issue is automation. It has been discussed whether human errors are eliminated by increased automation. Human errors are at a higher functional level, as the role of operator is shifted to a higher level. Introduction of new technology is coupled with new categories of human error. The ability of a pilot to stay ahead of an aircraft is lost by aircraft cockpit automation, if the pilot is not provided with sufficient information necessary to make decisions, or decisions are automatically made without providing the rationale to the pilot [99]. Modeling human action is one of the issues related to HRA. The effect of operator role shift on human performance and new types of error are not well understood. There is also limited understanding on the effects of new technologies on human performance. The nuclear industry has little experience with operator performance in ACRs. Error quantification is also a critical issue. There are few databases for quantification of human errors related to ACRs. A countermeasure is a simulation study, even though challenging issues exist. The effect of PSFs in simulators is different from that in the real world (*e.g.*, stress, noise, and distractions). Operators expect events which seldom occur in the real world to occur. Operator attention is aroused at initial detection of problems, meaning that underarousal, boredom, and lack of vigilance will not be significant. HRA methodology frequently depends on the judgment of SMEs to assist in human action modeling, development of base-case HEPs, and evaluation of importance and quantitative effects of PSFs. However, there are few human factor experts in the area of ACR design.

9.4.2 Role of Human Performance Evaluation for HRA

Human error is considered as a type of human performance. Human error is related to the results (product) of operator activities. Human performance includes the product and the process (how that result was achieved). The study of human performance provides the theoretical and empirical background for the study of human error. The study of human performance is designed so that the results of human performance evaluation are used for the study of human error.

9.5 Concluding Remarks

The HMI design in ACRs is validated by performance-based tests to determine if the design acceptably supports safe operation of the plant. Plant performance, personnel task performance, situation awareness, workload, teamwork, and anthropometric/physiological factors are considered for the human performance evaluation in HUPESS. Attention is paid to regulatory support, changed CR, practicality and efficiency, evaluation criteria, and new technologies for the

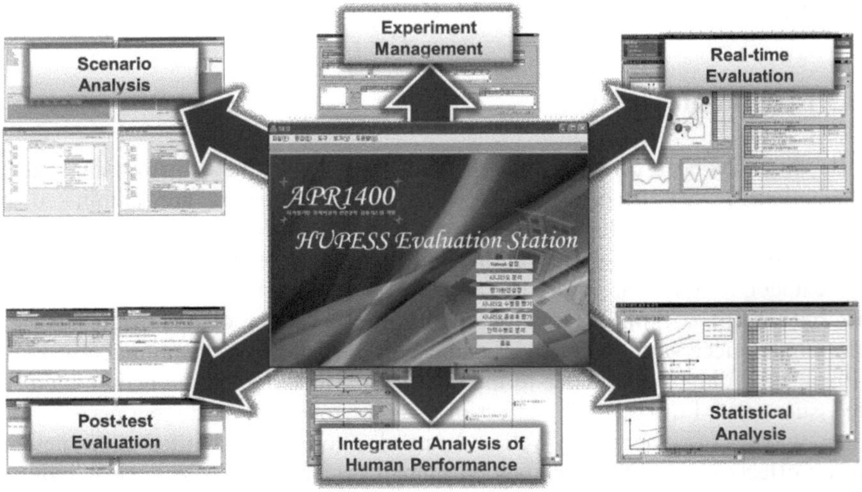

Figure 9.10. Main functions of HUPESS

development of HUPESS. Empirically proven measures used in various industries for the evaluation of human performance have been adopted with some modifications. This measure is called the main measure. Complementary measures are developed in order to overcome some of the limitations associated with main measures. The development of human performance measures is addressed on theoretical and empirical bases, considering the regulatory guidelines for the ISV, such as NUREG-0711 and NUREG/CR-6393. System configuration, including hardware and software, and methods such as integrated measurement, evaluation, and analysis are described. HUPESS provides functions, such as scenario analysis, experiment management, real-time evaluation, post-test evaluation, integrated analysis of human performance, and statistical analyses (Figure 9.10).

Issues related to HRA in ACRs are introduced and the role of human performance evaluation for HRA is discussed. Human performance is effectively measured, evaluated, and analyzed with HUPESS. HUPESS is an effective tool for the ISV in the ACR of Shin Kori 3 & 4 NPPs (APR-1400 type) which are under construction in the Republic of Korea. Further improvement and upgrade in HUPESS is necessary to cope with unexpected problems that are observed in the design of the ACR, as experience of human performance evaluation in the ACRs of Korean NPPs is accumulated.

References

[1] US Nuclear Regulatory Commission (1980) Functional criteria for emergency response facilities. NUREG-0696, Washington D.C.

[2] US Nuclear Regulatory Commission (1980) Clarification of TMI action plan requirements. NUREG-0737, Washington D.C.

[3] O'Hara JM, Brown WS, Lewis PM, Persensky JJ (2002) Human-system interface

design review guidelines. NUREG-0700, Rev.2, US NRC
[4] O'Hara JM, Higgins JC, Persensky JJ, Lewis PM, Bongarra JP (2004) Human factors engineering program review model, NUREG-0711, Rev.2, US NRC
[5] Barriere M, Bley D, Cooper S, Forester J, Kolaczkowski A, Luckas W, Parry G, Ramey-Smith A, Thompson C, Whitehead D, Wreathall J (2000) Technical basis and implementation guidelines for a technique for human event analysis (ATHEANA), Rev.01. NUREG-1624, US NRC
[6] Chang SH, Choi SS, Park JK, Heo G, Kim HG (1999) Development of an advanced human-machine interface for next generation nuclear power plants. Reliability Engineering and System Safety 64:109–126
[7] Kim IS (1994) Computerized systems for on-line management of failures: a state-of-the-art discussion of alarm systems and diagnostic systems applied in the nuclear industry. Reliability Engineering and Safety System 44:279–295
[8] Yoahikawa H, Nakagawa T, Nakatani Y, Furuta T, Hasegawa A (1997) Development of an analysis support system for man-machine system design information. Control Eng. Practice, 5-3:417–425
[9] Ohi T, Yoshikawa H, Kitamura M, Furuta K, Gofuku A, Itoh K, Wei W, Ozaki Y (2002) Development of an advanced human-machine interface system to enhanced operating availability of nuclear power plants. International Symposium on the Future I&C for NPP (ISOFIC2002), Seoul: 297–300
[10] Cho SJ et al. (2003) The evaluation of suitability for the design of soft control and safety console for APR1400. KHNP, TR. A02NS04.S2003.EN8, Daejeon, Republic of Korea
[11] Sheridan TB (1992) Telerobotics, automation, and human supervisory control. Cambridge, MA: MIT Press
[12] O'Hara JM, Stubler WF, Higgins JC, Brown WS (1997) Integrated system validation: methodology and review criteria. NUREG/CR-6393, US NRC
[13] Andresen G, Drøivoldsmo A (2000) Human performance assessment: methods and measures. HPR-353, OECD Halden Reactor Project
[14] Ha JS, Seong PH (2007) Development of human performance measures for human factors validation in the advanced MCR of APR-1400. IEEE Transactions on Nuclear Science 54-6:2687–2700
[15] Braarud PØ, Brendryen H (2001) Task demand, task management, and teamwork, HWR-657, OECD Halden Reactor Project
[16] Drøivoldsmo A et al. (1988) Continuous measure of situation awareness and workload. HWR-539, OECD Halden Reactor Project
[17] Moracho M (1998) Plant performance assessment system (PPAS) for crew performance evaluations. Lessons learned from an alarm study conducted in HAMMLAB. HWR-504, OECD Halden Reactor Project
[18] Skraning G jr. (1998) The operator performance assessment system (OPAS). HWR-538, OECD Halden Reactor Project
[19] Sim BS et al. (1996) The development of human factors technologies: the development of human factors experimental evaluation techniques. KAERI/RR-1693, Daejeon, Republic of Korea
[20] O'Hara JM, Hall RE (1992) Advanced control rooms and crew performance issues: implications for human reliability. IEEE Transactions on Nuclear Science 39-4:919–923
[21] Braarud PØ, Skraaning GJ (2006) Insights from a benchmark integrated system validation of a modernized NPP control room: performance measurement and the comparison to the benchmark system. NPIC&HMIT 2006: 12–16, Albuquerque, NM, November
[22] Saaty TL (1980) The analytic hierarchy process. McGraw-Hill

[23] Hollnagel E (1998) Cognitive reliability and error analysis method. Amsterdam: Elsevier
[24] Kemeny J (1979) The need for change: the legacy of TMI. Report of the President's Commission on the Accident at Three Miles Island, New York: Pergamon
[25] Adams MJ, Tenney YJ, Pew RW (1995) Situation awareness and cognitive management of complex system. Human Factors 37-1:85–104
[26] Durso FT, Gronlund S (1999) Situation awareness. In The handbook of applied cognition, Durso FT, Nickerson R, Schvaneveldt RW, Dumais ST, Lindsay DS, Chi MTH (Eds). Wiley, New York, 284–314
[27] Endsley MR, Garland DJ (2001) Situation awareness: analysis and measurement. Erlbaum, Mahwah, NJ
[28] Gibson CP, Garrett AJ (1990) Toward a future cockpit-the prototyping and pilot integration of the mission management aid (MMA). Paper presented at the Situational Awareness in Aerospace Operations, Copenhagen, Denmark
[29] Taylor RM (1990) Situational awareness rating technique (SART): the development of a tool for aircrew systems design. Paper presented at the Situational Awareness in Aerospace Operations, Copenhagen, Denmark
[30] Wesler MM, Marshak WP, Glumm MM (1998) Innovative measures of accuracy and situational awareness during landing navigation. Paper presented at the Human Factors and Ergonomics Society 42nd Annual Meeting
[31] Endsley MR (1995) Toward a theory of situation awareness in dynamic systems. Human Factors 37-1:32–64
[32] Lee DH, Lee HC (2000) A review on measurement and applications of situation awareness for an evaluation of Korea next generation reactor operator performance. IE Interface 13-4:751–758
[33] Nisbett RE, Wilson TD (1997) Telling more than we can know: verbal reports on mental process. Psychological Review 84:231–295
[34] Endsley MR, (2000) Direct measurement of situation awareness: validity and use of SAGAT. In Endsley MR, Garland DJ (Eds), Situation awareness analysis and measurement. Mahwah, NJ: Lawrence Erlbaum Associates
[35] Endsley MR, (1996) Situation awareness measurement in test and evaluation. In O'Brien TG, Charlton SG (Eds), Handbook of human factors testing and evaluation. Mahwah, NJ: Lawrence Erlbaum Associates
[36] Sarter NB, Woods DD (1991) Situation awareness: a critical but ill-defined phenomenon. The International Journal of Aviation Psychology 1-1:45-57
[37] Pew RW (2000) The state of situation awareness measurement: heading toward the next century. In Endsley MR, Garland DJ (Eds), Situation awareness analysis and measurement. Mahwah, NJ: Lawrence Erlbaum Associates
[38] Endsley MR (1990) A methodology for the objective measurement of situation awareness. In Situational Awareness in Aerospace Operations (AGARD-CP-478; pp. 1/1–1/9), Neuilly-Sur-Seine, France: NATO-AGARD
[39] Endsley MR (1995) The out-of-the-loop performance problem and level of control in automation. Human Factors 37-2:381–394
[40] Collier SG, Folleso K (1995) SACRI: A measure of situation awareness for nuclear power plant control rooms. In Garland DJ, Endsley MR (Eds), Experimental Analysis and Measurement of Situation Awareness. Daytona Beach, FL: Embri-Riddle University Press, 115–122
[41] Hogg DN, Follesø K, Volden FS, Torralba B (1995) Development of a situation awareness measure to evaluate advanced alarm systems in nuclear power plant control rooms. Ergonomics 38-11:2394–2413
[42] Fracker ML, Vidulich MA (1991) Measurement of situation awareness: A brief review. In Queinnec Y, Daniellou F (Eds), Designing for everyone, Proceedings of the

11th Congress of the International Ergonomics Association, London, Taylor & Francis 795–797
[43] Endsley MR (1995) Measurement of situation awareness in dynamic systems. Human Factors 37-1:65–84
[44] Wilson GF (2000) Strategies for psychophysiological assessment of situation awareness. In Endsley MR, Garland DJ (Eds), Situation awareness analysis and measurement. Mahwah, NJ: Lawrence Erlbaum Associates
[45] Taylor RM (1990) Situational awareness rating technique (SART): the development of a tool for aircrew systems design. In Situational Awareness in Aerospace Operations (AGARD-CP-478; pp. 3/1–3/17), Neuilly-Sur-Seine, France: NATO-AGARD
[46] Wickens CD, Hollands JG (2000) Engineering psychology and human performance, 3rd Edition. New Jersey, Prentice-Hall
[47] O'Hara JM, Higgins JC, Stubler WF, Kramer J (2002) Computer-based procedure systems: technical basis and human factors review guidance. NUREG/CR-6634, US NRC
[48] Kim MC, Seong PH (2006) A computational model for knowledge-driven monitoring of nuclear power plant operators based on information theory. Reliability Engineering and System Safety 91:283–291
[49] Ha JS, Seong PH (2005) An experimental study: EEG analysis with eye fixation data during complex diagnostic tasks in nuclear power plants. Proceedings of International Symposium On the Future I&C for NPPs (ISOFIC), Chungmu, Republic of Korea
[50] Wickens CD (1992) Workload and situation awareness: an analogy of history and implications. Insight 94
[51] Moray N (1979) Mental workload: its theory and measurement. Plenum Press, New York
[52] Hancock P, Meshkati N (1988) Human mental workload. North-Holland, New York
[53] O'Donnell RD, Eggemeier FT (1986) Workload assessment methodology. In Boff KR, Kaufman L, Thomas J (Eds), Handbook of perception and human performance: Vol. II. Cognitive Processes and Performance, John Wiley & Sons
[54] Norman DA, Bobrow DG (1975) On data-limited and resource-limited process. Cognitive Psychology 7:44–64
[55] Williges R, Wierwille WW (1979) Behavioral measures of aircrew mental workload. Human Factors 21:549–574
[56] Charlton SG (2002) Measurement of cognitive states in test and evaluation. In Charlton SG, O'Brien TG (Eds), Handbook of human factors testing and evaluation, Mahwah, NJ: Lawrence Erlbaum Associates
[57] Eggemeier FT, Wilson GF (1991) Subjective and performance-based assessment of workload in multi-task environments. In Damos D (Eds), Multiple task performance. London, Taylor & Francis
[58] Rubio S, Diaz E, Martin J, Puente JM (2004) Evaluation of subjective mental workload: a comparison of SWAT, NASA-TLX, and workload profile. Applied Psychology 53:61–86
[59] Wierwille WW, Rahimi M, Casali JG (1985) Evaluation of 16 measures of mental workload using a simulated flight task emphasizing mediational activity. Human Factors 27:489–502
[60] Johannsen G, Moray N, Pew R, Rasmussen J, Sanders A, Wickens CD (1979) Final report of the experimental psychology group. In Moray N (Eds), Mental workload: its theory and measurement. New York: Plenum
[61] Moray N (1982) Subjective mental workload. Human Factors 24:25–40
[62] Hill SG, Iavecchia HP, Byers JC, Bittier AC, Zaklad AL, Christ RE (1992) Comparison of four subjective workload rating scales. Human Factors 34:429–440

[63] Sterman B, Mann C (1995) Concepts and applications of EEG analysis in aviation performance evaluation. Biological Psychology 40:115–130
[64] Kramer AF, Sirevaag EJ, Braune R (1987) A psychophysiological assessment of operator workload during simulated flight missions. Human Factors 29-2:145–160
[65] Brookings J, Wilson GF, Swain C (1996) Psycho-physiological responses to changes in workload during simulated air traffic control. Biological Psychology 42:361–378
[66] Brookhuis KA, Waard DD (1993) The use of psychophysiology to assess driver status. Ergonomics 36:1099–1110
[67] Donchin E, Coles MGH (1988) Is the P300 component a manifestation of cognitive updating? Behavioral and Brain Science 11:357–427
[68] Boer LC, Veltman JA (1997) From workload assessment to system improvement. Paper presented at the NATO Workshop on Technologies in Human Engineering Testing and Evaluation, Brussels
[69] Roscoe AH (1975) Heart rate monitoring of pilots during steep gradient approaches. Aviation, Space and Environmental Medicine 46:1410–1415
[70] Rau R (1996) Psychophysiological assessment of human reliability in a simulated complex system. Biological Psychology 42:287–300
[71] Kramer AF, Weber T (2000) Application of psychophysiology to human factors. In Cacioppo JT et al. (Eds), Handbook of psychophysiology, Cambridge University Press 794–814
[72] Jorna PGAM (1992) Spectral analysis of heart rate and psychological state: a review of its validity as a workload index. Biological Psychology 34:237–257
[73] Mulder LJM (1992) Measurement and analysis methods of heart rate and respiration for use in applied environments. Biological Psychology 34:205–236
[74] Porges SW, Byrne EA (1992) Research methods for the measurement of heart rate and respiration. Biological Psychology 34:93–130
[75] Wilson GF (1992) Applied use of cardiac and respiration measure: practical considerations and precautions. Biological Psychology 34:163–178
[76] Lin Y, Zhang WJ, Watson LG (2003) Using eye movement parameters for evaluating human-machine interface frameworks under normal control operation and fault detection situations. International Journal of Human Computer Studies 59:837–873
[77] Veltman JA, Gaillard AWK (1996) Physiological indices of workload in a simulated flight task. Biological Psychology 42:323–342
[78] Bauer LO, Goldstein R, Stern JA (1987) Effects of information-processing demands on physiological response patterns. Human Factors 29:219–234
[79] Goldberg JH, Kotval XP (1998) Eye movement-based evaluation of the computer interface. In Kumar SK (Eds), Advances in occupational ergonomics and safety. IOS Press, Amsterdam
[80] Ha CH, Seong PH (2006) Investigation on relationship between information flow rate and mental workload of accident diagnosis tasks in NPPs. IEEE Transactions on Nuclear Science 53-3:1450–1459
[81] http://www.seeingmachines.com/
[82] http://www.smarteye.se/home.html
[83] Shively R, Battiste V, Matsumoto J, Pepiton D, Bortolussi M, Hart S (1987) In flight evaluation of pilot workload measures for rotorcraft research. Proceedings of the Fourth Symposium on Aviation Psychology: 637–643, Columbus, OH
[84] Battiste V, Bortolussi M (1988) Transport pilot workload: a comparison of two subjective techniques. Proceedings of the Human Factors Society Thirty-Second Annual Meeting: 150–154, Santa Monica, CA
[85] Nataupsky M, Abbott TS (1987) Comparison of workload measures on computer-generated primary flight displays. Proceedings of the Human Factors Society Thirty-First Annual Meeting: 548–552, Santa Monica, CA

[86] Tsang PS, Johnson WW (1989) Cognitive demand in automation. Aviation, Space, and Experimental Medicine 60:130–135
[87] Bittner AV, Byers JC, Hill SG, Zaklad AL, Christ RE (1989) Generic workload ratings of a mobile air defense system (LOS-F-H). Proceedings of the Human Factors Society Thirty-Third Annual Meeting: 1476–1480, Santa Monica, CA
[88] Hill SG, Byers JC, Zaklad AL, Christ RE (1988) Workload assessment of a mobile air defense system. Proceedings of the Human Factors Society Thirty-Second Annual Meeting: 1068–1072, Santa Monica, CA
[89] Byers JC, Bittner AV, Hill SG, Zaklad AL, Christ RE (1988) Workload assessment of a remotely piloted vehicle (RPV) system. Proceedings of the Human Factors Society Thirty-Second Annual Meeting: 1145–1149, Santa Monica, CA
[90] Sebok A (2000) Team performance in process control: influences of interface design and staffing. Ergonomics 43-8:1210–1236
[91] Byun SN, Choi SN (2002) An evaluation of the operator mental workload of advanced control facilities in Korea next generation reactor. Journal of the Korean Institute of Industrial Engineers 28-2:178–186
[92] Plott C, Engh,T, Bames V (2004) Technical basis for regulatory guidance for assessing exemption requests from the nuclear power plant licensed operator staffing requirements specified in 10 CFR 50.54, NUREG/CR-6838, US NRC
[93] Hart SG, Staveland LE (1988) Development of NASA-TLX (Task Load Index): Results of empirical and theoretical research. In Hancock PA, Meshkati N (Eds), Human mental workload, Amsterdam: North-Holland
[94] Stern JA, Walrath LC, Golodstein R (1984) The endogenous eyeblink. Psychophysiology 21:22–23
[95] Tanaka Y, Yamaoka K (1993) Blink activity and task difficulty. Perceptual Motor Skills 77:55–66
[96] Goldberg JH, Kotval XP (1998) Eye movement-based evaluation of the computer interface. In Kumar SK (Eds), Advances in occupational ergonomics and safety, IOS Press, Amsterdam
[97] Bellenkes AH, Wickens CD, Kramer AF (1997) Visual scanning and pilot expertise: the role of attentional flexibility and mental model development. Aviation, Space, and Environmental Medicine 68-7:569–579
[98] Roth EM, Mumaw RJ, Stubler WF (1993) Human factors evaluation issues for advanced control rooms: a research agenda. IEEE Conference Proceedings: 254–265
[99] Sexton G (1998) Cockpit-crew systems design and integration. In Wiener E, Nagel D (Eds), Human factors in aviation. Academic Press: 495–504

Part IV

Integrated System-related Issues and Countermeasures

10

Issues in Integrated Model of I&C Systems and Human Operators

Man Cheol Kim[1] and Poong Hyun Seong[2]

[1]Integrated Safety Assessment Division
Korea Atomic Energy Research Institute
1045 Daedeok-daero, Yuseong-gu, Daejeon, 305-353, Korea
charleskim@kaeri.re.kr

[2]Department of Nuclear and Quantum Engineering
Korea Advanced Institute of Science and Technology
373-1 Guseong-dong, Yuseong-gu, Daejeon, 305-701, Korea
phseong@kaist.ac.kr

Reliability issues and some countermeasures for hardware and software of I&C systems in large-scale systems are discussed in Chapters 1–6. Reliability issues and some countermeasures for human operators in large-scale systems are discussed in Chapters 7–9. Reliability issues and countermeasures when the I&C systems and human operators in large-scale systems are considered as a combined entity are in Chapters 10–12.

The conventional way of considering I&C systems and human operators as parts of large-scale systems is introduced in Section 10.1. Reliability issues in an integrated model of I&C systems and human operators in large-scale systems are summarized based on some insights from the accidents in large-scale systems in Sections 10.2 and 10.3. Concluding remarks are provided in Section 10.4.

10.1 Conventional Way of Considering I&C Systems and Human Operators

PRA is widely used for reliability and/or risk analysis of large-scale systems. PRA usually consists of the development of event trees and fault trees for describing possible scenarios after the occurrence of an initiating event and determining branch probabilities for event trees, respectively.

An example of how I&C systems and human operators are considered in conventional event-tree-and-fault-tree-based PRA models is shown in Figure 10.1 [1]: an event tree, including a low-pressure safety injection system (LPSIS) in an NPP, and a part of a fault tree for calculation of branch failure probability of safety injection by LPSIS. I&C systems are considered in the basic event for evaluated failure of safety injection actuation signal (SIAS) generating devices. Human operators are considered in the basic event for the failure of operator manually generating SIAS as part of the fault tree (Figure 10.1). I&C systems and human operators are not described in detail in conventional PRA models because PRA mainly focuses on hardware failures. I&C systems and human operators are considered to be independent in conventional PRA models (Figure 10.1).

10.2 Interdependency of I&C Systems and Human Operators

I&C systems and human operators are independently modeled in the conventional PRA (Section 10.1). Researchers in the field of quantitative safety assessment of large industrial systems consider the interdependency of I&C systems and human operators.

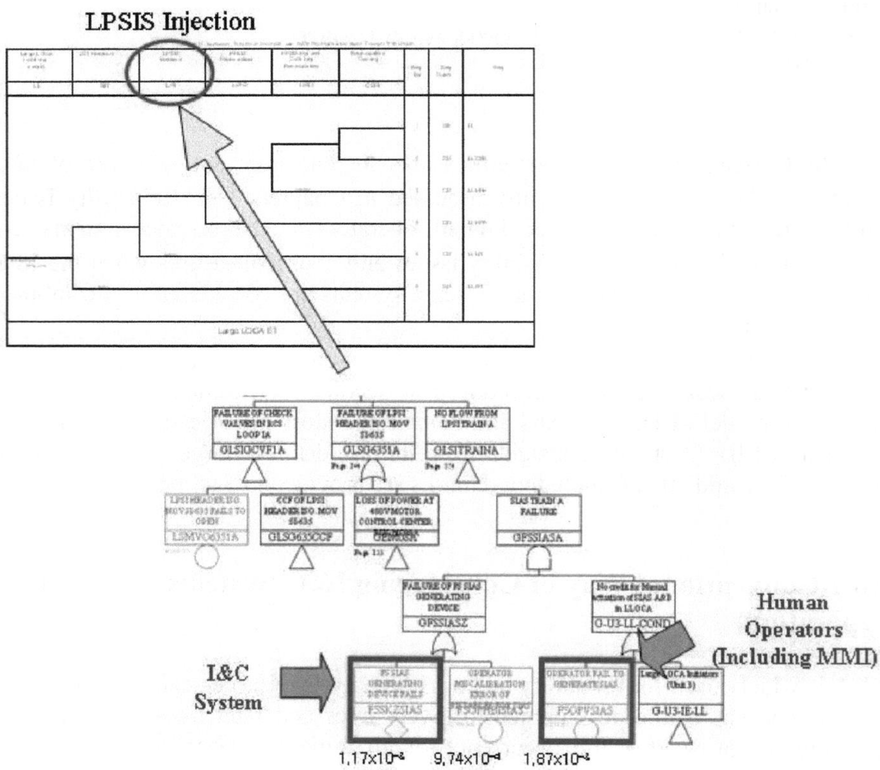

Figure 10.1. An example of how I&C systems and human operators are considered in conventional PRA models

10.2.1 Risk Concentration on I&C Systems

The basic concept of risk concentration in I&C systems is as follows. The plant parameters are measured using sensors and then displayed on indicators. Those signals are also transmitted to the plant protection system (PPS) as a large-scale digital control system. The PPS provides necessary signals to the engineered safety feature actuation system (ESFAS), and provides some alarms to human operators (Figure 10.2). The concept of risk concentration, if the PPS fails, is due to the possibility that ESFAS cannot generate automatic ESF actuation signals, and at the same time, operators cannot generate manual ESF actuation signals, because ESFAS cannot receive necessary signals from PPS and necessary alarms are not provided to human operators. The risk is thus concentrated in an I&C system.

The effect of risk concentration on I&C systems is limited considering that many control systems also provide general alarms to the human operator (Figure 10.2). For example, there can be a failure mode in the PPS of an NPP which prohibits PPS from generating a pressurizer pressure low reactor shutdown signal and ESFAS from generating a pressurizer pressure low safety injection (SI) actuation signal. These important alarms will not be given to human operators due to the failure of PPS. The pressurizer pressure control system will generate a "pressurizer pressure low & backup heater on" alarm, even in this situation. This alarm will draw operator attention, causing them to focus on trend.

One important insight from the concept of risk concentration on I&C systems is the possibility that the failure of I&C systems could deteriorate human operator performance (*i.e.*, there is a dependency between I&C systems and human operators in large-scale digital control systems).

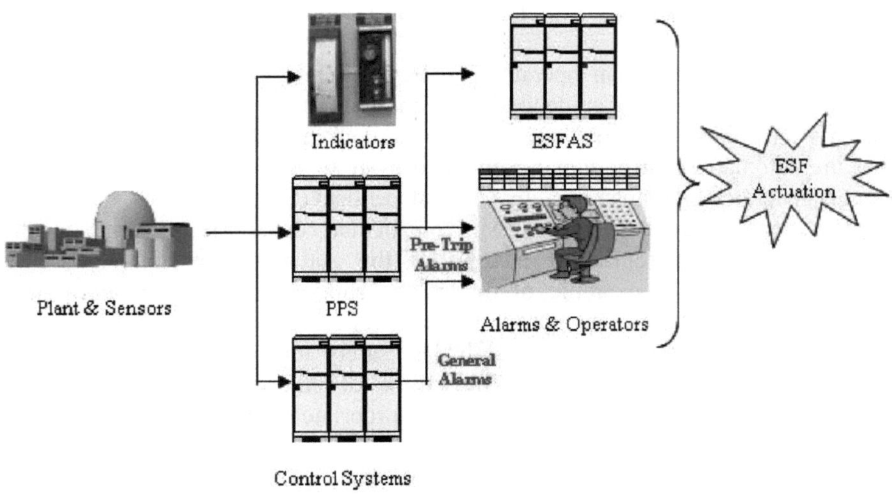

Figure 10.2. The concept of risk concentration of I&C systems

10.2.2 Effects of Instrument Faults on Human Operators

The effects of instrument faults on the safety of large-scale systems have also received a lot of attention. Instrument faults can affect the activation of safety features, such as emergency shutdown, not only by PPS and/or ESFAS, but also by human operators (Figure 10.2). An emphasis on unsafe actions due to instrument faults is found in many places of ATHEANA [2]. ATHEANA is a second generation HRA method developed by U.S. NRC. Appropriate selections from the ATHEANA handbook are:

- "There has been very little consideration of how instrument faults will affect the ability of the operators to understand the conditions within the plant and act appropriately." (p. 3–3)
- "As shown by EFCs for the Crystal River 3, Dresden 2, and Ft. Calhoun events, wrong situation models are frequently developed as a result of instrumentation problems, especially undiscovered hardware failures." (p.5–9)
- "Both tables also highlight the importance of correct instrument display and interpretation in operator performance." (p.5–14)
- "… unsafe actions are likely to be caused at least in part by actual instrumentation problems or misinterpretation of existing indications." (p.5–27)

The approach for analyzing errors of commissions proposed by Kim *et al.* [3] analyzes possibilities for NPP operators to be misled and make wrong situation assessments due to instrument faults, which will possibly result in unsafe actions.

10.2.3 Dependency of I&C Systems on Human Operators

The dependency of human operators on I&C systems is explained in Sections 10.2.1 and 10.2.2. Much evidence has been found for the dependency of I&C systems on human operators, even though the current PRA technology assumes that the automatic control signals generated by I&C systems and manual control signals generated by human operators are independent (Section 10.1).

The dependency of I&C systems on human operators is found in the appropriate or inappropriate bypassing of the automatic generation of ESF actuation signals. An example of this is the shut off of high-pressure safety injection flow by human operators in the TMI-2 accident. Recent incident reports also reveal that operators sometimes bypass safety functions when they cannot clearly understand the situation. The Office of Analysis and Evaluation of Operational Data (AEOD) identified 14 inappropriate bypasses of ESFs over 41 months [4]. The reliability of automatic control signals for mitigating accident situations generated by I&C systems is dependent on the situation assessment of human operators.

10.3 Important Factors in Situation Assessment of Human Operators

Human failure probabilities for correctly assessing the situation are assumed to be dominantly dependent on available time for human operators in conventional PRA technology. Time-reliability curves, which are determined mainly by expert consensus, are used to determine human failure probabilities in most conventional (first–generation) HRA methods, such as THERP [5], ASEP [6], and HCR [7]. Several second–generation HRA methods, such as ATHEANA [2] and CREAM [8], have been recently developed. Human failure probabilities are assumed to be dominantly dependent on contextual or environmental factors, in the case of CREAM. The relation between context factors and human failure probabilities are determined mainly by expert opinions.

Major accidents in large-scale industrial plants give insights about factors that should be considered when dealing with human operator situation assessment during abnormal and/or emergency situations in large-scale systems.

10.3.1 Possibilities of Providing Wrong Information to Human Operators

A brief illustration of the Bhopal accident is shown Figure 10.3. Human operators of the Bhopal plant could take mitigation actions, one of which was the transfer of methyl-isocyanate (MIC) in the main tank (Tank 610) to the spare tank (Tank 619), after the occurrence of the explosion and the release of toxic gas to the nearby environment. The level of the spare tank was indicated at about 20% full, even though the spare tank was actually almost empty (Figure 10.3). The wrong information prevented the human operators from immediately taking mitigation action. Several hours passed before mitigation action [9].

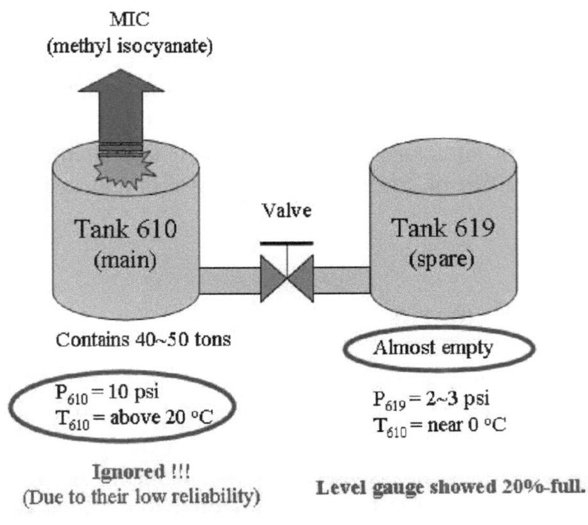

Figure 10.3. Some important aspects of the Bhopal accident

The information provided to the human operators of TMI-2 plant was that the pressure operated relief valve (PORV) solenoid was de-energized, even though the PORV was stuck open. This information was misinterpreted by the human operators of TMI-2 plant as a sign that the PORV was closed. About 2 hours were taken to recognize the main cause of the accident, which was the stuck open PORV [9].

Thus, the possibility of providing wrong information to human operators is an important factor that should be considered in the quantitative safety assessment of large-scale systems.

10.3.2 Operators' Trust on Instruments

Whether human operators trust information or not is completely another issue, even though human operators receive correct information. Human operators of the Bhopal plant did not trust the first indication of the accident because the sensors and indicators had often failed, even though they had received data about one hour and forty minutes before the accident.

Human failure probability is estimated to be about 10^{-4} if conventional HRA methods, such as THERP or ASEP, which are used in current PRA technology, are applied. One hour and forty minutes is a long time for the diagnosis of an abnormal situation.

The possibility of discarding information provided by I&C systems has not been considered in the safety analysis of large-scale systems, but is an important factor in the consideration of human operator response in such systems.

10.3.3 Different Difficulties in Correct Diagnosis of Different Accidents

Some accidents are easy to diagnose, and others are difficult to diagnose. Human operators are expected to easily diagnose an accident if the accident has unique symptoms. Human operators are expected to see difficulties in correctly diagnosing accidents if an accident has symptoms similar to other transients or accidents.

Current PRA technology provides a method to evaluate human failure probabilities in correctly diagnosing a situation without considering different difficulties of different situations. The development of a method for considering these different difficulties of correctly diagnosing different accident situations is required.

10.4 Concluding Remarks

Reliability and risk issues in the development of an integrated model of I&C systems and human operators in large-scale systems are reviewed in this chapter.

How I&C systems and human operators are generally considered in current PRA technology, and the basic assumption of independence between I&C systems and human operators is shown in Figure 10.4. Automatic control signals generated by control/protection systems, which are also a part of I&C systems, are modeled

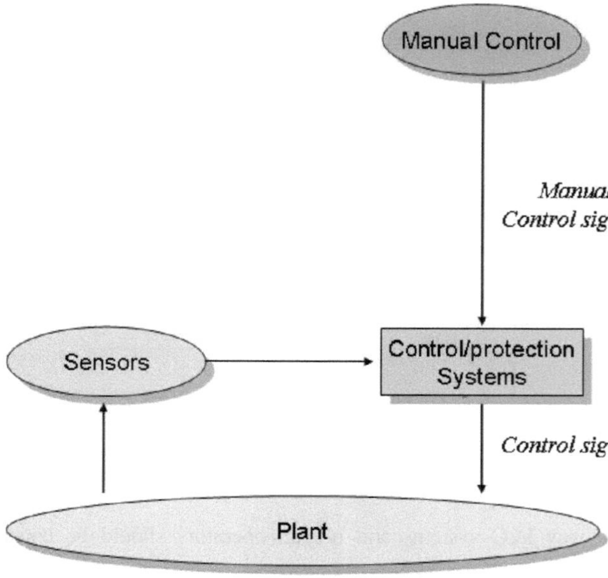

Figure 10.4. The way I&C systems and human operators are considered in current PRA technology

as independent from manual control signals generated by human operators. Manual control signals generated by human operators are also independent from information provided by I&C systems in current PRA technology (Figure 10.4).

There are interdependencies between I&C systems and human operators in large-scale systems. The concept of risk concentration in I&C systems shows the dependency of human operators on I&C systems. Concerns about possible effects of instrument faults on human operators are also indicated by advanced HRA methods [2]. The dependency of I&C systems on human operators has also been indicated in the inappropriate bypass of ESFs by human operators [4]. The development of an integrated model of I&C systems and human operators in large-scale systems is needed. An integrated model considering the interdependency of I&C systems and human operators is illustrated in Figure 10.5.

Important factors for an integrated model for I&C systems and human operators that can be used for reliability and/or risk analysis of large-scale systems are summarized in Section 10.3. These factors are: the possibility of an I&C system providing wrong information to human operators, human operator trust on the information provided by an I&C system, and different difficulties in correctly diagnosing different accident situations. Few of these factors have been considered in conventional PRA technology.

An integrated model of I&C systems and human operators in large-scale systems, which attempt to incorporate the interdependency of I&C system and human operators is described in Chapter 11. An integrated model (Chapter 11) provides the basic framework for future research.

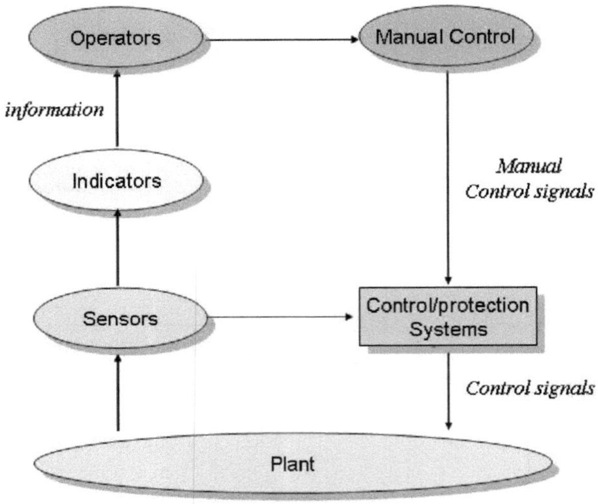

Figure 10.5. The way I&C systems and human operators should be considered in an integrated model

10.5 References

[1] KEPCO (1998) Full scope level 2 PSA for Ulchin unit 3&4: Internal event analysis, Korea Electric Power Corporation

[2] Barriere M, Bley D, Cooper S, Forester J, Kolaczkowski A, Luckas W, Parry G, Ramey-Smith A, Thompson C, Whitehead D, Wreathall J (2000) Technical Basis and Implementation Guideline for A Technique for Human Event Analysis (ATHEANA), NUREG-1624, Rev. 1, U.S. Nuclear Regulatory Commission, Washington D.C.

[3] Kim JW, Jung W, and Park J (2005) A systematic approach to analyzing errors of commission from diagnosis failure in accident progression, Reliability Engineering and System Safety, vol. 89, pp.137–150

[4] Office of Analysis and Evaluation of Operational Data (AEOD) (1995) Engineering evaluation – operating events with inappropriate bypass or defeat of engineered safety features, U.S. Nuclear Regulatory Commission

[5] Swain AD and Guttman HE (1983) Handbook of human reliability analysis with emphasis on nuclear power plant applications, NUREG/CR-1278, U. S. Nuclear Regulatory Commission

[6] Swain AD (1987) Accident sequence evaluation program: Human reliability analysis procedure, NUREG/CR-4772, U. S. Nuclear Regulatory Commission

[7] Hannaman GW et al. (1984) Human cognitive reliability model for PRA analysis, NUS-4531, Electric power Research Institute

[8] Hollnagel E (1998) Cognitive reliability and error analysis method, Elsevier

[9] Leveson NG (1995) SafeWare: system safety and computers, Addison-Wesley

11

Countermeasures in Integrated Model of I&C Systems and Human Operators

Man Cheol Kim[1] and Poong Hyun Seong[2]

[1]Integrated Safety Assessment Division
Korea Atomic Energy Research Institute
1045 Daedeok-daero, Yuseong-gu, Daejeon, 305-353, Korea
charleskim@kaeri.re.kr

[2]Department of Nuclear and Quantum Engineering
Korea Advanced Institute of Science and Technology
373-1 Guseong-dong, Yuseong-gu, Daejeon, 305-701, Korea
phseong@kaist.ac.kr

Reliability and risk issues related to the development of an integrated model of I&C systems and human operators in large-scale systems are discussed in Chapter 10. The development of an integrated model that address these issues is discussed in this chapter.

I&C systems and human operators are completely different entities by nature. However, both I&C systems and human operators process information. I&C systems gather information from the plant and processes it to provide automatic control signals to the plant. I&C systems also provide information to human operators in a form that human operators can understand. Human operators receive the information from I&C systems and process it to provide manual control signals for I&C systems.

The way information is processed in I&C systems is generally known, while the way information is processed by human operators is not well known. The development of an integrated model of I&C systems and human operators starts from the development of a model for how human operators process incoming information, especially during abnormal or emergency situations.

11.1 Human Operators' Situation Assessment Model

11.1.1 Situation Assessment and Situation Awareness

Operators try to understand what is going on in their plants when an abnormal or emergency situation occurs in large-scale systems. This process is called *situation assessment*. How operators correctly understand the situation is called *situation awareness*. Operators in large-scale systems, who are usually highly experienced, are expected to correctly understand the situation in most cases. Consequences can be catastrophic if the operators incorrectly understand the situation. The question is, "how often the operators are likely to incorrectly understand the situation?" Understanding how operators will process information they receive and what kind of conclusions they will make in various situations is necessary to answer this question.

Many situation awareness models have been developed, mostly including the process of situation assessment. The models for situation awareness are categorized [1] into three major approaches, the information-processing approach [2], the activity approach [3], and the ecological approach [4]. Situation awareness models describe basic principles and general features about how people process information or interact with the environment to accumulate their situation awareness. Those models have some limitations in helping predict what will be happening, due to their descriptive and qualitative nature, even though the models help to understand situation assessment when retrospectively analyzing events. Quantitative models are needed.

11.1.2 Description of Situation Assessment Process

Human operators usually work as field operators for more than five years before becoming main control room (MCR) operators of large-scale systems. Training courses in full-scope simulators are used to learn how to respond to various accident situations before becoming MCR operators and while working as MCR operators. Their experience as field and MCR operators are major sources of establishing the model for large-scale systems. Their expectations on how the systems will behave in various accident situations are established with their model for the systems and their experience in full-scope simulators. Example expectations for NPP operators are "when a LOCA occurs in an NPP, the pressurizer pressure and pressurizer level will decrease, and the containment radiation will increase," and "when a steam generator tube rupture (SGTR) accident occurs in an NPP, the pressurizer pressure and pressurizer level will decrease and the secondary radiation will increase." These expectations form rules on the dynamics of large-scale systems. The rules are used to understand the situation in abnormal and accident situations.

Human operators usually first recognize the occurrence of abnormal and accident situations by the onset of alarms. The major role of alarms is to draw the attention of operators to indicators relevant to the alarms. Operators will read the relevant indicators after receiving alarms. The operators might obtain level 1 SA, *perception of the elements in the environment*, among Ensley's three levels of SA [2], after this process. Operators try to understand what is going on in the plant after reading the indicators, and consider the possibility of sensor or indicator failures. Operators will read the relevant indicators when receiving other alarms. Operators may decide to monitor other indicators to explore the possibility of abnormal or accident situation occurrence, even though they do not receive other alarms. Observations from other indicators will change the current understanding of the situation, regardless of monitoring of other indicators.

Operators form rules on plant dynamics and use them to understand the situation. An example is "when pressurizer pressure and pressurizer level is decreasing in an NPP, the occurrence of a LOCA or SGTR accident is possible." The operator might obtain level 2 SA [2], *comprehension of the current situation*, using this reasoning process.

Operators will discard many possibilities and consider only a few based on the observations. Operators can predict what they will see in the future based on an understanding of the situation. An example is "The pressurizer pressure and pressurizer level will continue to decrease if a LOCA or SGTR accident occurs in an NPP. The increase in the containment radiation will be observed if a LOCA occurs in an NPP. The increase in the secondary radiation will be observed if an SGTR accident occurs in an NPP." The operator might obtain level 3 SA [2], *projection of future status*, using this prediction process. Predictions are expected to guide the active seeking of further information.

11.1.3 Modeling of Operators' Rules

The modeling of operator rules is needed to first model the situation assessment process. Two assumptions are made in establishing the model for operator rules:

1. Plant status is modeled as mutually exclusive. This assumption is for concentrating the modeling effort on a single accident or transient. The possibilities of simultaneous occurrence of more than one accident or transient are insignificant compared to the possibility of a single accident or transient. Transients do not include sensor and indicator failures. This assumption is similar to the first-order approximation in mathematics, which considers only first order of a small value ($<<1$) and ignores higher than second orders. This assumption is also supported by the fact that PRA for large-scale systems usually does not consider the simultaneous occurrence of more than one accident. Operators are usually confronting a single accident or transient, or a single accident or transient with sensor or indicator failures during the training of operators in full-scope simulators. This observation is also used to support the assumption. This assumption does not mean that the model described in this chapter cannot consider the simultaneous occurrence of more than one accident or transient. The simultaneous occurrence of more than one accident or transient is considered as another accident which is mutually exclusive to other accidents or transients.

2. Operators are assumed to use deterministic rules. Operator rules are formed from their understanding of plant dynamics and their training in simulators. Operator rules are deterministic because indicator behavior under clearly specified accident scenarios is clearly determined. This assumption is consistent with interview results of a retired operator. [5]

The model for operator rules is shown in Figure 11.1. X indicates the plant status (the situation), Y_i's (I = 1, 2, ..., m.) indicate various indicators, and Z_i's indicate various sensors. In mathematical form, X, Y_i's, and Z_i's are defined as:

$$X = \{x_1, x_2, \ldots, x_l\} \tag{11.1}$$

$$Y_i = \{y_{i1}, y_{i2}, \ldots, y_{in_i}\}, \text{ where } i = 1, 2, \ldots, m. \tag{11.2}$$

$$Z_i = \{z_{i1}, z_{i2}, \ldots, z_{in_i}\}, \text{ where } i = 1, 2, \ldots, m. \tag{11.3}$$

For example, "if the plant status is x_k, then the value or the trend of the indicator Y_i is expected to be y_{ij}." These rules can be collected from interviews with actual operators or simulator simulations, depending on the purpose of the model. Deterministic rules can be described using conditional probabilities, as:

$$P(y_{ij} \mid x_k) = \begin{cases} 1 & \text{if } y_{ij} \text{ is expected upon } x_k \\ 0 & \text{if } y_{ij} \text{ is not expected upon } x_k \end{cases} \tag{11.4}$$

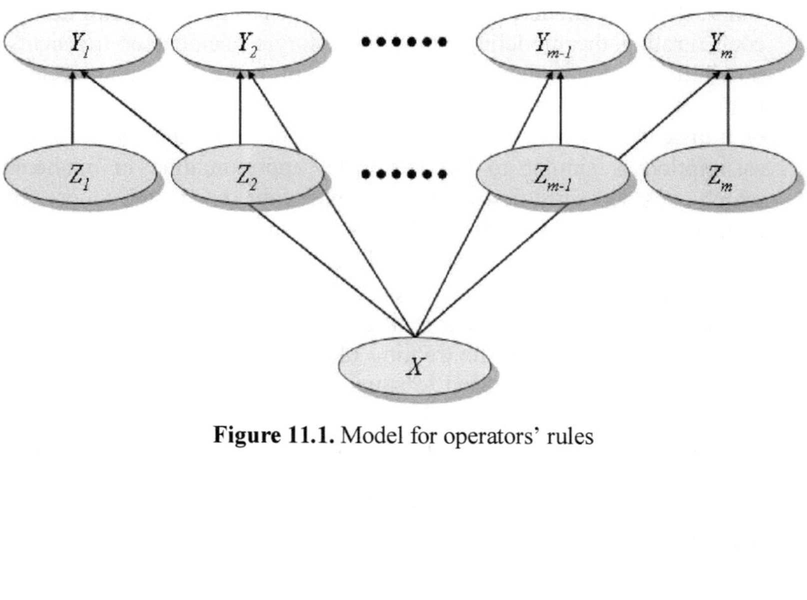

Figure 11.1. Model for operators' rules

11.1.4 Bayesian Inference

Normal people change their expectations upon observing events. For example, an employee can be assumed to have the following two rules:

1. If his boss is in his office, it is highly likely that boss's car is in the parking lot. If the boss is not in his office, it is highly likely that boss's car is not in the parking lot.
2. If the boss is in his office, it is highly likely that the boss will answer the office phone. If the boss is not in his office, it is almost impossible for him to answer the office phone.

The probability that his boss is in his office early in the morning is 0.5. The probability of his boss answering the office phone early in the morning is also about 0.5 without further observations or information. His boss is likely in his office and likely to answer the office phone if he observes the boss's car in the parking lot early in the morning.

This is the process of Bayesian inference and revision of probabilities. Human operators have a capability of Bayesian inference and revision of probabilities, even though the results are not as accurate as mathematical calculations.

Operators understand the situation by using their rules (Section 11.1.2). Operators will increase the probability of LOCA occurrence based on their rule "when a LOCA occurs in an NPP, the pressurizer pressure and pressurizer level will decrease, and the containment radiation will increase," if operators observe the pressurizer pressure is decreasing in an NPP. Based on their other rule "when an SGTR accident occurs in an NPP, the pressurizer pressure and pressurizer level will decrease and the secondary radiation will increase," the probability of the occurrence of an SGTR accident in the NPP will also be increased.

Mathematically, if the operators observe y_{ij} on the indicator Y_i, the probability of the plant status x_k can be revised as:

$$P(x_k \mid y_{ij}) = \frac{P(y_{ij} \mid x_k)P(x_k)}{\sum_{h=1}^{l} P(y_{ij} \mid x_h)P(x_h)} \tag{11.5}$$

The probability distributions for unobserved indicators are also revised following the revision of probability distribution for the plant status. The increase in containment radiation in the NPP is expected to be observed if the probability of the occurrence of a LOCA in an NPP is increased by the observation of the decrease of the pressurizer pressure in the NPP. This is related to level 3 SA, *projection of future status*.

11.1.5 Knowledge-driven Monitoring

Operators monitor relevant indicators and read values and trends (increase or decrease) of the indicators when operators receive alarms. This kind of monitoring is called *data-driven monitoring*. Operators establish their situation model and actively monitor other indicators to confirm or modify their situation model after monitoring relevant indicators. This kind of monitoring is called *knowledge-driven monitoring*. Monitoring is often knowledge driven [6].

Operators develop their situation model after data-driven monitoring. The situation model based on one or several observations is not clear. A great amount of uncertainty is associated with the initial situation model. Operators actively look for information to more clearly understand the plant status. Information refers to messages, data, or other evidence that reduces uncertainty about the true state of affairs in information theory [7]. Knowledge-driven monitoring is understood as the process of seeking information to reduce operator uncertainty about plant status. Operators are expected to further reduce uncertainty as they receive more information. Operators tend to get as much information as possible through the knowledge-driven monitoring process, if the intent is to reduce the uncertainty about plant status. The tendency of operators is assumed based on this speculation, as follows:"Operators have a tendency to select one of the most informative indicators, which means the indicators that provide the most information to operators, as the next indicator to monitor."

A quantitative measure of expected information from each indicator is needed to determine the most informative indicators. The amount of information transmitted from the observation of y_{ij} on the indicator Y_i to the plant status x_k is defined from the information theory as:

$$I(x_k;y_{ij}) = \log_2 \frac{P(x_k \mid y_{ij})}{p(x_k)} \text{ bits} \tag{11.6}$$

The expected information from the observation of the indicator Y_i is defined based on Equation 11.6 as:

$$\begin{aligned} T(X;Y_i) &= \sum_{k=1}^{l}\sum_{j=1}^{n_i} p(x_k,y_{ij}) I(x_k;y_{ij}) \\ &= \sum_{k=1}^{l}\sum_{j=1}^{n_i} p(x_k,y_{ij}) \log_2 \frac{P(x_k \mid y_{ij})}{p(x_k)} \end{aligned} \tag{11.7}$$

Equation (11.7) can be rewritten as:

$$T(X;Y_i) = H(X) + H(Y_i) - H(X,Y_i) \tag{11.8}$$

where

$$H(W) = \sum_i p(w_i)\log_2 \frac{1}{p(w_i)}. \tag{11.9}$$

11.1.6 Ideal Operators *Versus* Real Human Operators

A quantitative model for situation assessment of human operators in large-scale systems in abnormal or accident situations is described in Subsections 11.1.4 and 11.1.5. The model assumes that the situation assessment process is guided by mathematical calculations. Operators do not calculate revised probabilities and expected information from each indicator in their mind, even though human operators can do Bayesian inference in general and wisely determine informative indicators. Instead, operators do simple estimations that are not as accurate as mathematical calculations.

The mathematical model is regarded as a situation assessment model for *ideal operators*, who can correctly revise probabilities after observations and correctly calculate the expected information from each indicator. Ideal operators have the following characteristics:

1. Awareness of deterministic rules on the dynamics of the large-scale systems. For example, ideal operators have a rule that pressurizer pressure and pressurizer level will decrease in the case of a LOCA or SGTR accident in NPPs.
2. Full awareness of the failure modes and failure probabilities of various sensors and indicators.
3. Ability to correctly revise the probability distribution for the plant status based on Bayesian inference after observations of various indicators.
4. Ability to correctly calculate the expected information from each indicator, and select next most informative indicator to monitor.

The situation assessment process and changes in their understanding of plant status is explained by the assumptions of ideal operators. Can the model be applied to explain the situation assessment of real human operators? A degree of consistency between the situation assessment process of ideal operators and real human operators is believed to exist, even though real human operators have limited ability for mathematical calculations, and may behave illogically. The situation assessment model of real human operators is developed on the basis of the situation assessment model of ideal operators by identifying the limitations and characteristics of real human operators. The situation assessment model of ideal operators is used as the starting point for the development of the situation assessment model of real human operators.

11.2 An Integrated Model of I&C Systems and Human Operators

11.2.1 A Mathematical Model for I&C Systems and Human Operators

The mathematical model for I&C systems and human operators when the interdependency between I&C systems and human operators are considered is similar to the mathematical model (Section 11.1). The structure of the model and definitions of the variables are summarized in Figure 11.2. W indicates the plant status (the situation), Z_i (i = 1, 2, ..., m) indicates sensors, and Y_i indicates indicators. X indicates the operator situation model, V indicates the manual control signal, and U indicates the control signal. In mathematical form, the variables are defined as:

$$W = \{w_1, w_2, \cdots, w_l\} \tag{11.10}$$

$$V = \{v_1, v_2, \cdots, v_f\} \tag{11.11}$$

$$U = \{u_1, u_2, \cdots, u_g\} \tag{11.12}$$

Manual control is assumed to be determined by the situation model of human operators. The relations between the situation model of human operators and manual control are assumed to also be deterministic. Human operators are assumed to always decide to actuate the manual reactor trip if human operators believe that a LOCA is occurring in an NPP, but the automatic reactor trip is not actuated.

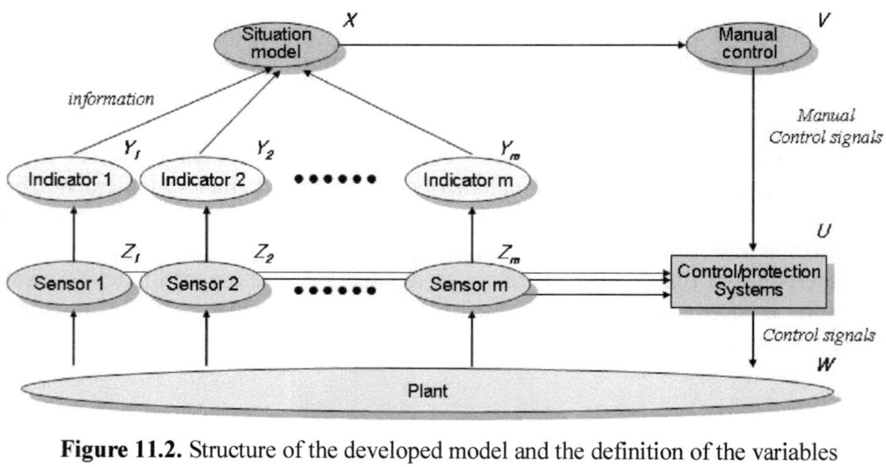

Figure 11.2. Structure of the developed model and the definition of the variables

The possibilities of action errors (*e.g.*, pushing the wrong button) in manual control are considered, and the conditional probability, $P(v_i \mid x_k)$ (i = 1, 2, ..., f and k = 1, 2, ..., l) in the mathematical model for manual control are determined. The estimation of the action error probabilities, which determine $P(v_i \mid x_k)$'s, follows conventional HRA methods such as ASEP, THERP, and HCR.

The reliabilities of I&C systems, which consist of sensors and control/protection systems (Figure 11.2) are calculated using fault tree analysis. The analysis is easier when the reliability graph with general gates (RGGG) method [8] is used. Reliabilities of I&C systems are used to determine conditional probabilities, $P(u_k \mid v_i, z_{1j} \dots, z_{mj})$ (k = 1, 2, ..., g, i = 1, 2, ..., f and k = 1, 2, ..., n).

The probability distribution for the control signal U is determined based on the plant status W and conditional probabilities.

11.3 An Application to an Accident in an NPP

11.3.1 Description on the Example Situation

The developed method is applied to an example situation. The example situation is a LOCA with CCF of pressurizer pressure sensors in a pressurized water reactor (PWR) plant. The Compact Nuclear Simulator (CNS) which was originally developed by KAERI and Studsvik Inc. in 1986, and recently renewed by KAERI, was used as the simulator. The reference plants of CNS are Kori 3&4 NPPs which are Westinghouse 900MWe 3-loop type PWR.

The parameter for pressurizer pressure measured by sensors for a normal operation value was fixed by modifying the simulation code. The physical meaning of the modification is the occurrence of the CCF of four pressurizer pressure sensors. A small LOCA with a break size of 5 cm^2 was induced. The trends of various plant parameters are shown in Figure 11.3, including the trends of actual pressurizer pressure and of pressurizer pressure measured by failed sensors. The measured pressurizer pressure does not change, while the actual pressurizer pressure continuously decreases. The automatic reactor trip signal by low pressurizer pressure criterion is not generated by RPS because the measured pressurizer pressure does not change. The generation of an automatic reactor trip signal by the over-temperature delta-T (OTΔT) criterion is expected in this situation. The above-right trend graph in Figure 11.3 shows the calculated OTΔT and its trip setpoint. This indicates that the calculated OTΔT is always below the setpoint, which means that an automatic reactor trip signal by the OTΔT criterion will not be generated. The CCF of pressurizer pressure sensors also prohibits the RPS from generating the reactor trip signal by the OTΔT criterion because the measured pressurizer pressure is also used to calculate the OTΔT. The automatic reactor trip will not occur and the accident continues to proceed due to the failure of generating a reactor trip signal by the two criteria. The decrease in pressurizer level and the setpoint for low pressurizer level alarm generation is shown in the

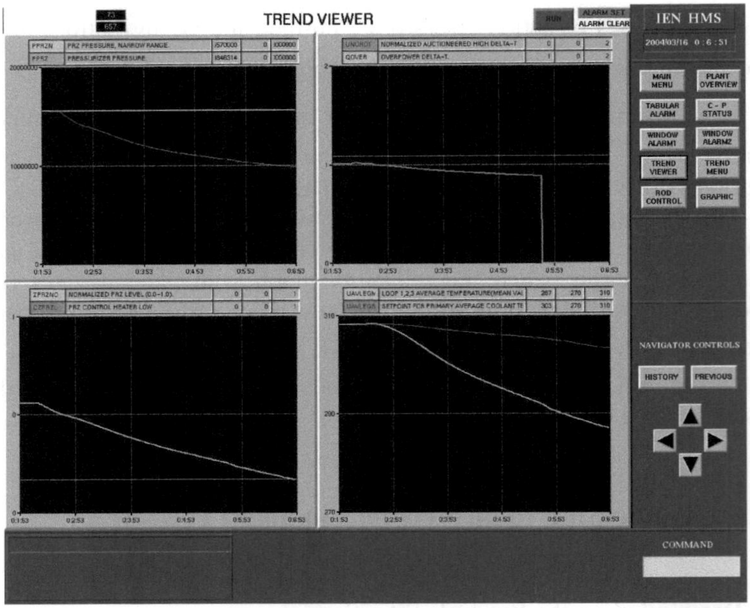

Figure 11.3. Trends of various plant parameters by CNS for the example situation

below-left trend graph of Figure 11.3. The average temperature of the reactor coolant system, T_{avg}, and the reference temperature determined by turbine load, T_{ref}, is shown in the below-right trend graph of Figure 11.3.

The SI signal for low pressurizer pressure will not be generated by the ESFAS due to the CCF of pressurizer pressure sensors. The CCF of pressurizer pressure sensors can simultaneously prohibit the RPS from generating the reactor trip signal and the ESFAS from generating the SI signal. Operators will see several alarms generated by control systems and alarm systems, which inform the operators of the occurrence of an abnormal situation. The generated alarms are shown in Figure 11.4. The role of the operator in this situation is to correctly recognize the occurrence of an accident and generate the manual reactor trip and SI actuation signals, and follow emergency operation procedures (EOPs). Whether operators can correctly recognize the occurrence of an accident is unknown, even though there are several symptoms that indicate the occurrence of a LOCA such as the decrease in the pressurizer level and the increase in the containment radiation. What is important is the probability that human operators correctly recognize the occurrence of an accident and generate the manual reactor trip signal or the SI actuation signal in the viewpoint of PRA. Conventional HRA methods cannot provide appropriate probabilities for these since only the allowable time or the type of tasks (skill-based, rule-based or knowledge-based) are considered.

Figure 11.4. Generated alarms by CNS for the example situation (the LOCA occurs at 3 minutes)

11.3.2 A Probable Scenario for the Example Situation

Operators may think that the plant is in normal operation before the occurrence of the accident is recognized. Human operators receive a containment radiation high alarm at 49 seconds after the accident (Figure 11.4). Operators will move to the containment radiation indicator and observe that containment radiation is increasing. This is an example of data-driven monitoring. Two possibilities are considered by operators in this situation: the failure of containment radiation sensors and indicators in normal operation, or the occurrence of a LOCA. Other indicators are monitored so that operators more clearly understand the situation. This is the process of knowledge-driven monitoring. The situation is understood as the failure of containment radiation sensors or indicators in normal operation if operators observe that pressurizer pressure does not change, due to the CCF of pressurizer pressure sensors. Other indicators need to be observed to make sure of their clear understanding of the situation. The possibility of the LOCA occurrence is considered if a decrease in reactor power is observed. The occurrence of a LOCA cannot be certain at this point, even though human operators think there is a possibility of a LOCA. Human operators are expected to monitor more indicators to clearly understand the situation.

11.3.3 Quantitative Analysis for the Scenario

A quantitative model for the example situation is shown in Figure 11.5. Four kinds of plant status, normal operation, LOCA, SGTR, and steam line break (SLB) are assumed, and seven indicators, reactor power, generator power, pressurizer pressure, pressurizer level, steam/feedwater deviation, containment radiation, and secondary radiation, are modeled. Each indicator has three states, increase, no change and decrease:

$$W = \{\text{normal operation, LOCA, SGTR, SLB}\} \quad (11.13)$$

$$Y_i = \{\text{increase, no change, decrease}\} \text{ where } i = 1, 2, ..., 7 \quad (11.14)$$

The plant is assumed to be in normal operation when human operators are unaware of the occurrence of the accident. Operator understanding of plant status is assumed to be:

$$P(X) = \{0.9997, 0.0001, 0.0001, 0.0001\} \quad (11.15)$$

Human operators receive a containment radiation high alarm at 49 seconds after the occurrence of the LOCA, when the increase in containment radiation is observed. This observation changes operator understanding of plant status, which is calculated using Equation 11.5. The Bayesian network model provides an equivalent result (Figure 11.6).

$$P(X) = \{\ 0.50055, 0.49935, 0.00005, 0.00005\ \} \quad (11.16)$$

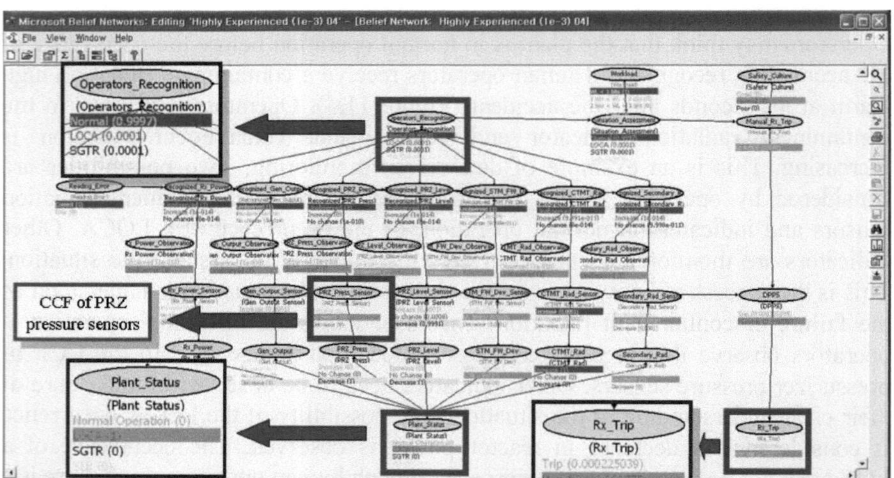

Figure 11.5. Bayesian network model for the example situation when the operators are unaware of the occurrence of the accident

Operators consider two possibilities: sensor failure in normal operation or a LOCA, with almost equal probabilities. The failure probability of the manual reactor trip is about 0.5 (Figure 11.6).

Operator understanding of plant status is changed if human operators monitor the pressurizer pressure indicators after monitoring the containment radiation indicators, and then observe that the pressurizer pressure does not change, due to the CCF of pressurizer pressure sensors, as:

$$P(X) = \{0.999102, 0.000798, 8\times10^{-8}, 0.0001\} \tag{11.17}$$

Human operators are likely to understand the situation as the failure of containment radiation sensors in normal operation if human operators do not monitor other indicators, and therefore, do not immediately actuate the manual reactor trip. Human operators are likely to observe that reactor power is decreasing, if operators monitor the reactor power indicators after monitoring the pressurizer pressure indicators. The possibility of observing no change in reactor power or that power is increasing, due to sensor or indicator failures, also have to be considered. Operator understanding of plant status is changed when all these possibilities, with corresponding probabilities, are considered, as:

$$P(X) = \{0.112777, 0.887072, 8.9\times10^{-5}, 6.1\times10^{-5}\} \tag{11.18}$$

Human operators put more belief on the occurrence of a LOCA, and therefore are more likely to actuate the manual reactor trip.

The change in operator understanding of plant status matches the description of the scenario in Section 11.3.2. The process for change in operator understanding of

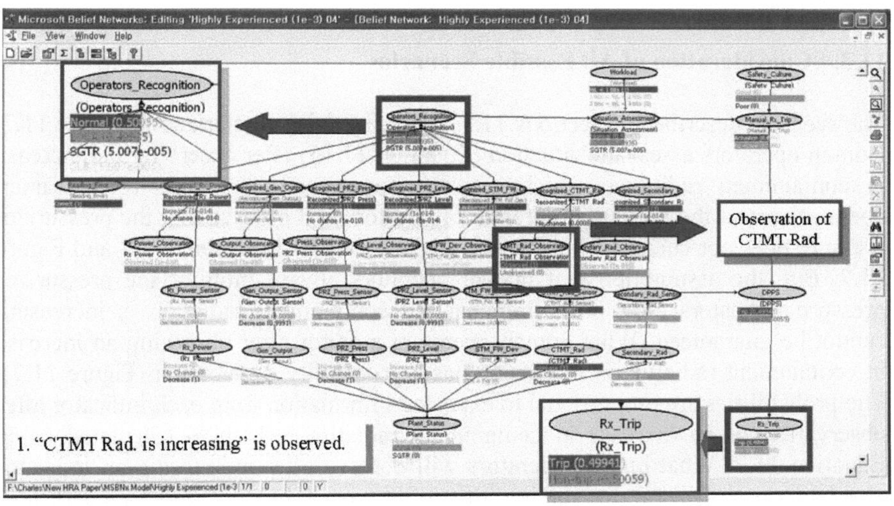

Figure 11.6. Bayesian network model for the example situation when the containment radiation is increasing is observed

Table 11.1. Change in operators' understanding of the plant status

	Indicator	Normal operation	LOCA	SGTR	SLB	Rx. trip failure probability
0	No observation	0.9997	0.0001	0.0001	0.0001	–
1	CTMT rad.	0.50055	0.49985	5.0×10^{-5}	5.0×10^{-5}	0.50095
2	PRZ press	0.999102	0.000798	8.0×10^{-8}	0.0001	0.99918
3	Rx. power	0.112777	0.887072	9.0×10^{-5}	6.0×10^{-5}	0.11284
4	PRZ level	0.001717	0.997416	0.000356	1.0×10^{-5}	0.001733
5	Gen. output	4.1×10^{-5}	0.998988	0.000947	2.4×10^{-5}	7.0×10^{-5}
6	STM/FW dev.	4.8×10^{-5}	0.999506	0.000446	2.7×10^{-7}	5.3×10^{-5}
7	2nd rad.	5.4×10^{-5}	0.999945	1.1×10^{-6}	2.9×10^{-7}	6.0×10^{-5}

plant status, and the failure probability of a manual reactor trip, as the human operators gradually monitor the indicators, is summarized in Table 11.1.

11.3.4 Consideration of All Possible Scenarios

The scenario described in Sections 11.3.2 and 11.3.3 is summarized in Figure 11.7. Human operators assess the situation (Equation 11.16) after observing the increase in containment radiation, which is also summarized in Figure 11.7. Human operators select the pressurizer pressure indicators and observe that the pressurizer pressure does not change. The situation is described by Equation 11.17 and Figure 11.7. But, the assumption that human operators always monitor the pressurizer pressure indicators after observing that the containment radiation is increasing cannot be guaranteed. What human operators monitor after observing an increase in containment radiation is increasing has a probability distribution (Figure 11.7). The probabilities are proportional to expected information from each indicator after observation of an increase in containment radiation, which is calculated using Equation 11.7. What human operators will observe after monitoring an indicator also has a probabilistic distribution, due to the possibilities of sensor or indicator failures. There are 18 possible observations, since there are six indicators and each indicator has three different kinds of states (Table 11.2). Probabilities of observations are all different. Each observation will produce a different operator understanding of the plant status. The observation of no change in pressurizer pressure and the resultant operator understanding of plant status is one possibility among 18 after observation of an increase in containment radiation (Table 11.2).

The expected probability distribution for operator understanding of plant status and the expected reactor trip failure probability are calculated (Table 11.2) based on the probabilities for 18 possibilities, the operator understanding of plant status for 18 possibilities, and the corresponding reactor trip failure probabilities. The change of operator understanding of plant status and the reactor trip failure probability can be calculated (Figure 11.8). As operators receive more information, their belief in the occurrence of a LOCA increases, and the failure probability of manual reactor trip decreases (Figure 11.9).

11.3.5 Consideration of the Effects of Context Factors

Quantitative effects of various context factors on the safety of large-scale systems is evaluated with developed framework by making quantitative assumptions. A brief summary of assumptions for the effect of context factors on the process of situation assessment of human operators is shown in Figure 11.10. The details of the quantitative assumptions for the example situation are summarized:

1. The sensor failure probabilities are assumed to be 0.001.
2. A quantitative assumption of the safety culture was made since the adequacy of organization affects the safety culture of the organization. The assumptions for the four levels of safety cultures are:

   - Very good: Manual reactor trip when any abnormal situation occurs.
   - Good: Manual reactor trip when LOCA, SGTR or SLB occurs.
   - Moderate: Manual reactor trip when LOCA or SGTR occurs.
   - Poor: Manual reactor trip when LOCA occurs.

   The safety culture is assumed to be moderate for quantitative analysis.
3. Quantitative assumptions for the three levels of working conditions are made since working conditions possibly affect error probabilities of indicator reading, as follows:

   - Advantageous: Indicator reading error = 0.0003
   - Compatible: Indicator reading error = 0.001
   - Incompatible: Indicator reading error = 0.003

   The working condition is assumed to be compatible for quantitative analysis.
4. Quantitative assumptions for the four levels of the adequacy of HMI are made since the adequacy of HMI affects the indicator reading time, as follows:

   - Supportive: Indicator reading time = 5 sec
   - Adequate: Indicator reading time = 10 sec

Table 11.2. Possible observations and resultant operator understanding of plant status after observing increased containment radiation

Observations		Probability	Normal operation	LOCA	SGTR	SLB	Rx. trip failure probability
Rx. power (p=0.249992)	Increase	2.5E-05	0.333456	0.333256	3.34E-05	0.333256	0.666696
	No change	0.0002	0.999201	0.000799	8.00E-08	8.00E-08	0.999176
	Decrease	0.249542	0.0001	0.9998	0.0001	1.00E-08	0.000106
Gen. output (p=0.249842)	Increase	2.5E-05	0.5001	0.4998	5E-05	5E-05	0.50014
	No change	0.0002	0.999201	0.000799	8.00E-08	8.00E-08	0.999176
	Decrease	0.249392	0.0001	0.9997	0.0001	0.0001	0.000206
PRZ press (p=0.249842)	Increase	0	0.5001	0.4998	5E-05	5E-05	0.50014
	No change	0.249617	0.999101	0.000799	8.00E-08	9.99E-05	0.999176
	Decrease	0	0.0001	0.9998	0.0001	1.00E-08	0.000106
PRZ level (p=0.249842)	Increase	2.5E-05	0.5001	0.4998	5E-05	5E-05	0.50014
	No change	0.0002	0.999101	0.000799	8.00E-08	9.99E-05	0.999176
	Decrease	0.249392	0.0001	0.9998	0.0001	1.00E-08	0.000106
STM/FW dev. (p=0.000319)	Increase	3.19E-08	0.250113	0.249962	0.249962	0.249962	0.500065
	No change	0.000319	0.50015	0.49985	4.00E-08	4.00E-08	0.50014
	Decrease	3.19E-08	0.5001	0.4998	5E-05	5E-05	0.50014
2nd rad. (p=0.000163)	Increase	1.63E-08	0.333456	0.333256	0.333256	3.34E-05	0.333484
	No change	0.000163	0.500125	0.499825	4.00E-08	5E-05	0.500165
	Decrease	1.63E-08	0.5001	0.4998	5E-05	5E-05	0.50014
Average		0.9991	0.250341	0.748626	7.49E-05	5.83E-05	0.250397

- Tolerable: Indicator reading time = 20 sec
- Inappropriate: Indicator reading time = 40 sec

The adequacy of HMI is assumed to be adequate for quantitative analysis.

5. The following quantitative assumptions were made since the time of day (circadian rhythm) affects the response time of the human operators:

   - Day-time (adjusted): Reading time + Situation assessment time
   - Night-time (unadjusted): (Reading time + Situation assessment time) × 1.5

   The time of day (circadian rhythm) is assumed to be day-time (adjusted) for the quantitative analysis.

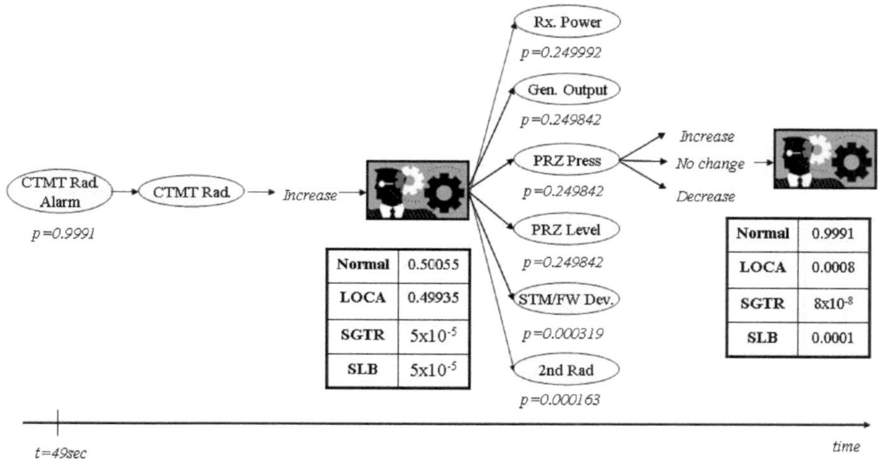

Figure 11.7. Change in operator understanding of plant status after observation of an increase in containment radiation

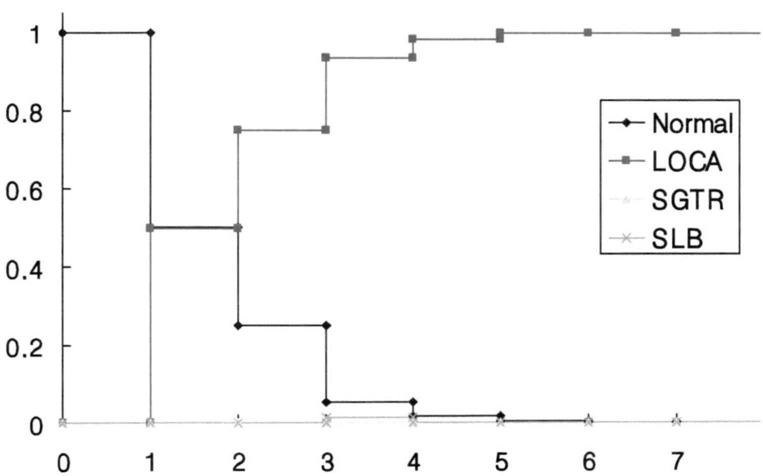

Figure 11.8. Change of operator understanding of plant status as operators monitor indicators

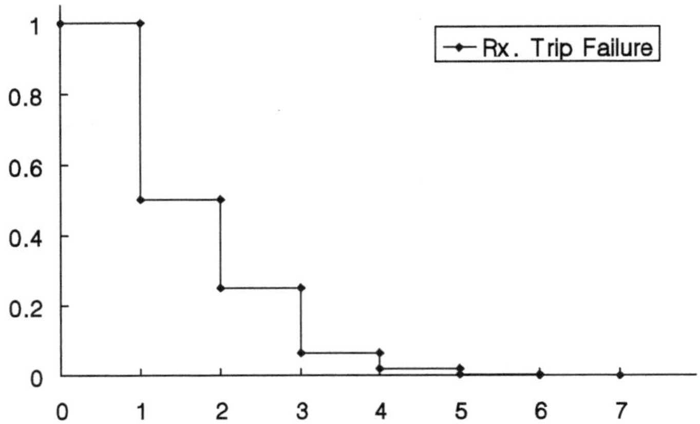

Figure 11.9. Change of reactor trip failure probability as operators monitor indicators.

6. The following quantitative assumptions are made since the adequacy of training is related to the knowledge or expertise of human operators:

   - Adequate, high experience: Know the behavior of all 7 indicators
   - Adequate, limited experience: Know the behavior of 6 indicators
   - Inadequate: Know the behavior of 5 indicators

   The adequacy of training is assumed to be adequate and human operators are assumed to be highly experienced for quantitative analysis,.
7. The following quantitative assumptions are made since the crew collaboration quality can affect verbal communication error probabilities:

   - Very efficient: Verbal communication error probability = 0.0003
   - Efficient: Verbal communication error probability = 0.001
   - Inefficient: Verbal communication error probability = 0.001
   - Deficient: Verbal communication error probability = 0.003

   Crew collaboration quality is assumed to be efficient for quantitative analysis.
8. Quantitative assumptions on the stress of human operators were made since the available time affects the stress of human operators. Quantitative assumptions for the four levels of the stress of human operators are:

   - Optimal: Inference error probability = 0
   - Moderately high: Inference error probability = 0.01
   - Very high: Inference error probability = 0.1
   - Extremely high: Inference error probability = 0.5

The stress is assumed to be optimal for quantitative analysis.

The effects of each context factor on the safety of large-scale systems are evaluated based on quantitative assumptions. Changes in reactor trip failure

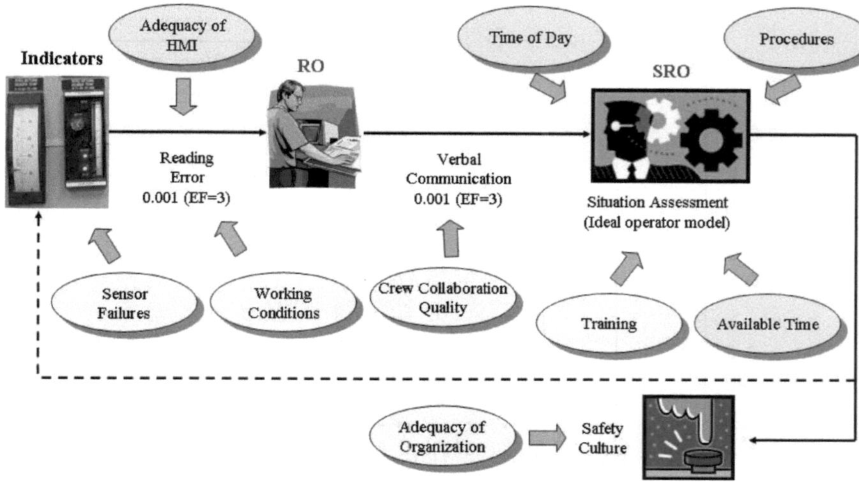

Figure 11.10. A brief summary of the assumptions for the effects of context factors on the process of situation assessment of human operators

probability as functions of time are shown in Figure 11.11, based upon the assumptions of four levels of the adequacy of organization (safety culture). The reactor trip failure probability is lowest when the safety culture is very good and highest when the safety culture is poor (Figure 11.11). The calculated reactor trip failure probabilities at 150 seconds based on the assumptions of four levels of the adequacy of organization (safety culture) are summarized in Table 11.3. The effects of context factors on the calculated reactor trip failure probabilities are summarized in Tables 11.4 to 11.9. The effect of the adequacy of HMI is shown in Figure 11.13. The effect of time of day (circadian rhythm) on the reactor trip failure probability is shown in Figure 11.14. The adequacy of procedure, available time, training/experience of human operators, and sensor failure probabilities are found to be relatively important compared to other factors, in this example situation.

11.4 Discussion

Several reliability issues for an integrated model of I&C systems and human operators in large-scale systems are discussed in Chapter 10. The integrated model, when applied to the example situation, suggests that faults in pressurizer pressure instruments can cause human operators to misunderstand the situation as a normal operation with the failure of containment instruments, even though a LOCA has occurred in the plant. The integrated model addresses the possible effects of instrument faults on human operators in the example application.

Signals from the control/protection systems are depicted as being dependent on the decisions of human operators (Figure 11.2). Signals from the digital plant protection system (DPPS) were modeled to be dependent on the decisions by

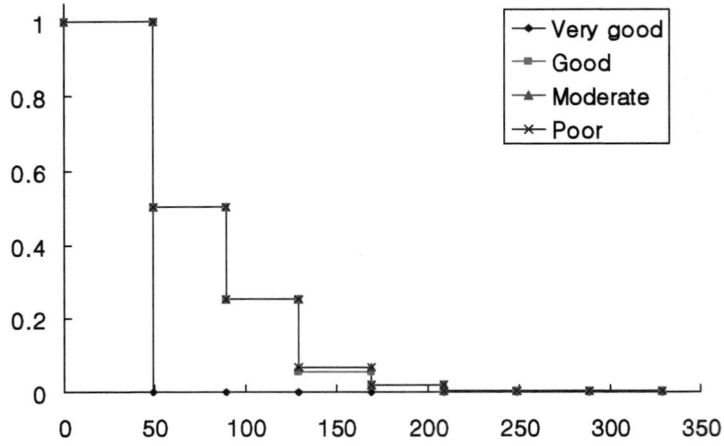

Figure 11.11. Changes of reactor trip failure probability as function of time (0 sec < Time < 500 sec)

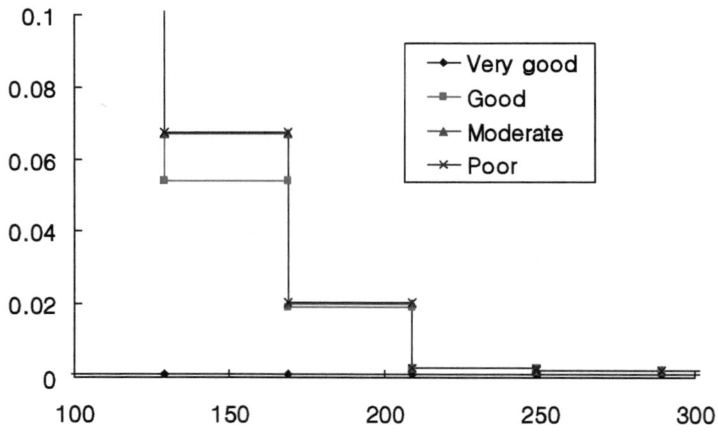

Figure 11.12. Changes in reactor trip failure probability as function of time (100 sec < Time < 500 sec)

Table 11.3. Effect of adequacy of organization (safety culture)

Very good	Good	Moderate	Poor
0.0009	0.54032	0.067228	0.067304

Table 11.4. Effect of working conditions

Advantageous	Compatible	Incompatible
0.066813	0.067228	0.068412

Table 11.5. Effect of crew collaboration quality

Very efficient	Efficient	Inefficient	Deficient
0.066813	0.067228	0.067228	0.068412

Table 11.6. Effect of adequacy of procedures

Appropriate	Compatible	Inappropriate
0.000907018	0.0028025	0.999725

Table 11.7. Effect of stress (available time)

Optimal	Moderately high	Very high	Extremely high
0.067228	0.239487	0.299788	0.452516

Table 11.8. Effect of training/experience

Highly experienced	Experienced	Inexperienced
0.067228	0.074113	0.1234

Table 11.9. Effect of sensor failure probability

$P = 0.0001$	$P = 0.0003$	$P = 0.001$	$P = 0.003$	$P = 0.01$
0.066036	0.0663	0.067228	0.069887	0.079306

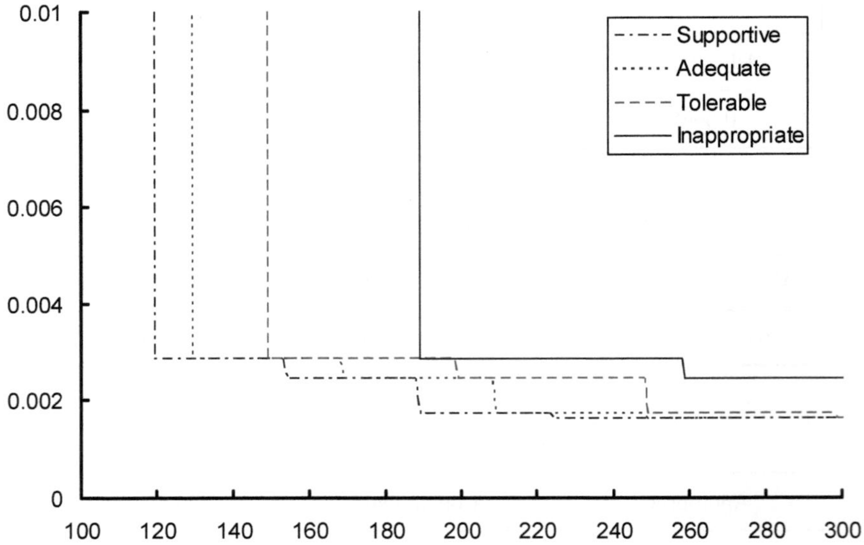

Figure 11.13. Effect of the adequacy of HMI

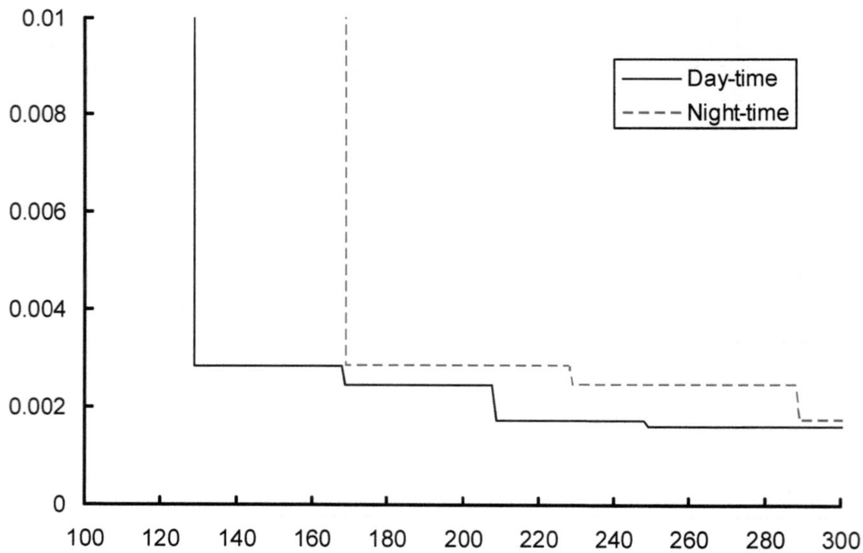

Figure 11.14. Effect of time of day (circadian rhythm)

human operators (Figure 11.5). The control signals are blocked by human operators if the operators believe that the plant is in a normal operation state. The integrated model models the dependency of I&C systems on human operators.

The integrated model models the effects of instrument faults on human operators. The integrated model can consider the possibilities of providing wrong information to human operators due to instrument faults, because the instrument faults provide wrong information to human operators. The integrated model considers the possibilities of providing wrong information by other operators due to communication errors, because the integrated model also considers the possibilities of failures in verbal communication (Figure 11.10).

Whether the human operators trust the provided information or not is a completely different issue, even though human operators receive information from instruments. Operator trust of instruments can be modeled in the integrated model when developing a model for operator rules (Figure 11.1). The information provided by the instrument has little effect on situation assessment of human operators if operators do not trust an instrument.

A reliability issue is the different difficulties of correctly diagnosing different accident situations (Chapter 10). The use of a human operator information-processing model, based on Bayesian inference in the integrated model, enables the integrated model to consider different difficulties in human operator situation assessment for different situations. An SLB accident in an NPP is easily diagnosed due to its unique symptom, an initial increase in the reactor power due to the insertion of positive reactivity. However, a LOCA and an SGTR accident in an NPP are not easily diagnosed due to similar symptoms (decrease in the reactor power, generator power, pressurizer pressure, and pressurizer level). Safety assessment based on the integrated model provides more realistic results compared to conventional safety assessment methods due to the ability to consider different difficulties in the diagnosis of different accident situations.

11.5 Concluding Remarks

The development of an integrated model of I&C systems and human operators in large-scale systems is described in this chapter. The integrated model was developed to address reliability issues described in Chapter 10, such as the effects of instrument faults on human operators, the dependency of I&C systems on human operators, possibilities of providing wrong information to human operators, operator trust in instruments, and different difficulties in correctly diagnosing different accident scenarios.

The application of the developed integrated method for example situation demonstrates that quantitative description by the integrated method for a probable scenario matches with qualitative description of the scenario. Available time, training/experience and sensor failure probability have important effects on reactor trip failure probabilities, according to this model. Application of the developed integrated method also demonstrated that it can probabilistically consider all possible scenarios. The method is used to quantitatively evaluate the effects of various context factors on the safety of large-scale systems.

The human operator information processing model used in the integrated model is not a mature model but an initial attempt to quantitatively model information processing of human operators. The integrated model described in this chapter is considered only as a basis for the development of a more advanced integrated model of I&C systems and human operators, and should consider controversies over the use of Bayesian inference to model the information processing of human operators. The development of a more advanced quantitative model for information processing of human operators will continue.

References

[1] Stanton NA, Chambers PRG, and Piggott J (2001) Situational awareness and safety, Safety Science, vol. 39, pp.189–204

[2] Endsley MR (1995) Toward a theory of situation awareness in dynamic systems, Human Factors, vol. 37, pp. 32–64

[3] Bendy G and Meister D (1999) Theory of activity and situation awareness, International Journal of Cognitive Ergonomics, vol. 3, pp. 63–72,

[4] Adams MJ, Tenney YJ, and Pew RW (1995) Situation awareness and the cognitive management of complex systems. Human Factors, vol. 37, pp. 85–104

[5] Park JC (2004) Private communication

[6] Barriere M., Bley D, Cooper S, Forester J, Kolaczkowski A, Luckas W, Parry G, Ramey-Smith A, Thompson C, Whitehead D, Wreathall J (2000) Technical Basis and Implementation Guideline for A Technique for Human Event Analysis (ATHEANA), NUREG-1624, Rev. 1, U.S. Nuclear Regulatory Commission, Washington D.C.

[7] Seridan TB and Ferrel WR (1981) Man-machine systems, MIT Press, Cambridge

[8] Kim MC and Seong PH (2002) Reliability Graph with General Gates: An Intuitive and Practical Method for System Reliability Analysis, Reliability Engineering and System Safety, vol. 78, pp. 239–246

12

INDESCO: Integrated Decision Support System to Aid the Cognitive Activities of Operators

Seung Jun Lee[1], Man Cheol Kim[2] and Poong Hyun Seong[3]

[1]Integrated Safety Assessment Division
Korea Atomic Energy Research Institute
1045 Daedeok-daero, Yuseong-gu, Daejeon, 305-353, Korea
sjlee@kaeri.re.kr

[2]Integrated Safety Assessment Division
Korea Atomic Energy Research Institute
1045 Daedeok-daero, Yuseong-gu, Daejeon, 305-353, Korea
charleskim@kaeri.re.kr

[3]Department of Nuclear and Quantum Engineering
Korea Advanced Institute of Science and Technology
373-1 Guseong-dong, Yuseong-gu, Daejeon, 305-701, Korea
phseong@kaist.ac.kr

The possibility of human failure or human error has a significant impact on the safety or reliability of large-scale systems. Most analysis results of accidents, including Chernobyl and TMI-2 accidents indicate that human error is one of the main causes of accidents. Forty-eight percent of incidents in an analysis of 180 significant NPP events occurring in the United States were attributed to failures in human factors [1]. Human factors are analyzed to prevent human errors and are considered in performing a more reliable safety assessment of a system. HRA and human factors are described in Chapter 7 and Chapter 8. An approach to assessing the safety of a system, including human operators, is introduced in Chapter 11, which suggests an integrated safety model that includes both digital control systems and human operators. The adequacy of procedures, stress (available time), training/experience of human operators, and sensor failure probabilities are found to be relatively important compared to other factors in a sensitivity analysis described in Section 11.3.5. The safety of a system is more affected by these four factors. Efficient improvement in the safety of a system is achieved by improving them in the system. Such factors related to humans have been becoming more important than other factors related to hardware and software because only highly reliable hardware and software components can be used in safety-critical systems, such as NPPs.

There are two approaches for preventing human errors. One is to improve the capabilities of humans and the other is to improve systems that assist humans. Good education and training which are included in the first approach are important. The improvement of HMI design and development of automated systems and COSSs help humans operate a system more easily, which results in the reduction of human errors and the improvement of system safety. The latter approach has negative effects in some situations, especially in safety critical systems.

Several problems which occur by adapting automated systems and COSSs to actual systems have been discussed: the automation surprise, adaptive system versus adaptable system problem, authority distribution, MABA–MABA (Men Are Better At – Machines Are Better At), complacency, the reduction of situation awareness, and skill degradation [2–4]. One of the most serious problems for adapting automated systems or COSSs is to establish whether human operators or the system should be the final decision maker [5]. Human operators are able to detect the failure and override the decisions of automated systems or COSSs when those systems fail to respond correctly. Some tasks need to be retained by human operators in order to reserve such backup capabilities of human operators. The problem of losing backup capabilities of human operators due to excessive automation is called "out-of-the-loop unfamiliarity" [6]. The automated system or COSSs that do not manage a particular problem will degrade the performance of human operators [7]. The concept of human-centered automation is considered for more efficient automation as the level of automation of an advanced MCR increases [8]. A moderate level of automation is important for maintaining the SA of human operators [2]. A fully automated system is more efficient for simple tasks, while a COSS is more efficient for managing complex tasks that operators need to comprehend and analyze, since high levels of automation may reduce the operator's awareness of the system dynamics. Operators in the MCR of an NPP must correctly comprehend a given situation in real time so that human operators and not systems are the final decision makers. COSSs are necessary for MCR operators to help in efficient and convenient operations, leaving operators as the final decision makers.

A COSS for MCR operators in NPPs, the integrated decision support system to aid cognitive process of operators (INDESCO) [9], is discussed in this chapter.

12.1 Main Control Room Environment

MCR operators in large-scale systems, such as NPPs, have a supervisory role for information gathering, planning, and decision making. These operation tasks are very complex and mentally taxing. The importance of human errors in NPPs has been a considerable concern since the early 1980s. Prevention of human errors occurs by improving MCR interface designs and developing support systems that allow more convenient operation and maintenance for large-scale systems. Systems and their interfaces have been designed by considering many characteristics of human operators.

The design of I&C systems of NPPs is rapidly moving toward full digitalization, with an increased proportion of automation [10]. The trend is moving toward the application of modern computer techniques in the design of advanced MCRs (modernized MCRs) for NPPs as the processing and information presentation capabilities of modern computers increase [11]. Advanced MCRs adapting modern computer techniques are much simplified by using LDPs and LCD displays, instead of conventional analog indicators, hand switches, and alarm tiles. These computerized systems are aimed at improving the performance of human operators by filtering or integrating raw process data, interpreting plant state, prioritizing goals, and providing advice. Human operators focus their attention on the most relevant data and highest priority problems and dynamically handle changing situations more easily using computerized systems. Computerized support of operational performance is needed to assist human operators, particularly in coping with plant anomalies, so that any failures of complex dynamic processes can be managed as quickly as possible with minimal adverse consequences [12].

Various kinds of automated and support systems are applied in advanced MCRs for safer and more stable operation. The roles of an HMI and decision support systems (DSSs) are shown in Figure 12.1. An independent DSS used in conventional MCRs is shown in the left diagram in Figure 12.1 and an HMI including a DSS for advanced MCRs is shown in the right diagram. DSSs are included as a part of an HMI because advanced MCRs are computer-based systems.

Combining HMI and DSSs into one system is more efficient in computer-based systems. There are various support systems at work for operators of large-scale systems, aiding with surveillance, diagnostics, and prevention of human errors. Some of these, such as early fault detection systems [13], are capable of doing tasks which are difficult for human operators. Operation validation systems are intended to prevent human errors [14]. More support systems will be adapted as MCRs evolve. A support system does not guarantee an increase in the performance of human operators, according to several published results [15]. Some support systems will degrade the SA capability of human operators and may increase the mental workload of human operators. The information provided by DSSs is not basic information but supplemental information. Using the information provided by DSSs may reduce their potential for human errors, although operators can operate a system without that additional information. Some support systems may generate helpful information for operators. Operators can identify and comprehend plant conditions more easily and may be able to recognize potential

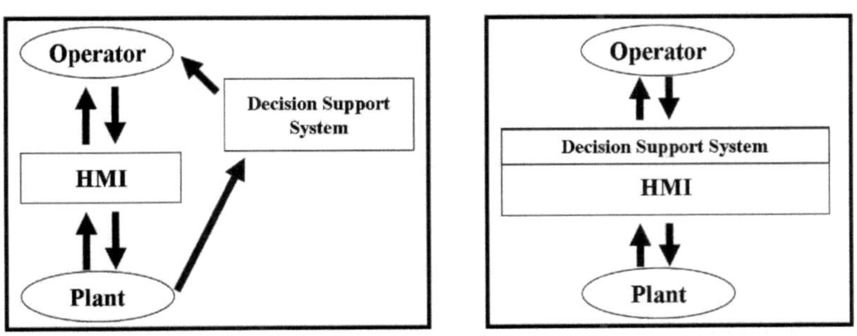

Figure 12.1. Independent support system and combined support system [9]

human errors before they make those errors using these systems. The performance of human operators can thus be enhanced using well-designed DSSs. However, some of these systems may generate unnecessary information. Operators seldom use or want to use overly informative systems. Information overload result from unnecessary information, and overly informative systems have adverse effects on the performance of human operators. Moreover, even if a DSS was proved to be generally efficient, the efficiency of the system will vary according to specific situational or environmental factors.

DSSs should be designed with consideration of two points. The first is to provide correct information and the second is to provide convenient and easy-to-use information. Most research, however, focuses on only the first point. The information from DSSs is useless in some situations, even if the information is perfectly correct. An adverse effect of a fault diagnosis system designed improperly has been shown in an experiment [15]. A type of fault diagnosis system provides only possible faults without their expected symptoms or causes in the experiment. Operators have to infer expected symptoms and compare them to plant parameters in order to confirm the results; this results in decreased performance. The use of a fault diagnosis system providing expected symptoms leads to increased performance of human operators. The performance of human operators is improved by the provision of not only accurate but also easy-to-use information to human operators.

12.2 Cognitive Process Model for Operators in NPPs

12.2.1 Human Cognitive Process Model

INDESCO is designed with a consideration of major cognitive activities for NPP operations underlying "A technique for human error analysis (ATHEANA) [16]" [9]. Understanding the basic cognitive processes associated with monitoring, decision-making, and control of a plant and how these processes can lead to human errors is important [17]. Appropriate DSSs are suggested through an analysis of the cognitive processes during operations.

Major cognitive activities for NPP operations underlying ATHEANA are: (1) monitoring/detection, (2) situation assessment, (3) response planning, and (4) response implementation [17]:

1. *Monitoring/detection*: This refers to the activities involved in extracting information from the environment.

2. *Situation assessment*: Humans actively try to construct a coherent, logical explanation to account for their observations when confronted with indications of the occurrence of an abnormal situation. This process is referred to as the situation assessment.

3. *Response planning*: This refers to the process of making a decision about which actions to take. For many cases in NPPs, when written procedures are available and deemed appropriate to the ongoing situation, the need to generate a response plan in real time may be essentially eliminated. However, operators still need to (1) identify appropriate goals based on their own situation assessment, (2) select appropriate procedures, (3) evaluate whether the actions described in the procedures are sufficient to achieve those goals, and (4) use the procedures for the ongoing situation as necessary.

4. *Response implementation*: This refers to taking specific control actions required to perform a task. Taking discrete actions or continuous control actions are involved in this activity.

12.2.2 Cognitive Process Model for NPP Operators

Human operators in an MCR monitor and control an NPP according to the human cognitive process. The relations among a human, an HMI, I&C systems, and a plant are shown in Figure 12.2 [18]. All HMIs in MCRs have display and implementation systems for monitoring and controlling the plant. Human operators obtain plant information through the display system in the HMI layer and assess the ongoing situation using the obtained information. Human operators select the operations corresponding to the assessed situation. Finally, the operations are implemented using the implementation system. The operation process of human operators is represented in this way using the human cognitive process. DSSs, in general, make use of one of the following two approaches to improve the

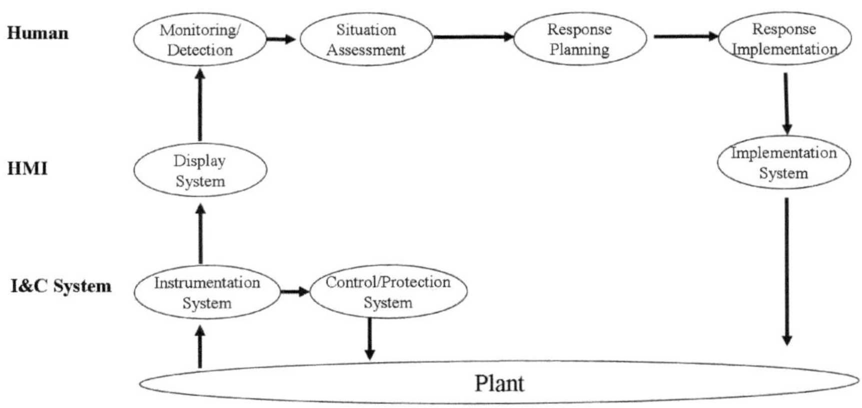

Figure 12.2. The operation process of human operators in large-scale systems [18]

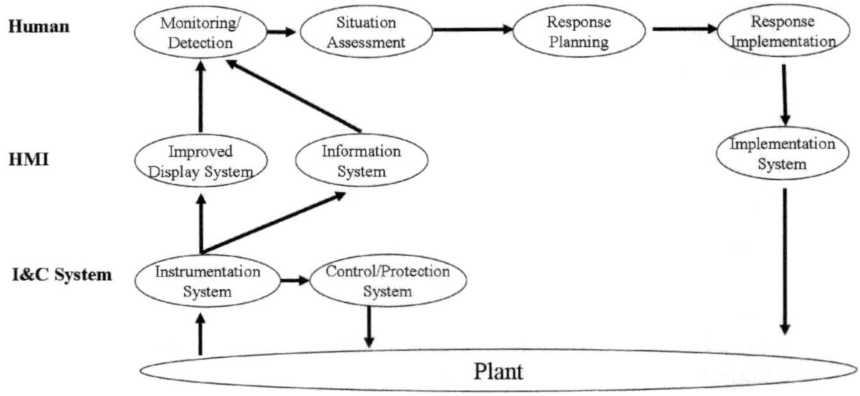

Figure 12.3. The operation process of a large-scale system with indirect support systems [9]

performance of human operators [19]. One approach is the improvement of MCR displays, which are considered as "indirect support." The indirect support system is described in Figure 12.3. Improved display systems using integrated graphic displays, configurable displays, and ecological interface designs and information systems, such as an alarm system, are examples of indirect support systems. These systems improve operator perceptual and awareness abilities. Operators perceive the plant status more easily and quickly using the information provided by the improved display system and using digested data from the information system. Indirect support systems improve the performance of the monitoring/detection activities in the cognitive process of human operators.

The second approach is the development of DSSs, which are called "direct support." Direct support systems include intelligent advisors, computer-based procedures, fault diagnostic systems, and computerized DSSs. Several direct support systems, such as a fault diagnosis system and a CPS, are added as part of the advanced HMI (Figure 12.4). The fault diagnosis system assists and supports

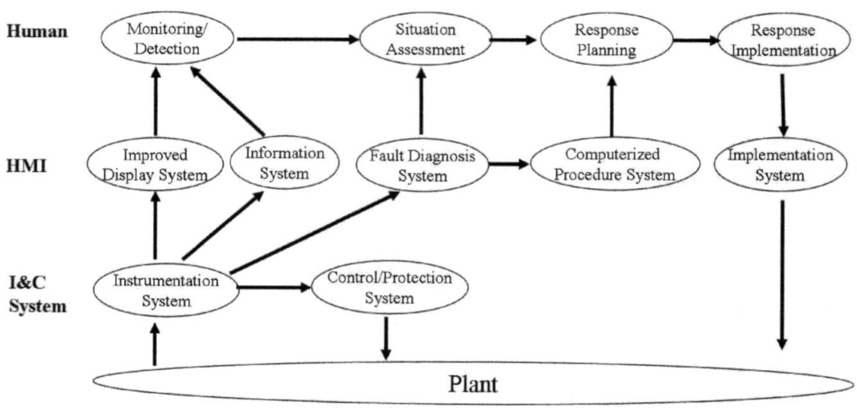

Figure 12.4. The operation process of a large-scale system with direct and indirect support systems [9]

situation assessment activities of the human operator cognitive process. Response planning activities are supported by a CPS in a similar way. The relationships among an operator, an HMI, I&C systems, and a plant even if the design and components of an HMI are changed are represented by the model in Figure 12.4. The model shows which cognitive activity an added support system relates to and supports. Indirect support systems mainly support monitoring/detection activities, which is the first of the four major cognitive activities. Several kinds of direct support systems support other cognitive activities (Figure 12.4). Support systems necessary to support specific cognitive activities are suggested and selected based on this model.

12.3 Integrated Decision Support System to Aid Cognitive Activities of Operators (INDESCO)

12.3.1 Architecture of INDESCO

Various support systems for the development of INDESCO are separately developed to aid major cognitive activities in the human cognitive process. Those support systems are integrated into one system to maximize efficiency. Each support system in INDESCO supports its corresponding cognitive activities. INDESCO supports all major cognitive activities in an integrated manner. The conceptual architecture of INDESCO is shown in Figure 12.5. The integrated HMI including the DSSs and the included DSSs support four major cognitive activities (Figure 12.5).

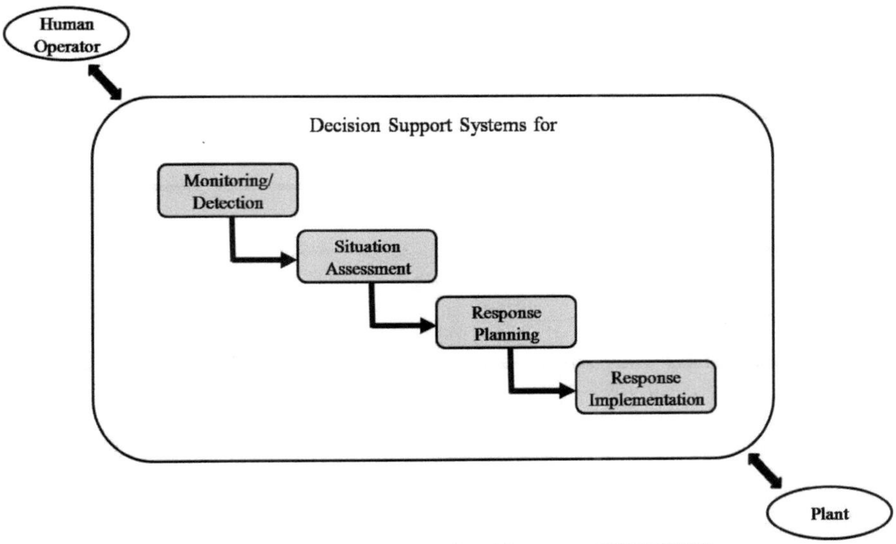

Figure 12.5. The conceptual architecture of INDESCO

12.3.2 Decision Support Systems for Cognitive Process

Various indirect or direct support systems are added to the HMIs to support cognitive process activities. The most appropriate support systems are selected based on the cognitive process of human operators to enhance the operational efficiency. Several kinds of support systems and related cognitive activities are selected (Figure 12.6). A display system, which is an indirect support system, supports the monitoring/detection activities. A fault diagnosis system, a CPS, and an operation validation system are direct support systems and support three other cognitive activities. Several sub-systems can be added, such as an alarm prioritization system, an alarm analysis system, a corresponding procedure suggestion system, and an adequate operation suggestion system, which also support cognitive activities. The former four systems are classified as main support systems since the latter four systems are implemented as sub-systems. These support systems facilitate the whole operation process of human operators, which include monitoring plant parameters, diagnosing the current situation, selecting corresponding actions for the identified situation, and performing the actions.

12.3.2.1 Support Systems for the Monitoring/detection Activities

Monitoring/detection activities access a high volume of information from a large-scale system in order to detect abnormal situations. This kind of activity is performed by instruments and alarms in MCRs. Operators always monitor the instruments and alarms in order to detect variations in instrument values, changes of color or sounding of alarms. Operators proceed to situation assessment upon detecting an abnormal situation. There are many instruments that indicate the status of the large-scale system such as NPPs. The sheer number of instruments makes it impossible for operators to individually examine each. An analysis of all instruments is the best way to ensure correct detection and diagnosis. Operators

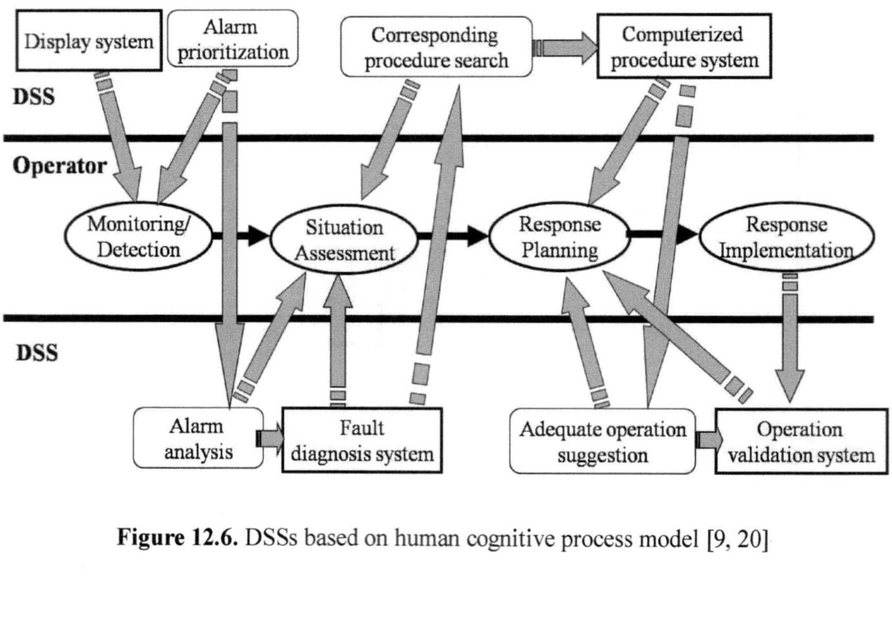

Figure 12.6. DSSs based on human cognitive process model [9, 20]

have to consider too many instruments and an operation will take too much time if there is no alarm which serves as a major information source for detecting process deviations. A slow reaction of the operator could result in accidents with serious consequences. Alarms help operators to make quick detections by reducing the number of instruments that must be considered. Alarms are helpful, but there are too many of them. A typical MCR in an NPP has more than a thousand alarms. Hundreds of lights turn on or off within the first minute in emergency situations such as LOCA or SGTR in an NPP. To have many alarms that repeatedly turn on and off causes confusion for human operators.

There are two approaches to supporting monitoring/detection activities. The first approach is to improve the interface of an MCR, and the second approach is the development of an advanced alarm system.

Advanced MCRs have been designed as fully digitalized and computer-based systems with LDP and LCD displays. These display devices are used for more efficient display, but have several disadvantages. A more flexible information display is possible by using LDPs and computerized display systems. Human operators select and monitor only necessary information. However, the space of computerized display devices is limited. Human operators must navigate screens in order to find necessary information. Excess information on a system increases the number of the necessary navigations. The system becomes inefficient if too many navigations are required to manipulate a device or to read an indicator. A key support for monitoring/detection activities is the efficient display of information.

An advanced alarm system supports monitoring/detection activities. Conventional hardwired alarm systems, characterized by one sensor–one indicator, may confuse operators with avalanching alarms during plant transients. Conventional alarm systems possess several common problems, including too many nuisance alarms and annunciating too many conditions [21]. Advanced alarm systems feature general alarm-processing functions, such as categorization, filtering, suppression, and prioritization. Such systems also use different colors and sounds to represent alarm characteristics. These functions allow operators to focus on the most important alarms.

12.3.2.2 Support Systems for Situation Assessment Activities

Operators analyze the situation at hand, make a situation model, and generate appropriate explanations for the situation during situation assessment activities. Systems which analyze the information related to the ongoing situation and generate estimated faults and expected symptoms are useful for supporting situation assessment activities. Fault diagnosis systems and alarm analysis systems are two examples.

Human operators make operation plans based on operating procedures which are categorized into two types: event-based procedures and symptom-based procedures. Different support systems are assigned to situation assessment activities on the basis of these different procedure types. Human operators start to execute procedural operations after identifying a situation in case of event-based procedures. Thus, fault diagnosis systems offering expected faults are useful for quick and easy situation assessment. Human operators using a symptom-based procedure do not begin by diagnosing a situation. Instead, they determine appropriate procedures by comparing the procedure entry conditions with ongoing parameters, and then act according to the selected procedure. A system to suggest appropriate procedures for a given situation is more useful than a fault diagnosis system for operators using such a method.

A critical issue for the support of situation assessment activities is the reliability of the support system. Human operators will not trust the information provided by support systems without a high degree of reliability. The support system will be rendered ineffective if human operators must always consider the possibility of incorrect results. There have been research projects using knowledge bases, neural networks, genetic algorithms, and other means to develop more reliable fault diagnosis systems [22–25].

12.3.2.3 Support Systems for Response Planning Activities

Response planning activities involve the situation model of human operators for the plant state to identify goals, generate alternative response plans, evaluate response plans, and select the most appropriate response plans relevant to the situation model. One or more steps may be skipped or modified in a particular situation [17]. When written operating procedures are available and judged appropriate to the situation, human operators handle the situation according to those procedures (Section 12.2.1). Errors arising from the omission of a step or selection of an incorrect step are of particular concern in such cases. Written operating procedures are designed to avoid such errors, and procedures intended to avert emergent situations are designed with more strict and formal linguistic formats. NPP EOPs intended to handle most serious accidents consists mainly of IF-THEN-ELSE statements.

There is the potential for human errors, though human operators may be provided with well-written procedures. The information can sometimes be overwhelming, making it difficult to continuously manage the requisite steps, since the content of the paper-based operating procedure is written in a fixed format in natural languages. CPSs have been developed and implemented since the 1980s due to deficiencies in paper-based operating procedures [26, 27]. Information about procedures and steps, relations between the procedures and steps, and the parameters needed to operate the plant are displayed in CPSs. Such systems also provide functions, such as a function of checkoff provisions, to prevent human errors, such as omitting a step or selecting an incorrect step. System functions, such as the provision of a list of candidate operations, may help human operators determine which operation should be performed next.

12.3.2.4 Support Systems for Response Implementation Activities

Response implementation activities are those activities which execute the selected operation after planning a response (*e.g.*, flipping a switch or closing a valve). Simple errors rather than decision-making errors are of concern in this step. Operators can commit an unsuitable operation even though they correctly assessed a situation and made an appropriate plan. Accidents caused by such commission errors have been reported.

Support systems to response implementation activities, such as an operation validation system, have been proposed to prevent commission errors [28]. The objective of the operation validation system is to detect inadequate operations and to warn operators about them in order to allow a chance to double-check operations which might result in commission errors. One of the most important considerations in the design of an operation validation system is to optimize the system-initiated interruptions. Such a system allows operators to do as they prefer when operators follow operation rules and procedures [14]. Too many interruptions may result in excessive operation validation time, although a validation system should interrupt all operations which may go wrong. Repeated interruptions may make operators insensitive to them. The double check loses its original significance if operators are frequently required to double-check their operations. The objective cannot be accomplished by a validation system with a too liberal validation filter. Therefore, an optimized and efficient filtering algorithm is needed to validate operations are necessary.

12.4 Quantitative Effect Estimation of Decision Support Systems

Several kinds of DSSs were introduced in Section 12.3.2. The systems are independent or an integrated system to support all cognitive activities of human operators. The effects are estimated in cases where no DSS is used, one or two DSSs are used, or all DSSs that aid complete cognitive activities are used. Evaluations are performed using the evaluation model introduced in Chapter 11. Several nodes are added to the model for operator situation assessment in order to consider the effects of DSSs. HRA event trees are used to define additional nodes and their relations pertaining to DSSs. Several PSFs are considered in order to create a model that considers human operators. Operator expertise and stress level are used as PSFs. Several assumptions are made to perform evaluations because the values about the reliability and the estimated effects of the DSSs are not clearly defined. Evaluations are performed for two scenarios based on the implemented model under certain assumptions [20].

12.4.1 Target System of the Evaluation

The target system of the evaluation is an application of INDESCO (Figure 12.7). Each main system supports each major cognitive activity for the operation of an NPP, based on the cognitive model used in ATHEANA, as explained in the previous section. The application has four main systems, two sub-systems, and four databases. All of the systems are connected and work together (Figure 12.7). The suggested architecture is just one application of INDESCO. INDESCO is also redesigned to adopt such systems whenever more efficient and useful support systems are developed. Only four main support systems are considered in order to simplify the evaluation; these are the display system, the fault diagnosis system, the CPS, and the operation validation system.

12.4.2 HRA Event Trees

The BBN model (Chapter 11) is modified by adding nodes related to DSSs. HRA event trees are used to define the relations among those nodes in the modified model. The basic HRA event tree (Figure 12.8) does not include any DSS. The final operation result is correct only if all tasks over the four steps are correct. a_c and a_w indicate the probabilities that a human operator reads an analog indicator correctly and incorrectly, respectively. In the same way, b_c and b_w indicate the probabilities of correct and incorrect situation assessment by a human operator; c_c and c_w indicate the probabilities of right and wrong operation selection by a human operator without checkoff provisions; and d_c and d_w indicate the probabilities as to whether a human operator performs an action correctly and incorrect, respectively. The HEP in reading digital indicators is used instead of analog indicators, when digital indicators are used instead of analog indicators. The structure of the basic HRA event tree is not changed in this case. e_w indicates the HEP in reading digital indicators. The HEP for omission error is changed to an HEP that considers checkoff provision if a function for checkoff provision is provided by the CPS. The structure of the basic HRA event tree is also not changed in this case. g_w indicates the HEP for omission error when a function for checkoff provision is provided.

New branches are added to the basic HRA event tree when a fault diagnosis system or an operation validation system which detects erroneous decision making and provides an additional opportunity to correct such errors is used. The HEP event tree for those cases is shown in Figure 12.9. f_c and f_w indicates the probabilities whether the fault diagnosis system generates correct and incorrect results and h_c and h_w indicates the probabilities whether the operation validation system detects operator wrong actions or fails to detect, respectively. Three

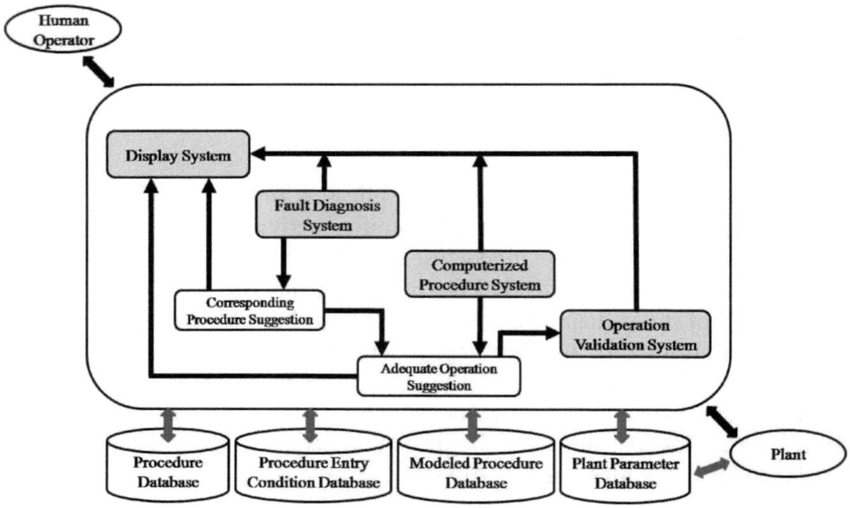

Figure 12.7. The architecture of an application [9]

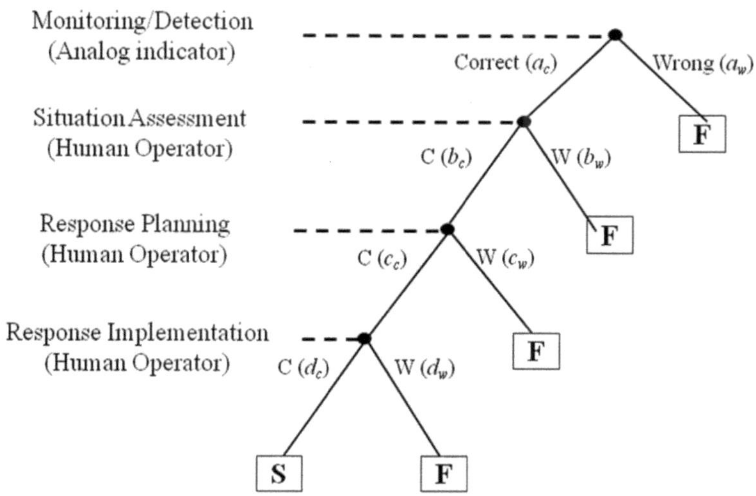

Figure 12.8. HRA event tree in the case of no DSS [20]

parameters are considered with regard to recovery probabilities. These parameters represent the cases where the decision of human operators is different from that of the DSSs. The whole HRA event tree that considers these parameters is shown in Figure 12.9. The recovery probability q means that the human operators do not change their correct decision even if the fault diagnosis system generates incorrect results. Operators are capable of identifying inappropriate recommendations from the fault diagnosis system, based on their knowledge and experience, because the

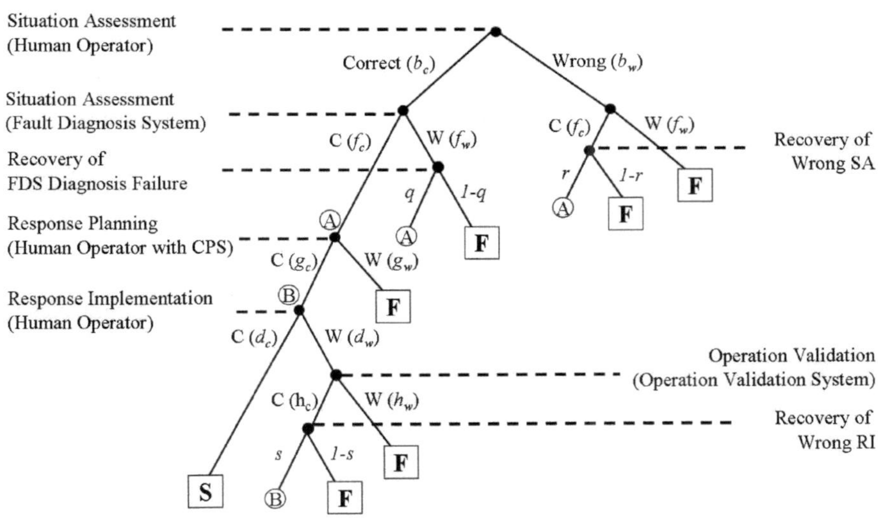

Figure 12.9. HRA event tree when all DSSs are used [20]

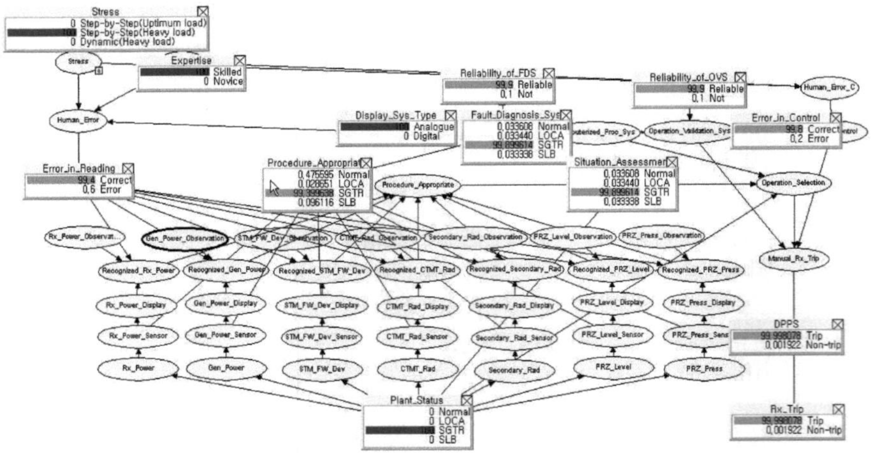

Figure 12.10. BBN model for the evaluation [20]

fault diagnosis system provides a list of possible faults and their expected causes. r represents the probability that human operators recognizes wrong diagnosis results from the fault diagnosis system. r indicates the recovery probability that human operators change their decision according to correct results of the fault diagnosis system, when they assess the current situation incorrectly. Human operator faults are corrected by consulting the correct diagnosis results of the fault diagnosis system, even if human operators assess the current situation incorrectly. r represents the probability of those cases. The probability of a correct situation assessment is defined in mathematical form when a fault diagnosis system is considered:

$$\text{Probability of correct situation assessment} = b_c(f_c + f_w q) + b_w f_c r \quad (12.1)$$

where:

q: human operators' recovery probability from diagnosis failures of the fault diagnosis system

r: recovery probability of fault diagnosis systems from human operators' wrong situation assessment

The objective of the operation validation system is to detect operator commission errors, which are inappropriate actions for the situation of the plant. s indicates the case where human operators recognize their mistake via the warnings of the operation validation system. The probability of correct response implementation is defined in mathematical form when the operation validation system is used:

$$\text{Probability of correct response implementation} = d_c + d_w g_c s \quad (12.2)$$

where:

s: recovery probability of the operation validation system from human operators' wrong response implementation

The modified model from the model described in Chapter 11 is constructed by adding nodes for the DSSs based on the implemented HRA event trees (Figure 12.10).

12.4.3 Assumptions for Evaluations

Assumptions were made for the evaluations. DSSs, such as fault diagnosis systems and operation validation systems, are in the development phase, and do not have widely accepted HEP values. The objective of this evaluation is not to analyze the impact of certain specific systems that have already been developed, but to estimate the effect of the integrated DSS supporting human cognitive activities. Therefore, values of several parameters pertaining to DSSs are assumed.

The software tool, HUGIN, is used as a tool for the analysis of Bayesian networks [29, 30]. The evaluation model (Figure 12.10) was developed based on the following conditions and assumptions:

1. Only four representative states of an NPP (normal operation, LOCA, SGTR, and SLB) are considered for simplicity in the evaluations. Therefore,
   X = {normal operation, LOCA, SGTR, SLB}
2. Only fifteen sensors and indicators, reactor power, generator output, pressurizer pressure, pressurizer level, steam/feedwater deviation, containment radiation, secondary radiation, wid-erange water levels of steam generator (SG) A and B, steam pressures of SG A and B, feedwater flowrates of SG A and B, and steam flowrates of SG A and B are considered in the evaluations for simplicity. NPP operators are assumed to know that each indicator has three different states: increase, no change, and decrease. Therefore, Y_i = {increase, no change, decrease}, where i = 1,2,...15.
3. The possibilities of sensor failures are considered. NPP operators are assumed to believe that all fifteen sensors have an equal unavailability, 0.001, and that each sensor has three failure modes, fail-high, stuck-at-steady-state, and fail-low for simplicity. Therefore, Z_is are given as follows: Z_i = {normal operation, fail-high, stuck-at-steady-state, fail-low}, where i = 1,2,...,15.
4. NPP operators are assumed to believe that the probability distribution for Z_is, *i.e.*, $p(Z_i)$s, are given as: $p(Z_i)$ = {0.999, 0.0001, 0.0008, 0.0001}.
5. The initial probability distribution for the plant state without any observation is assumed to be: $P(x)$ = {0.9997, 0.0001, 0.0001, 0.0001}.
6. The four plant states and fifteen indicators are selected because the combination of four states and fifteen indicators is useful for demonstration of the model. NPP operators are assumed to consider all fifteen indicators in order to assess the ongoing situation.

7. Two PSFs, operator expertise and operator's stress level, are considered. The PSFs are mainly used in THERP since the HEPs used in the evaluations are from THERP [31]. Operator expertise has two states, a novice group and a skilled group. The stress level changes according to the task load, and the task load factor is assumed to have three states, a step-by-step task with an optimum load, a step-by-step task with a heavy load, and a dynamic task with a heavy load.
8. Indicators are classified into two types: analog and digital. HEPs for reading indicators in THERP are used (Table 12.1). Three factors are considered for the HEPs in reading of indicators; task load, expertise, and type of indicator.

Table 12.1. HEPs for the reading of indicators [20]

Task load	Step-by-step (optimum load)		Step-by-step (heavy load)		Dynamic (heavy load)	
Expertise	Skilled	Novice	Skilled	Novice	Skilled	Novice
Analog indicator	0.003	0.003	0.006	0.012	0.015	0.030
Digital indicator	0.001	0.001	0.002	0.004	0.005	0.010

9. NPP operators are assumed not to use checkoff provisions without a CPS and to be provided a function for checkoff provisions with a CPS. The values used as the HEPs for omission errors are shown in Table 12.2. The length of the target list and usage of a checkoff provision are considered for the HEPs for omission of an item. The target operating procedure is assumed to include more than ten steps because almost all emergency operating procedures have more than ten steps.

Table 12.2. HEPs for omission per item of instruction when the use of written procedures is specified [20]

		Median joint HEP	EF
Without checkoff provisions	Short list (<= 10 items)	0.003	3
	Long list (> 10 items)	0.01	3
With checkoff provisions	Short list (<= 10 items)	0.001	3
	Long list (> 10 items)	0.003	3

10. The possibility of action errors (*e.g.*, pushing the wrong button) in the manual control are considered. There may be no commission error or negligible error if there is just one control switch, such as a reactor trip switch. However, NPP operators are assumed to be able to commit a commission error by similar control switches, such as an SG A isolation switch or an SG B isolation switch. HEPs for such commission error may depend on interfaces, and THERP provides HEPs considering these factors (Table 12.3). The SG isolation switches are assumed to be identified by labels only.

Table 12.3. HEPs for commission errors in operating manual controls [20]

Select wrong control on a panel from an array of similar-appearing controls	Median joint HEP	EF
Identified by labels only	0.002	3
Arranged in well-delineated functional groups	0.001	3
Which are part of a well-defined mimic layout	0.0005	10

11. Three reliability levels are assumed for the fault diagnosis system and operation validation system (95%, 99%, and 99.9%); the values concerning the reliability and the effect of these systems are not clearly estimated.
12. Group operations are not considered in the evaluation model. In fact, a group of operators consisting of more than one operator operates a plant in real MCRs. However, operation processes of one operator are considered for simplicity.
13. NPP operators are assumed to be able to detect wrong results of the fault diagnosis system based on their knowledge and experience and to correct their wrong decisions by receiving appropriate advices from DSSs. Skilled operators are assumed to have more capabilities against those cases than novice operators. Skilled operators are assumed to be able to detect a wrong result in the fault diagnosis system with a probability of 50%; novice operators are assumed to be able to detect with a probability of 30% (the recovery probability q). Skilled operators are assumed to be able to correct their wrong decision according to the correct diagnosis of the fault diagnosis system with a probability of 50%; novice operators are assumed to be able to correct with a probability of 30% (the recovery probability r). The skilled operator is also assumed to be able to recognize their wrong action by considering the advice of the operation validation system with a probability of 70%; novice operators are assumed to be able to correct with a probability of 50% (the recovery probability s).

12.4.4 Evaluation Scenarios

The evaluation scenario comprises the occurrence of an SGTR with the CCF of pressurizer pressure sensors in a Westinghouse 900MWe-type pressurized water reactor NPP, which describes Kori Unit 3 and 4 and Yonggwang Unit 1 and 2 NPPs in the Republic of Korea. The simulator used in the evaluations is the CNS [32] introduced in Chapter 11. The PPS is assumed to not generate an automatic reactor trip signal and the ESFAS does not generate an automatic safety injection actuation signal due to the CCF of pressurizer pressure sensors in the simulation. Operators have to correctly understand the state of the plant as well as manually actuate reactor trip and safety injection.

Operators are required to perform two operation tasks against two evaluations in the scenario. The operation task in the first evaluation is to trip the reactor manually. The operation task in the second evaluation is to isolate the failed SG. The failed pressurizer pressure sensors cause the PPS to fail to trip the reactor automatically under these conditions. Operators have to diagnose the current status correctly and trip the reactor manually. Operators also have to identify the failed SG and isolate it.

Evaluations are performed for the following seven cases:

- Case 1: No DSS is used and the indicator type is analog.
- Case 2: The indicator type is digital.
- Case 3: The indicator type is analog and the fault diagnosis system is used.
- Case 4: The indicator type is digital and the fault diagnosis system is used.
- Case 5: The indicator type is analog and a CPS is used.
- Case 6: The indicator type is digital, and the fault diagnosis system and the CPS are used.
- Case 7: The indicator type is digital, and the fault diagnosis system, the CPS, and the operation validation system are used.

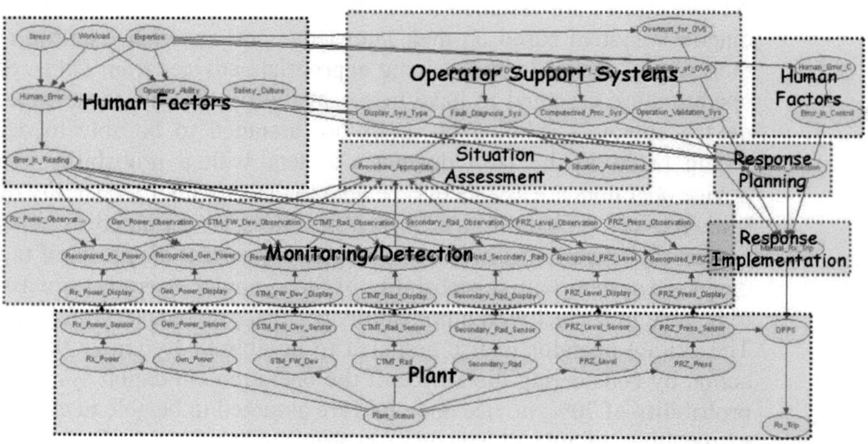

Figure 12.11. BBN model for Case 7 [20]

HRA event trees are made for all cases (Figures 12.8 and 12.9). BBN models for seven cases are constructed based on HRA event trees. Numerous nodes represent factors for humans, cognitive processes and DSSs. Their relationships are represented by arcs among the nodes. The BBN model for Case 7 is shown in Figure 12.11. There are nodes representing the plant and sensors at the bottom of the figure. Nodes for PSFs are in the upper left and upper right sides. There are also nodes for DSSs and major cognitive activities.

12.4.5 Evaluation Results

The results of the evaluations are obtained using the implemented BBN models (Tables 12.4 and 12.5). The values represent the failure probabilities of the tasks. The probability of situation assessment for a skilled operator without a DSS in the first evaluation, $P(X)$, is shown in Equation 12.3, and the BBN model for this case is shown in Figure 12.12. The final result is 0.017444, which represents the probability that a skilled operator fails to trip the reactor in the SGTR situation with the CCF of pressurizer pressure sensors.

$$P(X) = \{0.005403, 0.001466, 0.992490, 0.000641\} \quad (12.3)$$

The operation validation system is not considered in the first evaluation because no commission error is considered. This explains why the result values for Case 6 and Case 7 are identical in the first evaluation. The effect of the operation validation system is reflected in the result of the second evaluation; the result value for Case 7 is less than that for Case 6.

DSSs are shown to be helpful for reducing the failure probabilities of operators. The failure probability of the reactor trip operation is 0.017444 when a DSS is not used. The probability, however, is decreased to 0.004988 when four DSSs supporting major cognitive activities are used. The reliabilities of the fault diagnosis system and the operation validation system are both 99.9%. The failure probability is reduced by 71.4%. The failure probability of a novice operator without a DSS is 0.023344, but with all DSSs having 99.9% reliabilities, the failure probability is 0.006990. Here, the failure probability is reduced by 70.1%. The failure probability of a skilled operator without a DSS is 0.022820 for the failed SG isolation operation, and that of a skilled operator with all DSSs having 99.9% reliabilities is 0.006651. The failure probability is also reduced by 70.9% in this case. The failure probability of a novice operator without a DSS is 0.028994; with all DSSs having 99.9% reliabilities, it is 0.010370. The failure probability is reduced by 64.2%.

The DSSs yield good results when the fault diagnosis system and the operation validation system have 99% reliabilities. The failure probability of a skilled operator is reduced by 45.7% and that of a novice operator is reduced by 51.1% in

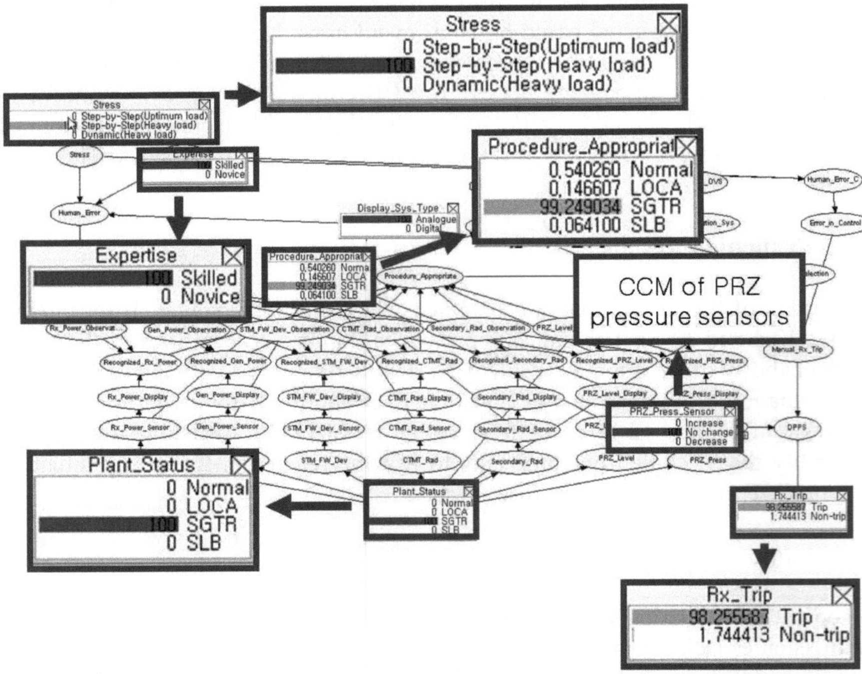

Figure 12.12. BBN model for Case 1 [20]

the first evaluation for the reactor trip operation. The failure probability of a skilled operator is reduced by 43.2% and that of a novice operator is reduced by 42.6% in the second evaluation for the failed SG isolation operation. However, adverse effects of DSSs result, if the reliabilities go down to 95%. In this case, the integrated DSS increases failure probabilities in almost all cases. The reliability of a DSS is very important in terms of enhancing the performance of human operators.

The results of both evaluations reflect good outcomes of the DSSs. The effect of the DSSs is greater for less-skilled operators than for highly skilled operators.

Table 12.4. Results of the first evaluation for the reactor trip operation [20]

Reliability of FDS and OVS	99.9%		99%		95%	
Case \ Expertise	Skilled	Novice	Skilled	Novice	Skilled	Novice
Case 1	0.017444	0.023344	0.017444	0.023344	0.017444	0.023344
Case 2	0.014512	0.018478	0.014512	0.018478	0.014512	0.018478
Case 3	0.013964	0.018106	0.016632	0.022121	0.029161	0.038133
Case 4	0.011974	0.013962	0.014180	0.018041	0.026337	0.034012
Case 5	0.013897	0.019738	0.013897	0.019738	0.013897	0.019738
Case 6	0.004988	0.006990	0.009477	0.013256	0.019453	0.027181
Case 7	0.004988	0.006990	0.009477	0.013256	0.019453	0.027181

Table 12.5. Results of the second evaluation for the failed SG isolation operation [20]

Reliability of FDS and OVS / Case \ Expertise	99.9%		99%		95%	
	Skilled	Novice	Skilled	Novice	Skilled	Novice
Case 1	0.022820	0.028994	0.022820	0.028994	0.022820	0.028994
Case 2	0.019854	0.024035	0.019854	0.024035	0.019854	0.024035
Case 3	0.017915	0.023238	0.020583	0.027431	0.030303	0.041194
Case 4	0.015202	0.018497	0.019656	0.024696	0.029555	0.038473
Case 5	0.017910	0.024128	0.017910	0.024128	0.017910	0.024128
Case 6	0.008239	0.011557	0.012725	0.017800	0.022693	0.031674
Case 7	0.006651	0.010370	0.011158	0.016631	0.021205	0.030568

The failure probability decrement by the DSSs with 99.9% reliability in the first evaluation is 0.012456 for skilled operators, and that for novice operators is 0.016354. The results from the second evaluation are similar.

12.5 Concluding Remarks

The impact of humans on the safety or reliability of systems becomes more important as technologies evolve. INDESCO is one of the countermeasures to reduce human errors. The operation processes of the human operator are analyzed with respect to cognitive processes. Various systems that support major activities of the human cognitive process are developed and integrated into one system. The integrated system, INDESCO, facilitates the entire operation process of human operators: monitoring plant parameters, diagnosing the ongoing situation, selecting corresponding actions for the identified situation, and performing those actions. A quantitative safety assessment to show the effect of INDESCO on the performance of human operators was performed by using the quantitative safety assessment method introduced in Chapter 11 and the HRA event tree method. The assessment results showed that INDESCO is helpful for reducing failure probabilities of human operators. DSSs with high reliability showed positive effects on the performance of human operators. However, the systems with low reliability showed negative effects. The effect of the system was greater for less-skilled operators than for highly skilled operators.

The models should be extended with consideration of group operations, more indicators, and more situations in order to obtain more reliable and credible evaluation results, even though several viable results were obtained from the evaluations. Both theoretical and experimental methods are necessary. The results of theoretical evaluations tend to be considerably affected by assumptions and data. Obtaining precise data, particularly data about human aspects of a system, is difficult. Evaluations using experimental methods, such as the method introduced in Chapter 9, are also needed compensate for the weakness of theoretical evaluations.

References

[1] Marsden P (1996) Procedures in the nuclear industry, In Stanton, N. (ed.). Human Factors in Nuclear Safety:99–116
[2] Miller CA, Funk HB, Goldman RP, Meisner J, Wu P (2005) Implications of adaptive vs. adaptable UIs on decision making. Human Computer Interaction International 2005
[3] Miller CA (2005) Trust in adaptive automation: The role of etiquette in tuning trust via analogical and affective methods. Human Computer Interaction International 2005
[4] Inagaki T, Furukawa H, Itoh M (2005) Human interaction with adaptive automaton: Strategies for trading of control under possibility of over-trust and complacency. Human Computer Interaction International 2005
[5] Kawai K, Takizawa Y, Watanabe S (1999) Advanced automation for power-generation plants-past, present and future. Control Engineering Practice 7:1405–1411
[6] Wickens CD (2000) Engineering psychology and human performance. New York: Harper Collins
[7] Perrow C (1984) Normal accidents. New York: Basic Books
[8] Green M (1999) Human machine interaction research at the OECD Halden reactor project. People in Control: An International Conference on Human Interfaces in Control Rooms, Cockpits and Command Centres:463
[9] Lee SJ, Seong PH (2007) Development of an integrated decision support system to aid cognitive activities of operators. Nuclear Engineering and Technology
[10] Ohi T, Yoshikawa H, Kitamura M, Furuta K, Gofuku A, Itoh K, Wei W, Ozaki Y (2002) Development of an advanced human-machine interface system to enhanced operating availability of nuclear power plants. International Symposium on the Future I&C for NPP (ISOFIC2002). Seoul:297-300
[11] Chang SH, Choi SS, Park JK, Heo G, Kim HG (1999) Development of an advanced human-machine interface for next generation nuclear power plants. Reliability Engineering and System Safety 64:109-126
[12] Kim IS (1994) Computerized systems for on-line management of failures: a state-of-the-art discussion of alarm systems and diagnostic systems applied in the nuclear industry. Reliability Engineering and Safety System 44:279–295
[13] Ruan D, Fantoni PF, et al. (2002) Power surveillance and diagnostics: Springer
[14] Gofuku A, Ozaki Y, Ito K (2004) A dynamic operation permission system for pressurized water reactor plants. International Symposium on the Future I&C for NPP (ISOFIC2004). Kyoto:360–365
[15] Kim JH, Seong PH (2007) The effect of information types on diagnostic strategies in the information aid. Reliability Engineering and System Safety. 92:171-186
[16] Barriere M, Bley D, Cooper S, Forester J, Kolaczkowski A, Luckas W, Parry G, Ramey-Smith A, Thompson C, Whitehead D, Wreathall J (2000) Technical basis and Implementation Guideline for A Technique for Human Event Analysis (ATHEANA), NUREG-1624, Rev. 1. U.S. Nuclear Regulatory Commission: Washington D.C.
[17] Thompson CM, Cooper SE, Bley DC, Forester JA, Wreathall J (1997) The application of ATHEANA: a technique for human error analysis. IEEE Sixth Annual Human Factors Meeting
[18] Kim MC, Seong PH (2004) A quantitative model of system-man interaction based on discrete function theory. Journal of the Korean Nuclear Society 36:430–450
[19] Niwa Y, Yoshikawa H (2003) The adaptation to main control room of a new human machine interface design. Human Computer Interaction International 2003:1406-1410
[20] Lee SJ, Kim MC, Seong PH (2007) An analytical approach to quantitative effect estimation of operation advisory system based on human cognitive process using the Bayesian belief network. Reliability Engineering and System Safety

[21] Kim JT, Kwon KC, Hwang IK, Lee DY, Park WM, Kim JS, Lee SJ (2001) Development of advanced I&C in nuclear power plants: ADIOS and ASICS. Nuclear Engineering and Design 207:105–119

[22] Lee SJ, Seong PH (2005) A dynamic neural network based accident diagnosis advisory system for nuclear power plants. Progress in Nuclear Energy 46:268–281

[23] Varde PV, Sankar S, Verma AK (1997) An operator support system for research reactor operations and fault diagnosis through a connectionist framework and PSA based knowledge based systems. Reliability Engineering and System Safety 60:53–69

[24] Yangping Z, Bingquan Z, DongXin W (2000) Application of genetic algorithms to fault diagnosis in nuclear power plants. Reliability Engineering and System Safety 67:153–160

[25] Mo K, Lee SJ, Seong PH (2007) A dynamic neural network aggregation model for transient diagnosis in nuclear power plants. Progress in Nuclear Energy 49-3:262–272

[26] Pirus D, Chambon Y (1997) The computerized procedures for the French N4 series. IEEE Transaction on Nuclear Science 8-13:6/3–6/9

[27] Converse SA, Perez P, Clay M, Meyer S (1992) Computerized procedures for nuclear power plants: evaluation of the computerized procedures manual (COPMA-II). IEEE Transactions on Nuclear Science 7-11:167–172

[28] Mo K, Lee SJ, Seong PH (2007) A neural network based operation guidance system for procedure presentation and operation validation in nuclear power plants. Annals of Nuclear Energy 34-10:813–823

[29] Jensen F (1994) Implementation aspects of various propagation algorithms in Hugin. Research Report R-94-2014, Department of Mathematics and Computer Science, Aalborg University, Denmark

[30] Jensen F., Andersen SK (1990) Approximations in Bayesian belief universes for knowledge-based systems. In Proceedings of the Sixth Conference on Uncertainty in Artificial Intelligence. Cambridge, Massachusetts:162–169

[31] Swain AD, Guttmann HE (1983) Handbook of human reliability analysis with emphasis on nuclear power plant applications, NUREG-1278. U.S. Nuclear Regulatory Commission: Washington D.C.

[32] Advanced compact nuclear simulator textbook. Nuclear Training Center in Korea Atomic Energy Research Institute (1990)

Acronyms and Abbreviations

ACR	Advanced control room
AECL	Atomic Energy of Canada Limited
AHP	Analytic hierarchy process
AI	Analog input
AOI	Area of interest
APR-1400	Advanced power reactor 1400
ATHEANA	A Technique for Human Event Analysis
AV	Audio/video
BARS	Behaviorally anchored rating scale
BBN	Bayesian belief network
BGA	Ball grid array
BP	Bistable processor
CBDTM	Cause-based decision tree methodology
CBHRA	Condition-based human reliability aanalysis
CCF	Common-cause failure
CCS	Calculus of communicating systems
CDF	Cumulative distribution function
CFP	Cognitive failure probabilities
CFR	Code of federal regulations
CNS	Compact nuclear simulator
COCOM	Contextual control model
COM	Consensus operator model
COSS	Computerized operator support system
COTS	Commercial-off-the-shelf
CP	Coincidence processor
CPC	Common performance condition
CPN	Colored petri nets
CPS	Computerized procedure system
CPT	Conditional probability table
CR	Control room
CREAM	Cognitive reliability and error analysis method
CRT	Cathode-ray tube
CSF	Critical safety function
DI	Digital input

DO	Digital output
DPPS	Digital plant protection system
DRAM	Dynamic random access memory
DSS	Decision support system
DURESS	Dual reservoir system
EEG	Electroencephalogram
EFC	Error forcing context
EID	Ecological interface design
EOC	Error of commission
EOO	Error of omission
EOP	Emergency operation procedure
EP	Evoked potential
EPC	Error-producing condition
EPROM	Erasable programmable read-only memory
ERP	Event-related potential
ESDE	Excessive steam demand event
ESDT	Extended structured decision tables
ESF	Engineered safety feature
ESFAS	Engineered safety feature actuation system
ETS	Eye-tracking system
FBD	Function block diagram
FDA	Food and drug administration
FIT	Failures-in-time
FLIM	Failure likelihood index methodology
FMEA	Failure mode and effect analysis
FOD	Function overview diagram
FPGA	Field programmable gate array
FTA	Fault tree analysis
GMTA	Goals–means task analysis
GTT	Generic task type
HAZOP	Hazard and operability studies
HCR	Human cognitive reliability
HCSS	High-capacity storage station
HE	Human error
HEART	Human error assessment and reduction technique
HEP	Human error probability
HFE PRM	Human factors engineering program review model
HFE	Human factors engineering
*HFE	Human failure event
HMI	Human–machine interface
HOL	Higher order logic
HR	Heart rate
HRA	Human reliability analysis
HRP	Halden reactor project
HRV	Heart rate variability
HS	HUPESS server
HTA	Hierarchical task analysis

HUPESS	Human performance evaluation support system
I&C	Instrumentation and control
IE	Integrated environment
IF	Instrumentation failure
INDESCO	Integrated decision support system to aid cognitive activities of operators
ISV	Integrated system validation
KAERI	Korea Atomic Energy Research Institute
KAIST	Korea Advanced Institute of Science and Technology
KSAX	Korean situation awareness index
LCD	Liquid crystal display
LDP	Large display panel
LOCA	Loss of coolant accident
LOTOS	Language of temporal ordering specification
LPSIS	Low-pressure safety injection system
MABA-MABA	Men are better at–machines are better at
MBU	Multi-bit upset
MCH	Modified Cooper–Harper
MCR	Main control room
MCS	Minimal cut sets
MDTA	Misdiagnosis tree analysis
MES	Mobile evaluation station
MIC	Methyl isocyanate
MRTF	Manual reactor trip failure
MTTF	Mean time to failure
NASA-TLX	National Aeronautic and Space Administration task load index
NEP	Nominal error probability
NFF	No fault found
NHPP	Non-homogeneous Poisson process
NPP	Nuclear power plant
NRC	Nuclear Regulatory Commission
NuFDS	N FBD-style design specification
NuSCM	Nuclear software configuration management
NuSCR	Nuclear software cost reduction
NuSDS	N software design specification and analysis
NuSEE	Nuclear software engineering environment
NuSISRT	Nuclear software inspection support and requirements traceability
NuSRS	Nuclear software requirements specification and analysis
OE	Operator error
OPAS	Operator performance assessment system
OTΔT	Over-temperature delta-T
OW	Overall workload
OVS	Operation validation system
PD	Plant dynamics
PDF	Probability density function

PDL	Primary heat transport low core differential pressure
PFS	Program functional specification
PLC	Programmable logic controller
PoF	Physics of failure
PORV	Pressure-operated relief valve
PPAS	Plant performance assessment system
PPS	Plant protection system
PRA	Probabilistic risk assessment
PSF	Performance shaping factor
PT	Post test
PVS	Prototype verification system
PWR	Pressurized water reactor
RBD	Reliability block diagram
RCS	Reactor coolant system
RGGG	Reliability graph with general gates
RIAC	Reliability information analysis center
ROM	Read-only memory
RPS	Reactor protection system
RTM	Requirements traceability matrix
S/W	Software
SA	Situation awareness
*SA	Software architecture
SACRI	Situation awareness control room inventory
SAGAT	Situational awareness global assessment technique
SART	Situational awareness rating technique
SBU	Single-bit upset
SC	Stacked capacitors
SCM	Software configuration management
SCR	Software cost reduction
SDS	Software design specification
SDT	Structured decision table
SEC	Single-error-correcting
SEFI	Single-event interrupt functional interrupt
SER	Soft-error rate
SES	Stationary evaluation station
SET	Single-event transient
SEU	Single-event upset
SFTA	Software fault tree analysis
SG	Steam generator
SGTR	Steam generator tube rupture
SI	Safety injection
SIAS	Safety injection actuation signal
SLB	Steam line break
SLI	Success likelihood index
SLIM	Success likelihood index methodology
SLOCA	Small loss of coolant accident
SME	Subject matter expert

SMoC	Simple model of cognition
SMV	Symbolic model verifier
SPDS	Safety parameter display system
SRAM	Static random access memory
SRK	Skills, rules, and knowledge
SRS	Software requirements specification
STA	Safety technical advisor
SWAT	Subjective workload assessment technique
TCP/IP	Transmission control protocol/internet protocol
TEC	Trenches with external charge
THERP	Technique for human error rate prediction
TIC	Trenches with internal charge
TMI	Three Mile Island
TSC	Technical support center
UA	Unsafe action
V&V	Verification and validation
VDM	Vienna development method
VDU	Visual display unit
VISA	Visual indicator of situation awareness
XML	Extensible markup language

Index

E

F

G

T

MIX
Papier aus verantwortungsvollen Quellen
Paper from responsible sources
FSC® C105338

If you have any concerns about our products,
you can contact us on
ProductSafety@springernature.com

In case Publisher is established outside the EU,
the EU authorized representative is:
Springer Nature Customer Service Center GmbH
Europaplatz 3, 69115 Heidelberg, Germany

Printed by Libri Plureos GmbH
in Hamburg, Germany